PNEUMATIC INSTRUMENTATION
Third Edition

Dale R. Patrick, Professor
and
Steven R. Patrick

Department of Industrial Education and Technology
College of Applied Arts and Technology
Eastern Kentucky University

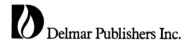 Delmar Publishers Inc.

NOTICE TO THE READER

Cover design by John DeSieno & Associates

Delmar Staff
Senior Administrative Editor: Vernon Anthony
Developmental Editor: Pat Konczeski
Project Editor: Carol Micheli
Production Coordinator: Mary Ellen Black
Art and Design Manager: Russell Schneck
Art Coordinator: Megan Keane DeSantis
Design Coordinator: Karen Kunz Kemp

For information, address Delmar Publishers Inc.
3 Columbia Circle, Box 15-015
Albany, New York 12212

COPYRIGHT 1993
BY DELMAR PUBLISHERS INC.

printed in the United States of America
published simultaneously in Canada
by Nelson Canada,
a division of The Thomson Corporation

  2   3   4   5   6   7   8   9   10   XXX   04   03

Library of Congress Cataloging-in-Publication Data

Patrick, Dale R.
    Pneumatic instrumentation / Dale R. Patrick and Steven R. Patrick.
    —3rd ed.
        p.   cm.
    Includes index.
    ISBN 0-8273-5482-7
    1. Pneumatic control. 2. Engineering instruments. I. Patrick,
Steven R. II. Title.
TJ219.P38   1993
629.8'045—dc20   92-31773
                                            CIP

CONTENTS ━━━━━━━━━━━━━━

PREFACE

Pneumatic instrumentation has changed significantly since the first edition of this book was published. At that time, industrial processes were largely controlled by a number of commonly used pneumatic instruments. The maintenance of these instruments was the responsibility of plant engineers. Technicians eventually assumed this responsibility. In-house technician training and upgrading was then a common practice. *Pneumatic Instruments* was developed as a text for this type of program.

The original intent of this material was to investigate pneumatic fundamentals through the operation of an instrument. The pneumatic instrument still serves as a vehicle for learning fundamental principles in the new edition. However, an attempt has been made to identify each instrument as a functioning system. The instrument is then reduced to a chain of usable components. These components are then used to apply fundamental principles. Through this approach the reader will be able to see how an instrument functions and how fundamental pneumatic principles apply to the operation of the instrument. Technicians are responsible for the maintenance of these instruments. A number of in-plant training programs, technology schools, apprentice training programs, and home study courses have found this material a valuable aid in technician training and upgrading.

The third edition, retitled *Pneumatic Instrumentation*, has a significant organizational change. This includes a change in reading level, chapter reorganization and format. Each chapter now has objectives, key terms for review, text, summary, suggested activities, and study questions. This has largely come about through the recommendations of previous users.

We would like to thank the numerous manufacturers that have supplied updated technical data, photographs, and suggestions in the presentation of the new edition of this material. We would also like to thank students, teachers, and users of previous editions for their countless corrections and suggestions.

Dale R. Patrick

Steven R. Patrick

ACKNOWLEDGEMENTS ▬▬▬▬▬▬▬▬

We would like to thank the following instructors for their excellent reviews:

Robert Allen
Boise State University College of Technology
Boise, ID 83725

Robert Dixon
Chemeketa Community College
Salem, OR 97305

David Ellingsworth
Rend Lake College
Ina, IL 62846

To
Courtney Grace Patrick,
daughter and grandaughter

1
Industrial Instrumentation

OBJECTIVES

Upon completion of this chapter, you will be able to:

- Draw a block diagram of a system.
- Explain the fundamental operation of each block of a system.
- Show how the system's concept applies to hydraulics and pneumatics.
- Explain the meaning of the term pneumatic instrument.

KEY TERMS

In the study of industrial instrumentation one frequently encounters a number of new and somewhat unusual terms. These terms play an important role in the presentation of this material. As a rule, it is helpful to review these terms before proceeding with the chapter.

Calibration. A procedure that determines, either by measurement or by comparison with a known standard, the correct value of a reading on an instrument or other device.

Compressor. A machine that causes a change in pressure mechanically altering the volume of gas or air in an enclosure.

Control. A basic function that affects or alters the operation of a system.

Control-full. An operation that causes the state of a system to be completely on or off.

Energy source. A system part that is responsible for supplying operational energy to the components.

Electronics.   A system or branch of science that deals with the emission, behavior, motion, and effects of electrically-charged particles.

Final control element.   The functional part of a system that is responsible for altering or working with a process variable.

Flow.   The movement or travel of liquid, gas, or solid material in response to an applied force.

Fluid.   A gas or liquid that is manipulated in the operation of a system.

Hydraulics.   A branch of science that deals with the transmission of energy through the flow of liquid which is placed in motion.

Indicator.   A system part that shows whether it is on or off or tells a specific operating value or quantity.

Load.   The part of a system that converts energy into a different form or performs the work function of a system.

Measurement.   The act of determining or comparing a standard value with an obtained value.

Mechanics.   A branch of science that deals with energy and forces and their effects on a body or machine.

Null.   An operating condition that represents a balanced or zero state.

Pneumatics.   A branch of science that deals with the mechanical properties of gas or air.

Pressure.   The force per unit area measured in pounds per square inch (psi), Newton's per square meter, or by the height of a column of water, mercury, or other fluids that it will support in feet, inches, or centimeters.

Process variable.   Any process parameter such as temperature, flow, liquid level, or pressure that changes its value during the operation of a system.

Setpoint.   A selected value adjustment or position setting of a process variable.

System.   An assembly of parts that are connected together to form a complete operational unit.

Transmission path.   The route or path that energy must flow through a system in order to make it operational.

Work.   The transferring or transforming of energy when a force is exerted to move something over a distance.

INTRODUCTION

The advancement of science and technology is responsible for a number of changes in the basic structure of industry. Most products are, for example, manufactured through some type of automatic processing equipment. This equipment is often complex and demands a variety of skilled personnel to keep it in operation. Technicians are called on to install equipment, evaluate operation, analyze problems, and make repairs. A wide range of experience is needed in order to cope with these situations.

At one time, most industrial equipment could be placed in operation with a few simple tools and some common sense. Today, however, a large part of our equipment contains numerous control devices that perform precise operations automatically. Technical personnel must, therefore, be concerned with evaluation procedures, calibration techniques, instrumentation, and troubleshooting procedures in order to keep equipment operational. Equipment breakdown often causes production shutdown, which tends to reduce operational efficiency and to increase production costs.

Equipment operation today relies heavily on measurement techniques, evaluation, and instrumentation. Technical personnel are constantly called on to evaluate performance by measuring specific conditions of operation. Instrumentation has, therefore, become a vital part of industry.

Instrumentation is a broad area of industry that deals with the measurement, evaluation, and control of process variables such as temperature, pressure, flow, fluid level, force, light intensity, pH, electrical conductivity, humidity, and others. These variables are usually involved in a manufacturing operation of some type that eventually leads to a finished industrial product. Instrumentation applications are widely utilized in automatic production operations.

The primary areas of concern in instrumentation are pneumatics, electronics, mechanics, and hydraulics. Each of these areas is unique, but all are similar in many respects. The systems concept is commonly used to show this relationship.

THE SYSTEMS CONCEPT

The systems concept is not particularly new to the study of industrial equipment or machinery. This approach represents a diagrammed method of showing the parts of a complex piece of equipment in some logical order. Through this approach, a "big picture" of the basic system is first presented. This diagram is then used to show the link between various system parts. Figure 1-1 shows the primary parts of a system in block-diagram form.

The role played by each block of a diagram represents the second step of the systems concept. This role, or function of each block, is much more meaningful when viewed in its composite form. The complete diagram can then be viewed as a general reference. This leads to the next step, which deals with component analysis of each block.

Component analysis of a system diagram represents the next area of study, often called the "nuts and bolts" of the presentation. The intricate workings of components are discussed in this area of study. A grouping of similar component operations should then become apparent, and a person should begin to see how individual pieces of a system begin to fit together. Therefore, the systems approach concept will serve as our general plan of procedure in the presentation of pneumatic instruments.

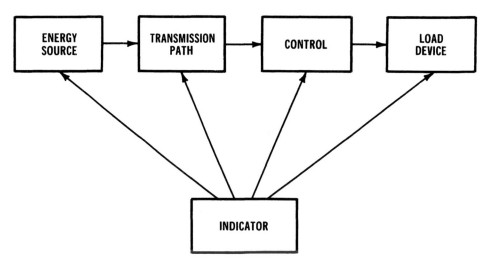

FIGURE 1-1   Block diagram showing the primary parts of a system.

SYSTEM PARTS

The essential parts of an operating system include an energy source, a transmission path, a control, a load device, and an optional indicator. Each part of the system has a specific role to play in the overall operation of the system. This role becomes extremely important when a detailed analysis of the system is to take place. Large numbers of components are sometimes needed to achieve a specific block function. Regardless of the complexity of the system, each block must still achieve its function in order for the system to be operational. Being familiar with these functions and being able to locate them within a complete system are very important steps in understanding the operation of the entire system.

The energy source of a system has the primary responsibility of changing energy from one form into another more useful form. Primary forms of energy include heat, light, sound, chemical, nuclear, and mechanical energy, among others. In its primary form, energy is not always useful to an operating system until changed into something more useful by the energy source.

The transmission path of a system is rather simplified when compared with other system parts. Its sole responsibility is the transference of energy. It provides a path for energy to flow from the source to the load device. In some systems, this function is achieved by pipes or flexible tubes that connect the energy source to the load. In complex systems, there may be a number of alternate conduction paths between different components.

The control section of a system is generally considered to be the most complex part of the entire system. In its simplest form, control is used to turn a system on or off. The terms full control, on/off, or two-state control are used to describe this operation. Control of this type may occur anywhere between the energy source and the load device. In addition, a system may include some type of partial control. This type of control usually causes a gradual change to occur somewhere in an operating system. Changes in time, pressure, and flow are typical of partial control.

The load of a system refers to a specific part, or number of parts, designed to achieve work. The term work, in this case, refers to an operation that occurs when energy changes form. Typically, heat, sound, and mechanical motion take place when work occurs. The load of a system generally consumes a large portion of the energy produced by the source. The load is generally the most recognizable part of system because of its obvious work function.

The indicator of a system is designed to display typical operating conditions at numerous points throughout a system. In some applications an indicator is optional, while in others it is an essential part. In the latter case, specific system operations are usually dependent upon indicator readings. The term operational indicator usually applies to this type of application. Test indicators are also used to measure specific values in maintenance and troubleshooting operations. Meters, gauges, and chart recorders are frequently used to perform this type of operation. To some extent, an indicator may add to the total load of an operating system.

FLUID POWER SYSTEMS ▰▰▰▰▰▰▰▰▰▰▰▰▰▰▰▰▰▰▰▰▰

The term fluid power is commonly used in industry today to describe those systems that employ liquid or gas as a means of controlling power. In ordinary usage, the terms fluid and liquid are used interchangeably. Technically, the term fluid applies to both liquids and gases. Hydraulics is used to describe liquid applications, while pneumatics applies to gaseous applications. The operating principles of the two systems are very similar in many respects.

Hydraulic Systems

In industry, hydraulic systems use liquid as a medium to transfer force between the source and the load device. This type of system is commonly found in equipment used in fabrication operations, robots, and material-handling processes. Hydraulic systems are simple to operate, are very reliable, and can be easily adapted to many applications. Figure 1-2 shows a double-acting hydraulic system used to control the ram of a punch-press.

The primary energy source of the hydraulic system is an electric-motor driven pump and reservoir. Rotary mechanical energy of the motor is changed directly into fluid energy through this device. The pump causes hydraulic fluid to be set into motion. Fluid entering the pump is set into motion and forced to leave through the outlet port. Each revolution of the pump rotor blade causes a fixed amount of fluid to be forced into the system. Fluid entering the system at this point encounters some resistance to its flow. This resistance is responsible for the development of hydraulic pressure.

The transmission path of Figure 1-2 may include solid metal tubes between some parts, with flexible tubing or hoses connected to others. As hydraulic fluid is forced to pass through the transmission path it builds system pressure. Typically, both high and low pressure are present in the system. Initial flow from the pump is higher than the pressure of the return line.

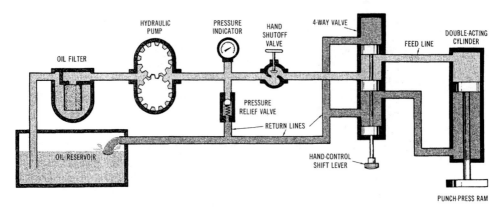

FIGURE 1-2   A double-acting hydraulic system.

A hydraulic system generally has both full and partial control of its fluid. The hand shutoff valve of Figure 1-2 is responsible for full control of fluid. It can turn it on or stop the flow from entering the system at this point. Pressure can also be altered by changing the motor speed of the pump.

The four-way valve is also a control device. It restricts flow, alters its direction, and can completely stop flow. The pressure-relief valve is a control that automatically releases system pressure. In general, a hydraulic system has many types of control.

The double-acting cylinder of Figure 1-2 serves as the load device of the hydraulic system. The primary function of this load is to change fluid flow into linear motion. The composite load of the entire system includes transmission-line resistance, cylinder resistance, and the outside work load. The load of this system will change to some extent according to the amount of outside work that it does.

The indicator of a hydraulic system is generally an optional item. In Figure 1-2, an indicator is used to monitor primary system pressure. It is placed near the pump. Indicators can also be placed at other locations. An indicator in the cylinder feed-line would be used to monitor the work function of the load. An indicator in the return line would monitor the low pressure response of the system. The location of the indicator in a system has a great deal to do with the measurement being observed.

Pneumatic Systems

Pneumatic systems are used in applications where air serves as a medium through which force is transferred between the source and the load device. This type of system is used to lift, move, rotate, position, and clamp products during machine operations. Pneumatic systems represent a unique form of fluid power that have open-ended return lines. Systems of this type derive air from the atmosphere, compress it to increase pressure, store it in a receiving tank, distribute it to do work, and ultimately return it to the atmosphere. A simple low-pressure pneumatic system is shown in Figure 1-3.

The energy source of a pneumatic system is the compressor/receiving tank assembly. The compressor is driven by an electric motor; the receiving tank is a container that holds

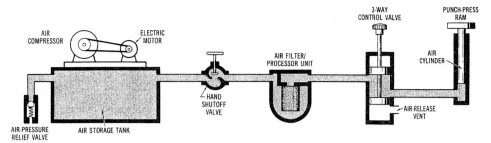

FIGURE 1-3   A simple low-pressure pneumatic system.

air. Through the action of the compressor, outside air is forced into the receiving tank under pressure. This air is stored and eventually released when the system calls for it.

After air has been compressed and placed in the receiving tank, it must then be conditioned before it can be used. Conditioning is responsible for the removal of foreign matter such as dirt, oil, and moisture. In some systems an air filter unit with a condensation trap is used. In other systems an air dryer may be used. Oil is one of the primary sources of contamination in a pneumatic system. In some systems, oilless conditioners may be placed in front of each device to provide added filtering. Conditioning may also include some method of regulating pressure.

The transmission path of a pneumatic system is composed of such things as pipes, tubing, and flexible hoses. This part of the system is primarily responsible for the distribution of air to each of the system components. A single transmission line is used to distribute air from the receiving tank to individual components. A unique feature of the pneumatic transmission path is its simple method of return. Air is dumped into the atmosphere instead of being returned to the receiving tank. Air distribution can therefore be accomplished by a single flow path without a return line.

The pneumatic system of Figure 1-3 has both full and partial control of air. The hand shutoff valve and pressure-relief valve both provide full control of air through the transmission path. Air flow is also changed partially through the air regulator and the three-way control valve. System control of this type may be achieved at numerous locations throughout the system.

The primary function of the pneumatic system load is to achieve some desired work function. In Figure 1-3, the load device is the air cylinder of a punch-press ram. This part of the system is designed to change the mechanical energy of air into a usable form of linear motion that drives a ram. The outside work being achieved by the ram also has some direct influence on the total performance of the entire system. All of this represents the composite load, which includes such things as transmission line resistance, the work being accomplished by the ram, and control resistance. Other pneumatic loads may be used to actuate a valve that controls another process or produces rotary motion.

Most of the pneumatic systems used in industry employ several indicators placed at different locations throughout the system. In general, system pressure is monitored by the indicator. The receiving tank and regulator of the system may employ pressure gauges as indicators. Any pressure loss in the feed line can also be monitored by an indicator. System maintenance and troubleshooting uses indicators to locate faulty components.

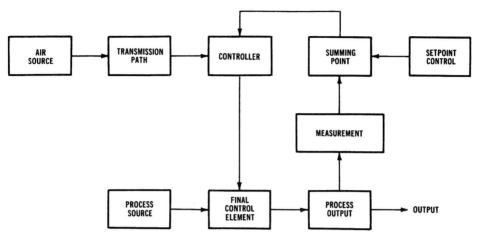

FIGURE 1-4   Block diagram of a pneumatic instrument control system.

PNEUMATIC INSTRUMENTS

Pneumatic instruments respond to air pressure as a medium that initiates some form of control in the operation of a system. Instruments of this type are used to measure and evaluate a condition of operation and to actuate a control element that alters a manufacturing process. The primary elements of this type of system are an expansion of the basic systems concept, which include such things as measuring instruments, controllers, and the final control element. In a strict sense, pneumatic instruments are used to initiate control that is ultimately used to alter the operation of a second system. Figure 1-4 shows a block diagram of this system application.

The pneumatic part of the system has an energy source, transmission path, controller, and load device. The final control element of the process system serves as the load device of the air system. This same component also serves as the control element of the process system. Measurement is achieved by an instrument attached to the output of the process system. The measured output is compared at the summing point with a manual setpoint adjustment. When these two values are equal, the controller output is in a null, or zero, state. When the values are not in agreement, the controller is energized to actuate the final control element. Through a constant change in balance and unbalance, the output of the process system is altered to agree with the setpoint control position. Automatic control of a process variable is commonly achieved through the use of pneumatic instruments today.

SUMMARY

Most of the products manufactured by industry today are produced through some type of automatic processing equipment. Equipment of this type relies heavily on measurement techniques and instrument analysis.

Instrumentation is a broad area of industry that deals with the measurement, evaluation, and control of process variables such as temperature, pressure, flow, fluid level, force, light intensity, pH, electrical conductivity, and humidity.

The primary areas of concern in instrumentation are pneumatics, electronics, mechanics, and hydraulics. The systems concept is used to show how these areas are related.

The systems concept is a diagrammed method of showing the parts of a piece of equipment in some logical order. A block diagram of system functions is represented as a "big picture" of the system. Typically, this includes an energy source, transmission path, control, a load device, and an optional indicator.

The energy source is responsible for changing energy from one form into another form that is more useful for system operation.

The transmission path of a system is responsible for transferring energy from the source to other system parts.

Control of a system may be either full or partial. Full control turns the system on or off while partial control is a variable condition.

The load of a system performs the work function. Energy from the source usually changes form when it produces work.

Indicators are designed to display operating conditions or to test system values in troubleshooting operations.

The energy source of a hydraulic system is a pump that circulates fluid through metal pipes or flexible hoses. Control is either full or partial through valves and regulators. A cylinder serves as the load of the hydraulic system, while indicators are gauges and meters.

Pneumatic systems are unique when compared with hydraulic systems. Air, for example, is dumped into the atmosphere instead of being returned. Compressors are used to develop air pressure, which is stored in a receiving tank until it is needed. Figure 5-1 shows the location of a motor-driven air compressor and receiving tank in a pneumatic packaging operation.

Pneumatic instruments respond to air pressure as a medium of control. These instruments are used to measure and evaluate the condition of an operation. In many applications, air is often used to control another process variable. Essentially, two interconnected systems are used to achieve this type of control.

ACTIVITIES

SYSTEM ANALYSIS ▰▰▰▰▰▰▰▰▰▰▰▰▰▰▰▰

1. Identify a system in the building where you are located. In a home or classroom facility this system can develop heat or distribute water.
2. Make a diagram of the system.
3. Identify where the source, path, control, load, and indicator are located.
4. Is the system self-contained or does it go beyond the building?
5. What are some of the distinguishing features of the system?

INDUSTRIAL SYSTEMS

1. Study the industrial pneumatic manufacturing system of Fig. 1-5.
2. Make a block diagram of the system.
3. Identify where the source, path, control, load, and indicator are located.
4. Explain the function of each system block.
5. What are some of the distinguishing features of the system?
6. What does the system do?

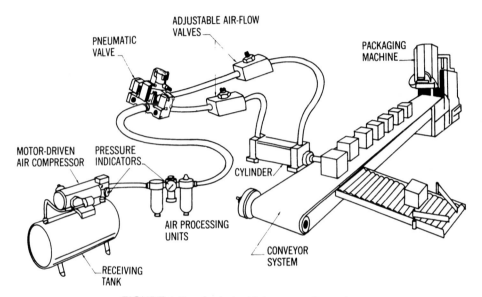

FIGURE 1-5   An industrial pneumatic system.

QUESTIONS

1. What are the basic parts of system?
2. What is meant by the term hydraulic?
3. What is meant by the term pneumatic?
4. Explain the function of the source of a system.
5. Explain the function of the transmission path.
6. What is the function of control in the operation of a system?
7. What does the load do in the operation of a system?
8. Explain the role of an indicator in the operation of a system.

2
Lever Mechanisms and Adjustments

OBJECTIVES

Upon completion of this chapter, you will be able to:

- Define the term lever.
- Distinguish between different classes of levers.
- Recognize a link and lever mechanism.
- Explain the operation of a link and lever mechanism.
- Explain how a link/lever mechanism is adjusted.
- Identify some special mechanisms used in an instrument.
- Define the role of a spring in an instrument.

KEY TERMS

Angularity. The quantity of being angular or having more than one angle.

Bellows. A pressure sensitive element consisting of a convoluted metal cylinder closed at one end. A pressure difference between the inside and outside of the bellows causes it to expand or contract.

Bourdon tube. A pressure sensing element that consists of a curved tube having a flattened elliptical cross-section that is closed on one end. A pressure difference between the inside and outside of the tube will cause it to change shape.

Controller. An instrument that alters or changes the operation of a system.

Effort force. An application of power that causes a push or pull to occur.

Flexure. The quantity of being turned, twisted or folded.

Force.   An effort that pushes or pulls and is measured in pounds, ounces, and grams.

Force balance.   A mechanism or instrument that detects or responds to a difference between applied forces.

Fulcrum.   A support around which a lever turns or pivots.

Integrator.   A device or function that continually totalizes, or adds up, the value of a quantity for a given time.

Lever.   A bar which is free to turn or pivot on a point, or fulcrum.

Linearity.   The degree to which the calibration curve of a device matches a straight line or in which a change in one property is directly proportional to a change in another property or quantity.

Link.   A connecting structure that attaches two or more items together.

Mechanical advantage.   A ratio of output force to applied force in a simple machine such as a lever.

Moment.   Another term meaning torque. The tendency to produce motion about a point or axis measured by the product of a force and the distance to that point or axis.

Moment balance.   A condition in which all the force-distance products occurring about a point is zero in their tendency to produce motion about the point.

Plane.   A surface of such nature that a straight line joining two of its points lies wholly on the same surface.

Positioner.   An instrument that alters or changes the location of a physical member according to signal data applied to its input.

Potentiometer.   A variable resistor used to control an electrical circuit. An unknown voltage-measuring circuit that determines values by comparing them with a known reference to produce a null condition.

Resistance force.   A force that opposes or is in opposition to the effort force applied to a lever.

Spiral.   A coil-like structure that winds around a center axis and gradually expands in diameter.

Standard cell.   A voltage cell that serves as a standard or reference in the measurement of an electromotive force.

Torsion.   The twisting or wrenching of a body by the exertion of forces tending to turn one end or part about an axis while the other is held fast or turned in the opposite direction.

LEVER BASICS

Almost every pneumatic instrument uses a lever in its operation. Levers are simple machines that make something work easier. Levers are used to multiply a force, increase speed, or cause a change in the direction of motion. In general, they deal with an effort force applied to an object that overcomes a form of resistance. Resistance is the weight of a component that is moved a distance.

A lever is a straight bar of material that is pivoted at one point. Levers have many shapes and sizes ranging from straight or bent forms to the wheel. A wheel is a number of levers acting at a common pivot point.

A lever is a piece of material that is free to turn about a fulcrum or pivot point. When force is applied to this piece of material, it produces motion. The resulting motion may be used to move a weight. This weight may be acting at another point along the piece of material. The resulting output of the lever is based on its classification. Levers are classified as first-, second-, or third-class.

First-Class Levers

A force-resistance diagram of a first-class lever is shown in Figure 2-1. Note that it has effort force, resistance force, and a fulcrum. The effort force must be great enough to overcome the resistance force. When this occurs the resistance force moves upward from the fulcrum. A child's seesaw and a measuring balance are examples of first-class levers.

An important concept of parallel forces can be applied to the measuring-balance lever. If the effort force equals the resistance force at the same time, balance occurs. Forces in this case act together and produce a single resulting force that will push downward. If effort force is greater than resistance force, movement will turn clockwise. Counter-clockwise turning will occur when resistance force is more than effort force.

The turning force or movement of a lever is called the moment of force or torque. The moment equals force multiplied by the distance it moves from the fulcrum. Moment-balance transmitters and moment-balance positioners utilize this principle. This application of the lever will be discussed in chapters 12 and 13.

Fulcrum location of a first-class lever is an important condition of operation. The fulcrum is located somewhere between the effort force and the resistance force. The mechanical advantage of this lever is quite variable. It is equal when balanced. It may be larger or smaller depending on the position of the fulcrum.

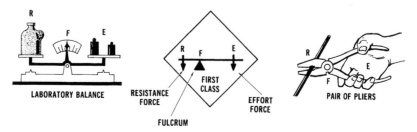

FIGURE 2-1   First-class levers.

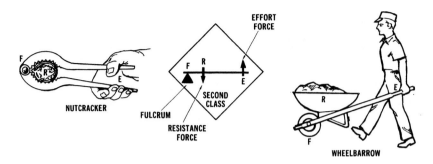

FIGURE 2-2   Second-class levers.

Second-Class Levers

A second-class lever has the resistance force located between the fulcrum and the effort force. An example of this lever would be a wheelbarrow or a nutcracker. Figure 2-2 shows an effort/resistance force diagram of a second-class lever. In this lever the effort force travels more than the resistance force. The resistance force is always larger than the effort force. The mechanical advantage of this lever is always greater than one.

Third-Class Levers

A third-class lever has the effort force between the resistance force and the fulcrum. The effort force is applied over a short distance and is always smaller than the resistance force. The mechanical advantage is less than one. Figure 2-3 shows an effort/resistance force diagram for a third-class lever. Note the position and direction of the force applied to the fulcrum.

   Third-class levers are designed to enable us to multiply distance at the expense of force. Common examples of the third-class lever are ice tongs, tweezers, and the human forearm.

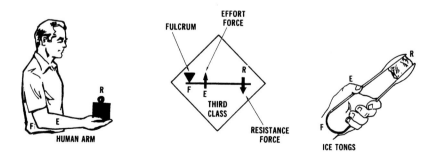

FIGURE 2-3   Third-class levers.

LINK AND LEVER MECHANISMS

Industrial instruments used for recording are all similar in principle. Except for some electronic instruments, almost all recorders use links and levers. A typical pneumatic recording instrument that employs a link/lever mechanism is shown in Figure 2-4.

Components

The mechanism of a typical instrument consists of:

1. An input lever.
2. An output lever.
3. A link connecting input and output levers together.

A lever is defined as a member that is free to pivot or rotate about a point called the fulcrum. Levers may have either one or two sides. The handle on a rotary wall-mounted can opener is an example of a one-sided lever. An example of a two-sided lever is a child's seesaw.

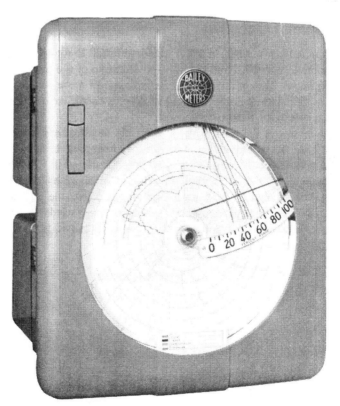

FIGURE 2-4   A typical pneumatic recording instrument. (*Courtesy of Bailey Controls Co.*)

In certain respects, a one-sided lever is similar to a wheel and axle mechanism. In instrument use, a one-sided lever travels less than 180° . An actual mechanism may only move 15° to 20°. This makes the mechanism more nearly a lever than a wheel.

Arrangements

The input lever may be one or two-sided. In a one-sided lever, the input is applied to a shaft about which the lever rotates. The input causes the shaft and lever to rotate with it. A link is attached to the lever, as shown in Figure 2-5A.

If a lever is two-sided, the input goes to only one side. This causes the lever to rotate about its pivot point, thus moving the alternate side. Figures 2-5B and 2-5C show two-sided lever construction.

The link attached to the input lever is connected to the output lever. The motion of the input lever is passed by the link to the output lever. The resulting action causes the output lever to rotate. This lever is then attached to a shaft that rotates with the lever. Attached to the shaft is the pen, or indicator. Also attached to the shaft are levers that feed controllers and alarm contacts. This type of mechanism is shown in Figure 2-6.

Links are usually straight. The levers connected by the link must be the same distance from the back or front of the instrument case.

Principle of Operation

Input applied to a mechanism causes the input shaft to rotate. This rotation is transferred to the output lever by a link mechanism. Rotation of the output lever causes the output shaft to rotate. This in turn causes the indicator to move. Other levers may be attached to the output shaft and may be used to drive a controller or alarm contacts, depending upon the function of the instrument.

Movement of the input lever is normally small compared to the movement of the output lever. To be usable, input lever movement must be multiplied. Assume that the angular rotation of the input lever is 10° and the output is 40°. This shows a multiplication factor of four. Motion multiplication is obtained by adjusting the ratio of lever length. The product of input angle and length equals the product of output angle and length. Suppose that the length of an output lever needs to be known. Assume that output lever moves 40° when a 3-inch input lever moves 30°. The length of the output lever is:

$$30 \times 3\text{-in} = 40° \times X$$
$$X = 90/40$$
$$X = 2\text{-}1/4\text{-in}$$

Figure 2-7 (page 19) shows a drawing and the calculations for this problem.

Physical, Actual, and Effective Levers

Operation of the link/lever mechanism is not as simple as it first appears. Two reasons for this are:

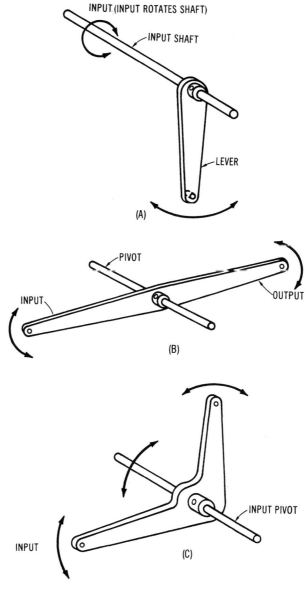

FIGURE 2-5   Input levers. (A) One-sided; (B) two-sided: 180°; (C) two-sided: 90°.

1.   The physical levers are not necessarily the actual levers.
2.   The actual levers are not necessarily the effective levers.

Examples of the actual, physical, and effective levers are shown in Figure 2-8.
    In this drawing, the actual lever is shown as a line that connects the point of rotation to the link. In many mechanisms the actual lever may fall outside of the physical lever. Location of the actual lever in a mechanism is important.

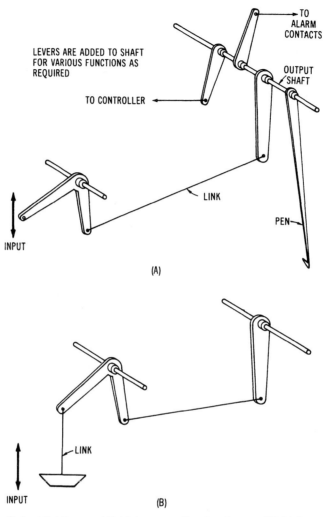

LEVERS ARE ADDED TO SHAFT
FOR VARIOUS FUNCTIONS AS
REQUIRED

TO CONTROLLER ←

→ TO
ALARM
CONTACTS

OUTPUT
SHAFT

LINK

PEN

INPUT

(A)

LINK

INPUT   (B)

FIGURE 2-6   Output link/levers. (A) Link connecting two levers; (B) Link connecting input to lever.

The effective lever is formed by a line starting at the point of rotation and connecting the link. When the two join, a 90° angle is formed. The effective lever changes length as the lever rotates, and the lengths of levers determine their travel. When the effective lever is not the actual lever, an angularity error occurs. Figure 2-9 shows an angularity error. The only point in this mechanism that is correct is where the actual and effective levers are similar. This point can be at any location on scale. It is usually either at midscale or at some critical point. If the critical point is not given, then a midscale point should be used.

PROBLEM:
FIND LENGTH OF OUTPUT LEVER IF:
1. INPUT ANGLE = 30°.
2. INPUT LEVER LENGTH = 3 in (7.6 cm).
3. OUTPUT ANGLE = 40°.

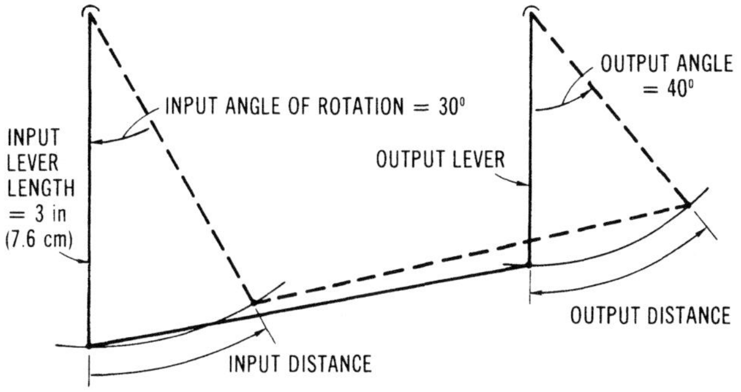

OUTPUT ANGLE = 40°

INPUT ANGLE OF ROTATION = 30°

INPUT LEVER LENGTH = 3 in (7.6 cm)

OUTPUT LEVER

OUTPUT DISTANCE

INPUT DISTANCE

INPUT DISTANCE = INPUT ANGLE × INPUT LEVER LENGTH
                = 30° × 3 in (7.6 cm)

OUTPUT DISTANCE = INPUT DISTANCE
                = 40° × OUTPUT LEVER LENGTH

THEN,

40° × OUTPUT LEVER LENGTH = 30° × 3 in (7.6 cm)

AND,

$$\text{OUTPUT LEVER LENGTH} = \frac{30° × 3 \text{ in (7.6 cm)}}{40°}$$

$$= 2\tfrac{1}{4} \text{ in (5.7 cm)}$$

NOTE: INPUT IS SET BY MEASUREMENT, OUTPUT IS SET BY PEN TRAVEL; THEREFORE, IF OUTPUT IS TO MATCH INPUT, LEVER LENGTHS MUST BE ADJUSTED.

FIGURE 2-7   Link/lever calculations.

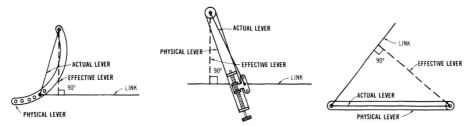

FIGURE 2-8   Examples of actual, physical, and effective levers.

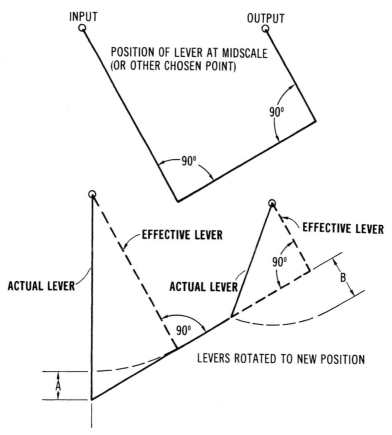

FIGURE 2-9   Angularity errors.

Angularity

An accurate relationship between input and output is obtained only under the most ideal conditions. This occurs when:

1.   The input lever is the same length as the output lever.
2.   Link length causes the input and output levers to be parallel.

In practice, these two conditions rarely occur. The introduction of different lengths of input and output levers results in an angularity error. Methods to minimize these errors will be discussed as link/lever adjustments. When these levers travel over different distances, they may not remain parallel. The link/lever mechanism of a recorder is shown in Figure 2-10.

When the input of a process variable is applied to a mechanism, it changes this value to an equivalent movement. This movement varies with the size of the variable. The amount of output causing pen travel has a fixed range of operation. The pen travels a

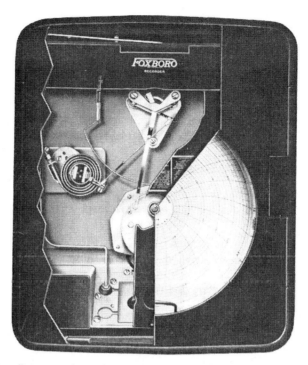

FIGURE 2-10   Cutaway view of a recording controller showing link/lever mechanism. (*Courtesy of the Foxboro Company*)

distance based on the width of the chart. This movement occurs regardless of the value of the variable or the specific output elements. To match measurements to different distances, input and output lever lengths must differ. The difference in lever length causes angularity errors. Lever lengths must be minimized to keep these errors within acceptable limits.

LINK/LEVER ADJUSTMENTS

Operation of a link/lever mechanism has input shaft rotation producing output shaft rotation. In this mechanism, input and output levers are usually of a different length. Correct readings can occur only when the levers are parallel and at 90° with the link. When the levers rotate they will not be parallel or form a 90° angle, which causes the mechanism to have an angularity error. Angularity errors can be minimized by an adjustment.

In a link/lever mechanism, rotation of the input shaft is controlled by the input signal. Rotation of the output shaft should be great enough to cause the pen to travel across the chart. To accomplish this, the lengths of the levers must be adjusted. This is called the span adjustment. The span is adjusted by altering the movement ratio of the output and input. This adjustment applies equally to mechanical, pneumatic, and electronic systems.

Final adjustment of the link/lever mechanism is used to match the input shaft to the output shaft, which occurs when the angle of rotation of the input and output correspond. Rotating the input shaft its first 10 percent should cause the output to rotate an equal 10 percent. This is called the span location, or zeroing, adjustment.

Angularity, span, and span location adjustments are all interrelated. One adjustment has some influence on the other two. As a rule, the adjustments are roughed in first. They are then refined for a more exact setting. The final adjustment is a fine-tuning operation. Manufacturers generally outline the adjustment procedure to follow for their instruments.

Adjustment Procedures

Examination of the link/lever mechanism is the first step of an adjustment procedure. This should indicate that the link/lever mechanism is correct when:

1.   The actual levers are parallel.
2.   The link forms a 90° angle with the actual lever.

Linearity (Angularity) Adjustment

Linearity or angularity is an adjustment that makes the levers parallel. This adjustment places the link at 90° to the levers. To make this adjustment, first change the length of the link, then rotate the levers relative to their shafts. It then becomes necessary to shift the rotational axis of the input lever.

Span Adjustment

The input must be capable of moving a lever through its angle of rotation. To observe this, first look at the output shaft. See if the pen will travel the width of the chart when the input is changed. Lever length should be adjusted if this does not occur. Changing the length of a lever will alter the ratio of input to output rotation. Span adjustments can also be corrected by changing spring constants in some instruments. Springs will be discussed near the end of this chapter.

Span Location Adjustment

Span location refers to the position setting of the pen mechanism. This adjustment is made by shifting the input and output levers relative to their shaft position. By shifting one lever or the other, the output shaft is made to assume its proper position. To make this adjustment, there must be a way to:

1.   Rotate the levers relative to their shafts.
2.   Shift the input relative to the input lever.

Some instruments may not have angularity, span, and span location adjustments. Some combination of the three is usually possible. Figure 2-11 shows a mechanism that has three adjustments. The linearity adjustment is connected to the measurement link,

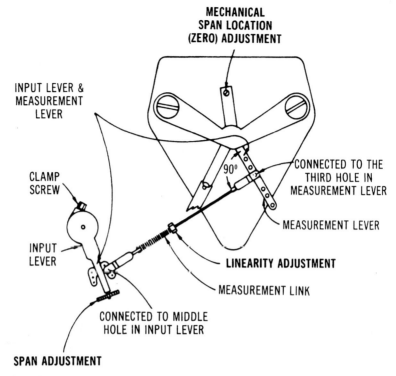

FIGURE 2-11   Span, zero and linearity adjustments in a link/lever system.

which is located near the center of the mechanism. Adjustments are made by changing the position of the nut/bolt assembly. The span adjustment is located in the lower left corner. Span adjustments are made by altering the position of a screw. Span location is adjusted by moving a mechanical arm located near the top center of the mechanism. These three adjustments all interact. Making one adjustment calls for additional adjustments of the other settings. When the mechanism has good linearity, there is no need for further adjustment. This occurs when the span and span location errors are minor.

MECHANICAL STOPS

In working with mechanisms, an extremely important consideration is the mechanical stops. Mechanical stops are any interference that makes it difficult for the mechanism to travel. For example, the pen should move from slightly below zero on the chart to slightly above 100 percent. Suppose that it can only move over 98 percent of the chart regardless of the input signal. This indicates that some mechanical part is restraining the pen. In other words, it is a mechanical stop.

Mechanical stops act at particular points in the low and high portions of mechanism movement. Investigation of the mechanism is first directed to the mechanical stops. This

FIGURE 2-12   Adjustable stops on a recorder. (*Courtesy of Bristol Babcock Inc.*)

should precede any mechanism adjustments. All too frequently, mechanism adjustments are made first to improve performance, when the first consideration in correcting a problem should instead be directed to the mechanical stops. Adjustments should be made only when stop problems have been corrected.

There are two types of mechanical stops: adjustable and fixed. Adjustable stops are arranged so that they can be easily moved. They make it possible to select points where the mechanism will stop. Examples of adjustable stops are those which limit the movement of a bellows or a Bourdon tube. These stops are mounted so that their position can be changed, which limits movement so that it corresponds to the measurement of interest. The adjustable stops of a recorder are shown in Figure 2-12.

Fixed stops are mechanical movement restraints that should not be changed. They may be due to interconnecting mechanisms. In a pen mechanism, travel is limited to an upper and lower movement extreme. The pen will strike these stops and not move beyond these positions. This type of stop usually serves as an overrange protection. It prevents damage to the mechanism. In most instruments, these stops are obvious and can be adjusted to some extent. In other instruments, they are neither adjustable nor obvious. In working with an instrument, one should recognize its mechanical stops.

Mechanical stops are used in a mechanism to limit its travel. For example, the drive mechanism of a recording pen needs to rotate 40° to make a mark on its chart. Most pen mechanisms are designed to rotate 44°. This permits the pen to have some overtravel. If the mechanism is out of adjustment, its overtravel could all be on one side. This could permit the pen to move excessively on the high side of the chart. It would probably not return to zero on the low side of the chart. In this case, mechanical stops would restrict the upper-side travel of the mechanism. This would prevent the pen from making a mark off of the chart. A stop on the lower side would also prevent the pen from moving off of the chart. Mechanical stops in this case prevent pen damage from excessive overtravel.

SPECIAL MECHANISMS

There are a number of special instrument mechanisms in addition to the link/lever. A major portion of these mechanisms fall into four categories:

1. Linearity adjustment mechanism.
2. Motion multiplying and dividing mechanism.
3. Antiarcing mechanism.
4. Nonlinear levers.

The first three categories are used to overcome the limitations of the link/lever mechanism. These mechanisms are refinements of the basic link/lever.

Linearity Adjustment Mechanism

In a link/lever mechanism, nonlinear operation arises because the output is a reproduction of the input. This occurs only when the actual lever is the same as the effective lever. As movement between each lever occurs, a succession of angles with the link are formed. In some mechanisms the link always forms a 90° angle with the lever. No linearity problems exist in these mechanisms, as shown in Figure 2-13. Suppose that a measurement signal causes the rack and gear segment to move. This would occur where the gear is fixed to the shaft. As the rack moves, so does the gear segment, which causes the shaft to rotate. The actual lever forms a 90° angle between the gear segment and the rack. For each rotation of the rack there is an equal incremental rotation of the shaft. A number of manufacturers use this design in their mechanisms. Operation is the same except that the drive is coupled to the gear segment with flat flexible straps.

Motion Multipliers

Suppose that instead of a gear segment and a rack, two gear sements are used, as shown in Figure 2-14A. Such an arrangement serves as an angularity compensating mechanism. Remember that rotation of the output depends on the ratio of input to output lever lengths. If the gears are of different diameters, then the ratio of input to output travel will depend on the gear ratio. This relationship can be stated as follows:

$$\frac{\text{angle of rotation of input}}{\text{angle of rotation of output}} = \frac{\text{diameter of output gear}}{\text{diameter of input gear}}$$

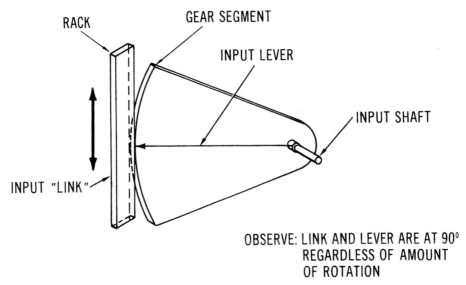

RACK

GEAR SEGMENT

INPUT LEVER

INPUT SHAFT

INPUT "LINK"

OBSERVE: LINK AND LEVER ARE AT 90⁰
REGARDLESS OF AMOUNT
OF ROTATION

FIGURE 2-13   Link/lever mechanism with no linearity (angularity) error.

A major advantage of the two-gear mechanism is its ability to attain wide variations of the input to output ratio. For example, with an input rotation of 10° it is possible to get an output rotation of 300°. Or, an input of 300° may cause an output of only 25°. An arrangement of any two gears may act as motion multipliers or dividers. The pressure gauge is an example of a motion multiplier. The pen-drive mechanism of a recorder is usually an example of a motion divider. These special mechanisms are shown in Figure 2-14.

Antiarcing Mechanisms

In large-case instruments, the pen arm is parallel to the chart surface. As the pen arm rotates, it "arcs." This means that the pen arm swings in a curve, with the axis of the pen arm serving as the radius of the arc. To compensate for the arcing, the time lines on the chart are also arced so that as the pen swings, it will be "on time." Such an arrangement is straightforward and presents no particular problem. Figure 2-15 shows an antiarcing mechanism and a representative chart.

Suppose that the motion of the pen arm is in a plane at a 90° angle to the chart surface. In this case, arcing of the pen would lift the pen off of the chart. It is necessary that a mechanism be developed in which the effect of arcing is eliminated. Figure 2-16 diagrams such a mechanism and is typical for small-case recorders. Other methods of dealing with this problem are used. One method obtains similar results by arcing the chart backup plate and the time lines on the chart.

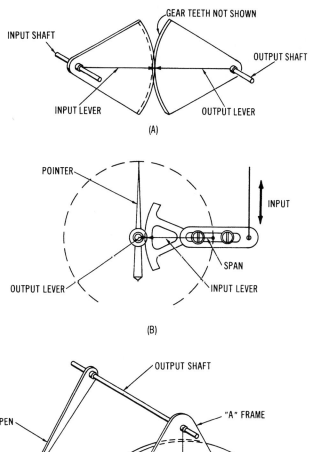

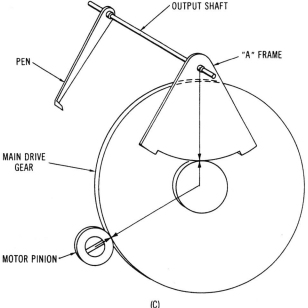

FIGURE 2-14   Special mechanisms. (A) Antiarcing mechanism; (B) Motion multiplier; (C) Motion divider.

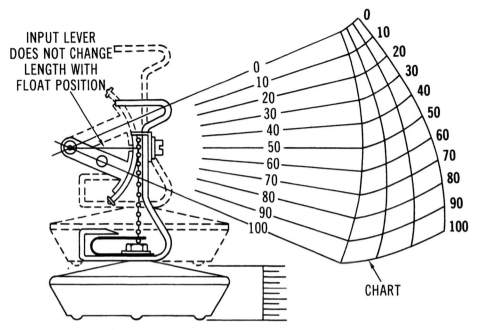

INPUT LEVER
DOES NOT CHANGE
LENGTH WITH
FLOAT POSITION

CHART

FIGURE 2-15   Antiarcing arrangement using curved chart.

Nonlinear Links

At times it is desirable to deliberately change the effect that the input has on the output, especially in flow measurement. When determining flow with an orifice, it is necessary to "take the square root" of the input signal. For example, if the input signal is 4, then the flow would be equal to the square-root of 4, which is 2. One way of doing this is by using a "nonlinear lever," such as the cam of Figure 2-17. This type of lever changes length as it rotates. Consequently, the ratio of input to output rotation changes. The shape of the cam also determines this ratio.

SPRINGS

A mechanism of great interest to industry is the common spring. Springs are used for many purposes. Examples are antibacklash mechanisms and force-balancing mechanisms in instruments, controllers, and control valves.

A notion that a spring becomes stiff when stretched or compressed is a misconception. This is not true so long as the spring is not overloaded. An essential property of a spring is its ability to deflect an equal amount for changes in the load. Suppose that an unloaded spring 2-in long stretches to 2 1/16-in when a 1-lb load is added. The spring has been stretched 1/16-in by the 1-lb load. Further suppose that the 1-lb load is replaced by a 12-lb load. This load stretches the spring 3/4-in from its original length. A load 12 times greater than the first load has produced a deflection 12 times greater than the first de-

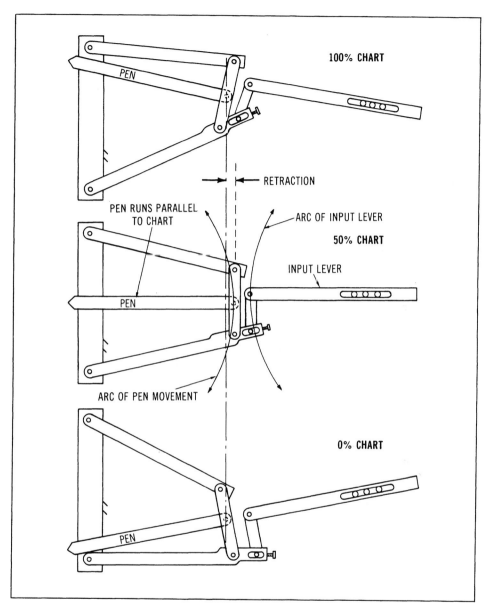

FIGURE 2-16   Antiarcing pen linkage.

flection. The relationship between load and deflection is called linear, meaning that equal changes in load produce equal changes in deflection.

The spring stiffness constant is fixed by its wire size, the material used, and the length. The heavier the wire, the stiffer the spring or the larger the spring constant. The elasticity of the spring material helps establish this constant. Spring steel is stiffer than

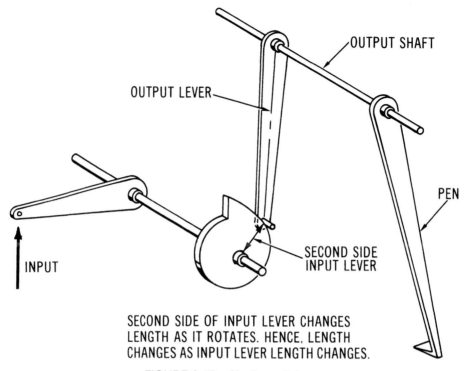

SECOND SIDE OF INPUT LEVER CHANGES
LENGTH AS IT ROTATES. HENCE, LENGTH
CHANGES AS INPUT LEVER LENGTH CHANGES.

FIGURE 2-17   Nonlinear linkage.

brass. When wire size and material are the same, a small diameter spring is stiffer than a large diameter spring.

Suppose two springs are similarly constructed except that one spring is 2-in long and the other is 3-in long. Suppose further, that a 12-lb load will stretch the 2-in spring 3/4-in. A 12-lb load will also stretch the 3-in spring 3/4-in plus 3/8-in. This means that a spring changes its deflection for a given load by changing its length. By selecting a specific spring for an application, the output can be changed for a given input. This is recognized as a change in span. Many instruments use a spring length adjustment to change the span.

SUMMARY

A lever is defined as a piece of material that rotates about a fulcrum, or pivot point. A link is explained as something that connects one or more levers together. These two basic mechanisms are used in almost all industrial recorders.

Transmitters, recorders, indictors, or controllers used in pneumatic instruments employ some type of lever in their operation. Basically, a lever is a simple machine that is used to multiply force, increase speed of movement, or cause a change of direction in the movement.

A first-class lever has the fulcrum near the center of its span with an effort force and resistance force at the ends. A second-class lever has the resistance force between the fulcrum and the effort force. A third-class lever has the effort force acting between the resistance force and the fulcrum.

Mechanical stops are extremely important considerations when working with instrument mechanisms. Proper stop position adjustment permits the mechanism to move through its travel range. Stops are either of the fixed or adjustable type. Adjustable stops are arranged so that they can be easily moved. Fixed stops are usually part of the mechanical structure and cannot be moved.

Springs are widely used in mechanisms today. A common misconception is that springs tend to become stiff when stretched. The elasticity of a spring depends upon the wire size, the material used, and the length of the spring. One use for the spring is to deflect an equal amount for an additional load. A ratio between the load applied to a spring and its deflection is normally equal.

ACTIVITIES

LEVER ANALYSIS

1. Refer to the link/lever mechanism of a pneumatic recording instrument.
2. Identify the function of the link/lever mechanism.
3. Describe the operation or movement of the link/lever mechanism.
4. Note where the fulcrum is located.
5. Does the lever mechanism have any linkage?
6. Can the lever mechanism be adjusted?

SPRING ANALYSIS

1. Refer to the internal structure of a pneumatic recording instrument.
2. Identify locations where springs are employed.
3. Note the function that the spring plays in the operation of the mechanism.
4. If more than one spring is involved, explain their relationship.

QUESTIONS

1. What is meant by the term lever?
2. What is a fulcrum?
3. Explain the meaning of resistive force and effort force?
4. What are the effort force, resistive force, and fulcrum relationships of a first-class lever?
5. What are the effort force, resistive force, and fulcrum relationships of a second-class lever?

6.   What are the effort force, resistive force, and fulcrum relationships of a third-class lever?

7.   Name the function of a link in the operation of an instrument?

8.   What are three adjustments of a link/lever mechanism?

9.   What are three types of spring structure?

3
Calibration

OBJECTIVES

Upon completion of this chapter, you will be able to:

- Define the term calibration.
- Explain the general procedure for calibrating an instrument.
- Identify several causes for instrument inaccuracy.
- Explain the role of a standard in instrument calibration.

KEY TERMS

Accuracy. A measurement that indicates the degree of exactness or correctness when compared to a standard value.

Calibration. An adjustment procedure for establishing the accuracy of a measuring instrument.

Calibration standard. A value determined by an authority that is used to establish an accurate measure of weight, quantity, or quality.

Dead space. A gap in indicator travel when the input is reversed or changed.

Friction. The resistance to sliding motion between two bodies in contact with each other.

Hysteresis. The physical property of a material that slows down an effect when the forces acting on a body are removed or changed.

Instrument. A device or mechanism used to determine the value of a quantity under observation.

Known inputs. A calibration standard value.

Sensitivity.   The minimum change in input value that an instrument can detect.

Standard.   An instrument or device having a recognized permanent value that is used as a reference or criterion in calibration.

INTRODUCTION

Recording instruments are mechanisms that indicate in some visible form the value of a process variable being measured. A primary concern of instrumentation is the accuracy of the measured values being obtained. To assure some degree of accuracy an instrument must be calibrated.

Calibration is defined as adjusting an instrument so that it can be used directly to determine the value of the measurement. A recorder is calibrated when its output is an accurate expression of the applied input value. Someone using an instrument should be able to determine the value of an applied input by looking at its output. The preciseness with which the output reproduces the input is called accuracy.

Some instruments do not require calibration. In particular, an uncalibrated instrument may be called on to perform a controlling function in addition to accepting and displaying a measurement. Such instruments are identical to recorders except that they do not display a measured value in visual form. The display is made in a way that the controller mechanism can understand what is being applied to the system. These instruments are generally considered to be blind. The accuracy of a blind instrument depends on the response of its internal components.

An important function of servicing deals with instrument calibration and accuracy determination. This chapter investigates these instrument functions.

CALIBRATION STANDARDS

Calibration makes it possible to determine the value of a measurement by examining the output of an instrument. For example, we can determine the temperature of a kettle of water by immersing a closed glass tube with a small bore partially filled with mercury. Heat from the water causes mercury to expand. By observing the height of a mercury column, we can tell water temperature if the glass tube is graduated. The mercury-in-a-glass device is recognized as a thermometer. The marks on a thermometer constitute the calibration that makes it possible to determine water temperature. The input is heat from water and the output is the length of a column of mercury.

In order to calibrate an instrument, the user needs to know a very precise value of the applied input. Knowing this, the user can then adjust the instrument so that it indicates the proper value.

Calibration standards make it possible to know precise input values. Some examples of these calibration standards are:

1.   The height of a column of water.
2.   The melting points of various materials.

3. A glass-stem thermometer calibrated by the Bureau of Standards.
4. A potentiometer using a standard cell.
5. The weight of disks of metal (the dead-weight tester).

Calibration standards make it possible to determine the value of the signal that is applied to the instrument. Such signals are called known inputs.

CALIBRATION PROCEDURES

To calibrate an instrument, inputs of known value are fed into the instrument. The response of the instrument to these known inputs is observed and the instrument is adjusted so that it will indicate the value of the input. Inputs are usually set at 10 percent, 50 percent, and 90 percent of the instrument range. The instrument indication is observed for these values.

Linearity Adjustment

In order to make a linearity adjustment, one must first look at of the levers and links when the input is at 50 percent. The levers should be parallel and the links should form 90° angles with the levers. If the levers and links are not formed like this, adjust the length of the link and/or the center of rotation of the input lever. The lever may need to be rotated relative to its pivot.

Span Adjustment

When making a span adjustment, one must cause the input to vary from 10 percent to 90 percent of the instrument range. During this time, the pen should travel 80 percent of the chart. If it travels than less 80 percent, increase the length of the input lever or decrease the length of output lever.

Span Location (Zero) Adjustment

When making a zero adjustment, first notice the position of the pen on the chart with a 50 percent input signal applied. The pen should fall on the 50 percent level of the chart. If it does not, shift a lever relative to its shaft until the pen position corresponds to the 50 percent input.

ACCURACY OF CALIBRATION

Accuracy defines how closely the pen (or indicator) agrees with the input value. Recorders and indicators can not be made any more accurate than the calibration standard. If the accuracy happens to be 102 percent, this would mean the calibration is off by 2 percent.

The instrument being calibrated is always a bit less accurate than the standard value. In general, it is unrealistic to try to attain better accuracy than that claimed by the manufacturer of an instrument.

CAUSES OF INACCURACY

The basic accuracy of an instrument is limited by the accuracy of the standard and the accuracy claimed by the manufacturer. Most instruments have an accuracy of 0.25 percent to 0.50 percent. Causes of inaccuracy are:

1.   Poor sensitivity.
2.   Dead space.
3.   Hysteresis.
4.   Friction.

Sensitivity is defined as the minimum change in input that the instrument can detect. The sensitivities of instruments expressed as a percentage of range are on the order of 0.01 percent to 0.1 percent. Little can be done to improve the sensitivity of link/lever instruments.

Dead space is defined as the gap in pen or indicator travel when the input is reversed. To detect dead space, slowly increase the input to some convenient value on the standard instrument and observe the pen position. Increase the input to a higher value, then slowly decrease the input to its original value and observe the reading. The difference is dead space. This dead space is due to necessary clearance in the instrument mechanism and should be so small as to be invisible to the eye.

Hysteresis is a physical property of materials. When a material is subjected to a force that deforms it, the material remains deformed when the force is taken away. When the force is reversed and then removed, the same material retains a slight deformation opposite in direction to the previous deformation.

An example of hysteresis might be a bellows acting against a spring. If air pressure is applied to the bellows, it expands and compresses the spring. The spring will not return exactly to its original position even though the air pressure is removed. Hysteresis is a major source of error in other mechanisms. In link/lever mechanisms, hysteresis errors can be neglected.

Friction is by far the most frequent cause of inaccuracies in link/lever instruments. Friction is the resistance to sliding motion between two bodies in contact with each other. Force is required to overcome this friction. Normally, the force available to move levers is fairly limited.

CALIBRATION OF A RECORDER

In the previous chapter, link/lever mechanisms were discussed and the adjustment of span, span location, and linearity were explained. The presence of mechanical stops were discussed. It was suggested that these stops be investigated first when working with an instrument.

Calibration was explained in this chapter. It was shown that calibration consists of making the instrument agree with a known standard. To accomplish this may require changing the span, span location, and linearity adjustments on the instrument.

The following paragraphs develop a calibration procedure that will apply to almost all link/lever mechanisms.

Equipment Required

To calibrate an instrument, the following equipment is required:

1. The calibration standard.
2. A means of feeding a signal to the calibration standard and the instrument being checked.
3. The materials (piping, wiring, etc.) to connect the calibration standard, the signal, and the instrument.
4. The tools needed to make the connections and to adjust the instrument.

One of the tests that help determine the causes of inaccuracy is the dead space test. Dead space is due to friction, slack in the linkage, clearance between holes and pivots, and sensitivity. Hysteresis can contribute to dead space, but in linkage mechanisms this rarely occurs. When encountering dead space, eliminate friction as a contributing factor by slowly and smoothly changing the input and observing the pen or indicator movement. If friction is present, movement will occur in small steps as the input signal is changed.

The slack in linkage can be felt. Slowly and carefully move the pen to see if it can be moved without changing the input element. If the element does move, the problem is slack in the linkage.

Preliminary Examination

1. Mechanical Stops: The mechanisms should be free to travel their range without hitting stops.
    a. Stroke the mechanism by changing the input signal.
    b. The pen should slightly above 100 percent.
    c. Adjust the pen micrometer, zero adjustment on levers slip levers on shaft, and shift input lever as required.
    d. Search for inadvertent stops which might be caused by misalignment or errors in the instrument assembly.
2. Accuracy: Is the mechanism (a) free of friction, (b) free of dead space, (c) free of hysteresis, and (d) sufficiently sensitive? Notice that this test is the test for dead space. If friction and slack in the linkage are eliminated as per (a) and (b) of the following, the dead space remaining is due to the sensitivity of the instrument.
    a. Examine for friction by slowing and steadily changing the input and observing pen reaction. The pen should move smoothly, or without hesitation. Examine for friction by feeling the resistance of the mechanism to change. Feel the clearances of the shafts and pivots. Feel the link connections. All connections should be free-running.
    b. To examine for slack in the linkage, with the pen at midscale, gently attempt to move the pen upscale and downscale without forcing the input. Feel for slack in the mechanism that would permit a slight pen movement without any

input movement. If slack is found, check for worn shafts, pivots, or lever/link connections and replace as required.

c. Hysteresis does not apply to link/lever instruments.

d. Examine for sensitivity. Slowly move the input upscale to a preselected point on the calibration standard. Observe the indicator. Then slowly move the input down scale to the same point on the calibration standard. The gap between the final indicator position should be 1/32-in. If it is greater than 1/32-in, recheck (a) and (b). There are no sensitivity adjustments on a link/lever instrument unless it is a "null-balance" instrument.

Calibration

Does the instrument reproduce the input signal as measured by the calibration standard?

a. Check linearity at midscale or at a critical point. Make levers parallel and cause the link to connect levers at 90°. Note: This adjustment is not critical and many times one must accept a compromise between levers which are nearly parallel and a 90° link/lever connection.

b. Check the span by feeding in a 10 percent signal, then a 90 percent signal. The pen or indicator should travel 80 percent of the chart. If it does not, adjust lever length.

c. Check zero by feeding in a 50 percent or critical point signal. Put the pen on 50 percent or some critical point by shifting the lever relative to the shaft or by shifting the input mechanism relative to the input lever.

d. Recheck (a) by putting a 10 percent, 50 percent, and 90 percent signal into the instrument. If the pen does not fall on all three points, and if the span and span location are correct, readjust the linearity.

e. Repeat (a) through (d) as required. Note: There is no requirement that the preceding sequence be followed as indicated. In general, adjusting linearity first, then span, and finally span location tends to give the best results.

Linearity

Check linearity by plotting the output obtained for inputs of 10 percent increments from 0 to 100 percent. The difference between the input values and the actual values obtained is the error. The allowable error varies, depending on the instrument. Allowable errors in general are less than 1 percent and greater than 1/4 of 1 percent. The linkage of an instrument can be calibrated so that no error exceeds ± 1/2 of 1 percent. That is, the output, or pen position, will be no farther away from the input than 1/2 of 1 percent.

The angle of the link relative to the lever determines the linearity of instrument linkage. To minimize linearity errors, the angles are adjusted. The specific type of adjustment needed to perform this operation depends on the type of linkage, the available adjustments, and how an error changes the uniformity.

A good understanding of the difference between actual and effective lever behavior is important in the adjustment of a mechanism. The effective lever should have maximum length at 90° while decreasing in length on either side of 90°. This indicates whether the angle should be increased or decreased.

SUMMARY

In practice, calibration standards are used to determine the accuracy of an instrument. As a practical matter, the instrument being calibrated is somewhat less accurate than its standard. Accuracy is generally expressed as a percentage of instrument range, or as a plus or minus value.

Calibration procedures first involve a preliminary examination of mechanical stops and mechanism movement. Second, the process involves adjustments for linearity and accuracy.

ACTIVITIES

INSTRUMENT EVALUATION

In this activity, a pneumatic instrument that will measure temperature will be evaluated to determine if it needs to be calibrated. The accuracy of this procedure depends largely on the accuracy of the input standard used. Pressure, flow, or liquid level can be used instead of temperature.

1. Prepare a pneumatic recording instrument for operation.
2. Place the temperature sensor or input probe of the instrument in a small container of water.
3. Start the recorder and note the temperature of the water sample.
4. Place a mercury-in-glass thermometer in the container of water.
5. After a few minutes measure and record the temperature of the water as indicated by the thermometer. This temperature value represents the calibration standard.
6. Compare the standard temperature value with the temperature value indicated by the recorder. If the standard and instrument values are similar, then the instrument does not need to be calibrated.
7. How does the accuracy of the standard compare with the recording instrument?
8. Place some crushed ice in the container. Repeat the evaluation procedure.
9. What are the limitations of this evaluation process?

INSTRUMENT CALIBRATION

In this activity we will look at the general calibration procedure of a link/lever pneumatic recording instrument. The calibration standard can be temperature, level, flow, or pressure. If the calibration documentation of a specific instrument is available, follow it for this procedure.

1. Identify the instrument according to its case size. A large case instrument is approximately one square foot in size.
2. Indicate the standard being used in this procedure.

3. Open the instrument door or remove the instrument from its housing.
4. Locate the link/lever mechanism.
5. Refer to the preliminary examination of an instrument discussed in this chapter. Perform a preliminary calibration examination of the recorder.
6. What is the status of the mechanical stops and accuracy of the recorder?
7. Follow the CALIBRATION OF A RECORDER procedure outlined in this chapter.
8. If the manufacturer's operational manual is available, follow the calibration procedure outlined in it.
9. Describe the calibration procedure performed on the instrument.

QUESTIONS

1. What is accomplished in the calibration of an instrument?
2. What is meant by the term input standard or calibration standard?
3. What is a mechanical stop?
4. What determines the dead space of an instrument?
5. How is the linearity of an instrument evaluated?
6. Explain how the linearity of a recorder is adjusted.
7. How are span and span location adjustments made?

4
Pressure

OBJECTIVES

Upon completion of this chapter, you will be able to:

- Define a number of commonly used terms such as atmospheric, absolute, gauge, and static pressure.
- Explain the meaning of head pressure.
- Describe the meaning of differential-pressure.
- Identify some of the variables expressed as head pressure.
- Use the vocabulary associated with pressure terms.

KEY TERMS

Absolute pressure. The pressure of a liquid or gas measured in relation to a vacuum or zero pressure.

Atmospheric pressure. The pressure exerted on a body by the air, equal at sea level to approximately 14.7-pounds per square inch.

Coherence. To hold together firmly parts of the same mass.

Density. Closeness of texture or consistency of particles within a given substance. The weight per unit volume.

Differential pressure. The difference in pressure between any two points or values of a system or component.

Force. A push or pull measured in units of weight such as pounds or Newtons.

Friction head. The loss of pressure or drop that occurs when a force drives fluid through a pipeline, also known as pressure drop.

Gas. A substance such as air that has mass but no definite shape and tends to expand indefinitely.

Gauge pressure.   The pressure of a liquid or gas that ignores atmospheric pressure. Its zero point is 14.7-psi absolute.

Head.   The height of a column or body of fluid above a given point expressed in linear units. Head is often used to indicate gauge pressure. Pressure that is equal to the height times the density of the fluid.

Newton (N).   The force exerted on an object that has a mass of 1-kilogram and a gravitational acceleration of 1-meter per square second.

Pascal (Pa).   The basic unit of pressure in the metric system of measurement. 1.0-Pa = 1.0-Newton per meter squared, or N/m^2.

Pitot tube.   A fluid velocity sensing probe with one end facing the fluid flow and the other end connected to a pressure gauge.

Pressure.   Weight or force distributed over an area. The driving force or cause of hydraulic or pneumatic flow in a system.

Pressure drop.   The difference in pressure between any two points of a pipe or device.

PSI.   Per square inch.

SI Unit System (Le Systeme International d'Unites).   The metric system of measurement adopted by most industries throughout the world.

Specific gravity.   A ratio between the density of liquid to the density of water, or the density of gas to the density of air.

Static pressure.   The pressure in a fluid at rest.

Vacuum.   Pressure less than atmospheric pressure that is expressed in absolute or gauge pressure.

Velocity head.   The speed of flow through a pipe or tube expressed as pressure.

Work.   The resulting energy used when a force is exerted through a definite distance expressed in foot-pounds.

INTRODUCTION

A major portion of all industrial measurement relates in some way to pressure. Flow, for example, is often measured by determining the pressure that exists at two different points in a system. In some instruments pressure changes are used to produce the mechanical motion of a recording stylus. The level of a container can be determined by a change in pressure at the bottom of the container. Pressure can also be used to measure temperature in a filled system through changes produced by the expansion or compression of liquid or gas in a closed tube sensing element. For these reasons, it is important to understand the meaning of pressure and its variations.

FORCE, PRESSURE, and WORK ■■■■■■■■

The force applied to an object of a system is defined as any cause which tends to produce or modify motion. To move a body or mass, an outside force must be applied to it. The amount of force needed to produce motion is primarily based on the inertia of the body. Force is normally expressed in units of weight. Weight is defined as the gravitational force exerted on a body (or mass) by the earth. Since the weight of a body is a force (not a mass), we must use units of force to express both weight and force. The basic unit of force in the English system is the pound (lb). In the Le Systeme International d'Unites (abbreviated SI), the basic unit of force is the Newton (N). A Newton represents the force exerted on an object having a mass of 1.0-kilogram, where the acceleration is 1.0-m/sec^2.

Pressure is a term used to describe the amount of force applied to a specific unit area. In the English system, pressure is expressed in pounds per square inch and is abbreviated psi or lb/in^2. An SI indication of pressure is expressed in Newtons per square meter (N/m^2). The name Pascal (Pa) has been assigned as the basic unit of pressure in the SI metric system. One Pascal is equal to one Newton per meter squared, or 1.0-Pa $= 1.0$-N/m^2. This unit is relatively small, which often leads to expressions such as 1,000-Pa (1-kPa) or 1,000,000-Pa (1-MPa) for larger values.

At sea level the pressure of the atmosphere on the surface of the earth is 14.7-lb/in^2. In industrial applications giant hydraulic presses are capable of squeezing metals with a pressure of 100,000,000-psi. Mathematically, pressure is expressed as:

$$P = F/A$$

where

P is the pressure in pounds per square inch or Pascals,

F is the force in pounds or Newtons,

A is the area in square inches or square meters.

Assume now that a force of 20-lbs is applied to a piston with an area of 2-square inches. The resulting pressure developed is:

$$P \ (\text{lb/in}^2) = F \ (\text{lb})/ \ A \ (\text{in}^2)$$
$$P = 20 \ \text{lb}/2\text{-in}^2$$
$$P = 10 \ \text{lb/in}^2$$

An important consideration about force and pressure is that they only represent a measure of effort. A measure of what is accomplished is called work. When an applied force causes an object to move a certain distance, work is accomplished. Work is commonly expressed in foot-pounds or Newton-meters. The mathematical formula for this relationship is:

$$W = F \times D$$

where

W is work in foot-pounds in the English system or Newton-meters in the SI system,

F is force in pounds or Newtons,

D is distance in feet or meters.

An example has a 10-pound weight lifted 10-feet. The amount of work done by this operation is:

$$W \text{ (ft-lbs)} = F\text{-(lbs)} \times D\text{-(ft)}$$
$$W = 10\text{-lbs} \times 10\text{-ft}$$
$$W = 100 \text{ ft-lbs}$$

GAS PRESSURE

A gas may be defined as a material in which there is no coherence or attraction between its individual particles or molecules. Any material in a gaseous form tends to fill the space into which it is placed. The application of heat to a gas "agitates" the individual molecules. As a result, the molecules move faster. The molecules bombard the internal surface of the enclosure. This bombardment constitutes gas pressure since the effect of all individual molecules is a force against the wall of the enclosure.

Pressure/Temperature Law

It has been said that heat causes the gas particles to move faster. This in turn, causes them to hit the container surfaces more frequently and to exert more pressure. The relationship between pressure and temperature has been expressed mathematically as:

$$P1/T1 = P2/T2$$

This expression means that the original pressure (P1) divided by the original temperature (T1) will equal the new pressure (P2) divided by the new temperature (T2). This relationship is true, provided the volume of the gas is kept constant.

A conclusion can be drawn showing that the pressure exerted by a fixed volume of gas depends on the temperature of the gas. This principle is widely used in a class of temperature-measuring instruments called fill systems. Gas-filled instruments are discussed in chapter 5 under Types of Filled Systems.

LIQUID PRESSURE

A liquid is a state of matter wherein there is a limited attraction between the molecules of which it is composed. The attraction is such that if a liquid is poured into a container, it will fill the container to a uniform level. Remember that a gas will completely fill the container, and a solid will retain its shape regardless of the container.

Liquids are fairly dense materials. The effect of gravity on liquids is quite extensive. A 2-ft^3 volume of water weighs 124.8-lbs. The density of water is determined by the weight of its mass divided by its size or volume. In this example, the density will equal 124.8-lbs divided by 2, or 62.4-lb/ft^3. The density of water is therefore 62.4-lb/ft^3. The density of mercury is 846-lb/ft^3, and the density of kerosene is 50.0-lb/ft^3.

If 1-ft^3 of water has the form of a cube, then the area of each side of the cube is 1-ft^2, or 144-in^2. In order to find the pressure exerted by this liquid, we must divide its weight (62.4-lb) by 144-in^2. This last figure represents the area of the bottom surface. The pressure is equal to 0.433-psi, which means that a column of water 1-ft high exerts a pressure of 0.433-psi.

It is common practice to express pressure in terms of water or mercury. For example, in a particular tank there may be a pressure of 10-ft of water. In other words, the tank has a head of 10-ft, which is shorthand for saying that the pressure in the tank is equal to the pressure exerted by a column of water 10-ft high.

Differential-pressure is a term that refers to the difference between two related pressures. Suppose that the pressure in a closed vessel partially full of water is to be measured. An example is shown in Figure 4-1A. The pressure at point 1 on the side of the tank depends on the height of the water above that point plus the air pressure on the top. Suppose the pressure was measured again at another location or at point 2. This pressure would be equal to the height of the water above this point plus the air pressure on the top of the water. The differential pressure would equal the head pressure at point 1 plus the air pressure minus the sum of the head and air pressures at point 2. Or,

$$\text{differential-pressure (PD)} = (H_1 + P_{air}) - (H_2 + P_{air})$$
$$= H_1 + P_{air} - H_2 - P_{air}$$
$$= H_1 - H_2$$

ATMOSPHERIC PRESSURE

We all are constantly subjected to the pressure of the earth's atmosphere. In fact, the earth is surrounded by atmosphere and at any point this atmosphere exerts a pressure on the earth's surface due to the weight of a column of air above a particular point. Sea level is used as a common reference point, where the air exerts a pressure of 14.7-psi. This atmospheric pressure will also cause a column of mercury to rise 29.92-in, or 760-mm. At 1000-ft above sea level the atmosphere is approximately 14.2-psi. This means that air pressure varies approximately 1/2-lb for each 1000-ft of altitude. In addition, air pressure varies according to other atmospheric conditions. Weather forecasters use pressure changes in the atmosphere to predict weather conditions. A barometer is an instrument used to measure atmospheric pressure.

ABSOLUTE PRESSURE

We all are constantly subjected to the pressure of the earth's atmosphere. It is often thought to be nonexistent because we are not aware of it's effect. If it were nonexistent,

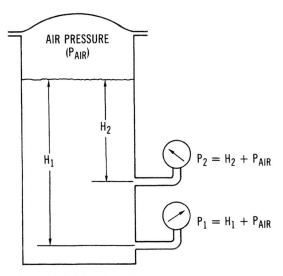

DIFFERENTIAL PRESSURE $= P_1 - P_2$

$$= H_1 + P_{AIR} - (H_2 + P_{AIR}) = H_1 - H_2)$$

(A)

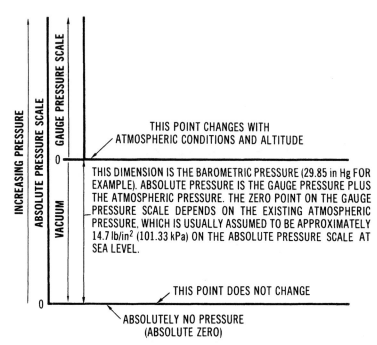

ABSOLUTE PRESSURES ARE MEASURED USING ABSOLUTELY NO PRESSURE AS A
STARTING POINT. GAUGE PRESSURES ARE MEASURED USING ATMOSPHERIC
PRESSURE AS A STARTING POINT AND CALLING THIS PRESSURE ZERO. THE GAUGE
ZERO SHIFTS WITH ATMOSPHERIC CONDITIONS AND ALTITUDE, BUT ABSOLUTE
(B)   ZERO DOES NOT. VACUUMS ARE GAUGED PRESSURES LESS THAN GAUGE ZERO.

FIGURE 4-1   Pressure and its variations. (A) Differential pressure; (B) Absolute and
gauge pressures.

the earth's surface would measure zero pressure. There is however, 14.7-psi on the surface of the earth at sea level. If we were to remove all of the air from a closed chamber, the inside pressure would be absolutely zero. We must be able to distinguish between these two "zero" pressure values. Pressure using the earth's atmosphere as a reference is called gauge pressure, or psig. The pressure in a chamber with the air removed is called absolute pressure. Absolute pressure is normally expressed in terms of pounds per square inch absolute, or psia. Absolute pressure is equal to gauge pressure plus atmospheric pressure. To find absolute pressure, add 14.7 to a pressure gauge reading.

A vacuum is gauge pressure that is less than atmospheric pressure. Figure 4-1B shows a comparison of different variations of pressure. It is common to refer to gauge pressures that are less than atmospheric in terms of inches or millimeters of mercury (Hg). For example, one might hear that a vacuum is 10-in Hg. This would mean that the pressure being measured has been reduced below atmospheric pressure by an amount equal to the pressure exerted by a column of mercury 10-in high. Since atmospheric pressure is 29.92-in of mercury, a vacuum of 10-in of mercury would be equivalent to a pressure of 19.92-in of mercury.

HEAD PRESSURE

The term head is commonly used in instrumentation to describe pressure in a number of different ways. For example, the term is used to describe a vacuum, friction, elevation, differential pressure, and velocity. This term also relates to liquid-level and flow measurement.

It has been shown that fluid exerts a pressure which depends upon:

1. The height of the fluid above a given point.
2. The density of the material.

To determine the pressure of a column of fluid, the height of the column is multiplied by the density of the fluid. An example is shown in Fig. 4-2. This shows 12 × 12, or 144, 1-in^2 columns of water at a height of 12-in. Each column will weigh:

$$62.4/144 = 0.433 \text{ lb.}$$

A column of water 12-in high will exert a pressure of 0.433 lb/in^2. A column of water 27.7-in high weighs 27.7/12, or 2.31 times as much as the 12-in column. Since 2.31 × 0.433 = 1, a column of water 27.7-in high will exert a pressure of 1-psi.

Suppose that someone needs to know what pressure a 30-in column of fluid exerts at the bottom of a tank. We would first need to know how much heavier the fluid is than water. One common fluid used to measure pressure is 2.97 times heavier than water. Therefore,

$$27.7/2.97 = 9.33\text{-in.}$$

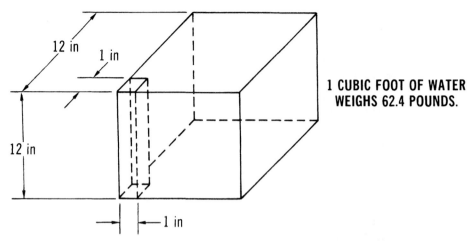

FIGURE 4-2   Determining the head pressure of a column of fluid.

This is the height of the fluid that will exert a pressure of 1-psi. Then,

$$30/9.33 = 3.2\text{-psi.}$$

The pressure exerted by a column of fluid of any density can then be determined by this method. It takes into account the known weight of the fluid in relation to an equivalent amount of water. This relationship is called the specific gravity of the fluid.

PRESSURE EXPRESSED AS HEAD

It has been shown that head of a fluid can be expressed as pressure, yet the reverse procedure is often desirable or necessary. For example, there could be a need to express the pressure exerted by a pump as a head. An example would be pressure expressed as 30-ft of water. If this is done, terms such as "pump head," "head pressure," or the "supply head" may be used. In this case, the pressure is converted to an equivalent column of water. Suppose the pump pressure is 30-psi, then

$$30\text{-psi} \times 27.7\text{-in/lb} /12\text{-in/ft} = 69.25\text{-ft.}$$

OTHER VARIABLES EXPRESSED AS HEAD

Some of the more common variables expressed as head are friction, velocity, and static pressure.

Static pressure is defined as the pressure of an inactive gas at rest. For example, the air or gas pressure on the surface of liquid in a tank is referred to as a static head. The pressure on a pump that is not operating is also referred to as a static head. A static head is pressure converted to equivalent inches of water.

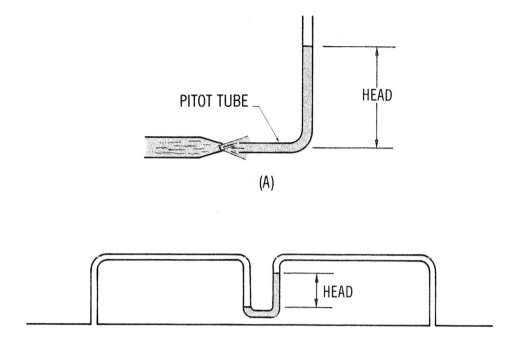

FIGURE 4-3   Velocity and friction heads. (A) Velocity head; (B) Friction.

Friction head describes the loss of pressure of a fluid flowing in a pipeline. For example, air may be flowing in a 1/4-in tube. At the inlet, the pressure might be 10-psi. At the outlet, the pressure might be 9-psi. In this case, there is a 1-psi drop in pressure, which is a pressure drop or loss due to the friction of fluid passing through the tube. A pressure drop expressed as inches of water is called the friction head.

A velocity head involves some moving material. The energy of this moving material is called kinetic energy. An example of this energy would be water forced through a fire hose. The water is forced through the hose at a fairly high pressure and leaves the nozzle at a fairly high velocity. This jet of water is under no pressure when it leaves the hose but has a force that can push a person down. Assume that this same stream of water is directed into a pipeline which has been bent to point upward. The water, in this case, would simply run up the pipe to a certain level and stop. This height (or head) is a measure of the velocity of the stream of water, called the velocity head. A device to measure velocity head is the Pitot tube. The concepts of friction and velocity heads are illustrated in Fig. 4-3.

The head concept, a mathematical technique that equates the physical properties of pressure, is accomplished by converting certain physical properties into equivalent inches of water. The advantage of this technique is that it enables a person to add, subtract, and otherwise manipulate different physical properties. This concept is important because many different instruments are used to determine head. The head concept also makes it easier to discuss and understand how and why these instruments are used.

SUMMARY

Pressure is the amount of force applied to a specific unit area. In instrumentation, a head is used to express pressure in terms of a column of mercury or water. Typical head pressure of a tank may be expressed as a height in feet or meters.

Differential pressure is an expression of the pressure between two different pressure values.

Atmospheric pressure is the standard weight that the atmosphere exerts on the surface of the earth. Common designations of atmospheric pressure are 14.7-psi or 29.2-in of mercury. Pressures that relate to the earth's atmosphere are called gauge pressures. Those values using zero pressure as a reference are called absolute pressure values.

Head is a general term that makes it possible to compare and work with different forms of energy. The term is used to describe pressure, vacuum friction, elevation, differential pressure, and velocity.

Head pressure depends on the height of fluid above a given point and the density of the material.

SUGGESTED ACTIVITIES

ATMOSPHERIC PRESSURE ▬▬▬▬▬▬▬▬▬▬▬▬▬▬▬▬▬

1. By means of a vacuum pump, remove the air from a flexible container.
2. When a considerable amount of air has been removed from the container, the atmospheric pressure will cause the container to collapse.
3. If the container is made of a flexible material, releasing the vacuum will cause the container to return to its original shape.
4. Try this procedure on another container made of non-flexible material.
5. Explain why the container collapses when pressure is removed internally.

BAROMETRIC PRESSURE ▬▬▬▬▬▬▬▬▬▬▬▬▬▬▬▬▬

1. Place a barometer in an upright position on a table or desk.
2. Observe the indicating scale of the instrument.
3. Record the value indicated by the instrument.
4. Barometers generally measure atmospheric pressure in inches of mercury. What does your instrument display?
5. What is a normal indication of atmospheric pressure in your area?
6. An increase in atmospheric pressure is an indication of what kind of weather conditions.
7. What does a drop in normal atmospheric pressure indicate?

VELOCITY HEAD

1. Connect a 10-ft or longer length of 1/4-in hose or tubing to the outlet of an air pressure regulator.
2. A pressure gauge should be attached to the outlet where the hose is connected.
3. The open end of the hose should be plugged to prevent air from escaping.
4. Turn on the air source and adjust the regulator to a pressure to 5-psi.
5. Remove the plug from the end of the hose or tube.
6. Measure and record the value indicated by the pressure gauge.
7. Turn off the pressure.
8. Replace the long length of hose or tubing with a length of 5-ft.
9. Repeat the test procedure.
10. Does the length of tubing have any effect on the pressure values observed?
11. What does this indicate?

QUESTIONS

1. How is pressure expressed?
2. What is meant by the term differential-pressure?
3. What effect does atmospheric pressure have on an object?
4. What are the similarities between atmospheric and absolute pressures?
5. Define the terms force, work and pressure.
6. If a force of 600-lbs is applied evenly over an area of 15-sq in, what is the pressure?
7. If a force of 10,000-Newtons is applied over an area of two square meters, what is the pressure in kPa?
8. If a force of 40-lbs is applied to an object and causes it to move 8-ft, how much work is being accomplished?
9. What is gauge pressure?
10. What is head pressure?
11. Determine the force on the surface of a piston with an area of 10-in^2 and under a pressure of 25-psi.
12. Determine the force on the head of a piston 645-mm^2 in area and under a pressure of 0.172-MPa.

5
Pressure Sensing Elements

OBJECTIVES

Upon completion of this chapter, you will be able to:

- Explain the operation of a Bourdon element.
- Describe the construction of a Bourdon tube.
- Identify different types of Bourdon elements.
- Describe how to measure temperature with a Bourdon element.
- Identify the construction of pressure-sensitive devices such as diaphragms, bellows, and manometers.
- Define and describe various types of filled systems.
- Explain how pressure is measured or sensed through the use of diaphragms, bellows, and manometers.

KEY TERMS

Ambient temperature.   The temperature surrounding any particular object.

Bimetallic element.   An element that is composed of two or more metal alloys and bends when heat is applied.

Bellows.   A pressure sensing element consisting of a convoluted metal cylinder closed at one end. The pressure difference between the outside and inside of the bellows causes it to expand or contract along its axis.

Bourdon force.   The force that is exerted on liquid inside a Bourdon tube by head pressure.

Bourdon tube. A pressure sensing element composed of a curved tube and having a flattened elliptical cross-section which is closed at one end.

Bulb volume. The amount of fluid that is contained in the bulb of a filled system.

Capillary. A small diameter tube having a very small bore or hollow center.

Compensation. A provision that equalizes or balances a condition by adjusting two or more things that are in opposition so that one balances the other.

Corrugated. To form or shape into wrinkles, folds, ridges, or grooves.

Diaphragm. A thin, flexible partitioning used to transmit pressure from one substance to another while keeping them from direct contact.

Feedback. Part of a closed loop system which provides information about a given condition for comparison with a desired condition.

Filled system. A closed tube network with a reservoir filled with gas or liquid. When the reservoir bulb is subjected to a temperature change it causes the pressure of the gas or liquid to change, thus causing the shape of the network to be altered.

Flange. A rib or rim of material used for strength, guiding, or for attachment to another object.

Flapper-nozzle. A combination of pneumatic elements used to control or regulate the flow of air or gas.

Gauge or gage. A device or instrument containing the primary elements for measuring a value.

Limp diaphragm. A diaphragm element composed of a material that has no definite shape and lacks firmness.

Manometer. An instrument for measuring the pressure of gases and vapors. It employs a hollow tube that balances liquid and compares the difference in a positional change.

Meniscus. The curved upper surface of a liquid column that is concave when the containing walls are wetted by the liquid and convex when not wetted.

Overrange. The amount by which a changing process variable exceeds the desired operational limit of an instrument.

Pinion gear. A gear with a small number of teeth designed to mesh with a larger wheel or rack.

Pulsation. A quantity that varies in value by rising and falling at a rate.

Siphon. A tube bent to form two legs of unequal length by which liquid can be transferred to a lower level by the force of atmospheric pressure.

Slurry. A watery mixture of insoluble matter such as mud, tomato pulp, or raw sewage.

Transducer.   A device that changes some physical quantity into a different physical quantity.

Vapor.   A substance in a gaseous state that is distinguished from the liquid or solid state of matter.

Vaporize.   To convert matter from a liquid state to a gaseous state.

INTRODUCTION

Pressure-sensitive devices play an important role in the overall evaluation of pneumatic and hydraulic systems. These devices may be a part of a gauge, controller, transmitter, or a recording instrument. A wide range of pressures must be measured today. A vacuum or negative pressure as low as 2×10^{-5} lb/in^2 (1.379×10^{-1} Pa) up to positive pressures as high as 1×10^6 lb/in^2 (6.895 GPa) must be measured in industrial equipment. This wide range of measurement requires a variety of pressure-sensing devices.

In this chapter, we shall discuss several of the common pressure-sensing devices that are used to measure, detect, or record system pressure. These devices essentially change shape or alter the position of a component when subjected to some form of pressure variance. Six devices will be discussed. Each device has a specific role in pressure measurement. It is therefore important for a technician to understand the role of a particular device. A manometer is the simplest device to use. The Bourdon element is by far the most common pressure-sensitive device in use today. The others have specific applications and reasons for their use. A listing of pressure-sensing devices includes:

1.   Limp diaphragm elements
2.   Bellows
3.   Diaphragms
4.   Manometers
5.   Bell displacer manometers
6.   Bourdon elements

The pressure-sensing element utilized for a specific measuring application is usually determined by the pressure range or differential pressure being measured. Pressure ranges, and the type of device recommended for each range, are shown in Table 5-1.

For differential pressure measurements, the diaphragm and the manometer are widely used. The diaphragm requires a pneumatic feedback circuit. The manometer can be arranged to record mechanically, or it can be designed to transmit a signal.

Limp Diaphragm Elements

As the name implies, the primary component of the limp diaphragm device is the diaphragm element. A limp diaphragm is constructed of a nonresilient material, such as a coated fabric, and is attached by linkage to an indicator. An example of a limp diaphragm is shown in Figure 5-1.

TABLE 5-1.   Pressure Measuring Devices

Pressure	Span	Device
Low	5 lb/in² (34.5 kPa) or less	Limp diaphragm. Manometer. Bell manometer.
Medium	5-25 lb/in² (34.5–172.4kPa)	Bellows. Manometer. Diaphragm in conjunction with pneumatic circuit. Bourdon element, if extreme sensitivity, is not needed.
High	25 lb/in² (172.4 kPa) and up	Bourdon element.

Opposite the diaphragm is a flat flexure spring. The diaphragm is sealed within two compartments. Pressure applied to an appropriate compartment makes it possible to indicate pressures as low as a few hundredths of an inch, or a few millimeters of mercury.

A limp diaphragm element, does not ordinarily have sufficient power to drive a pen mechanism. Because of this, limp diaphragms are generally used as indicators or transmitters. If the diaphragm is made large enough and if the sensitivity is not too great, this element can then be used to drive a pen.

Bell Displacer Manometers

The bell manometer of Figure 5-2 employs an inverted bell-shaped unit that floats in a pool of mercury. The open end of the bell sinks in the mercury, forming two compartments. One compartment is inside of the bell. High pressure is applied to this chamber through a standpipe. The standpipe extends above the pool of mercury. The other compartment occupies space outside of the bell and is inside of the meter body. Low pressure is applied to this chamber at the top of the unit. Mercury forms a seal between the two compartments. A differential pressure measurement has high pressure inside the bell and low pressure applied to the outside compartment. The pen input shaft is attached to the top of the bell. The input shaft runs through a pressure-tight bearing. The input shaft moves when pressure lifts the bell.

Differential pressure is a measure of the value difference between inside and outside chamber pressures. The bell displacer can be used to measure differential pressure. Pressure inside and outside of the bell balances to form a difference in the two pressures.

Bellows Pressure Elements

The main component of the bellows pressure device is the bellows itself. The bellows, shown in Figure 5-3A, is a cylindrical, thin-walled tube. This tube is "corrugated" so that it compresses or extends rapidly.

The simplest arrangement is to have the bellows closed at one end and admit pressure at the other end. The bellows expansion is opposed by a spring. As the pressure inside

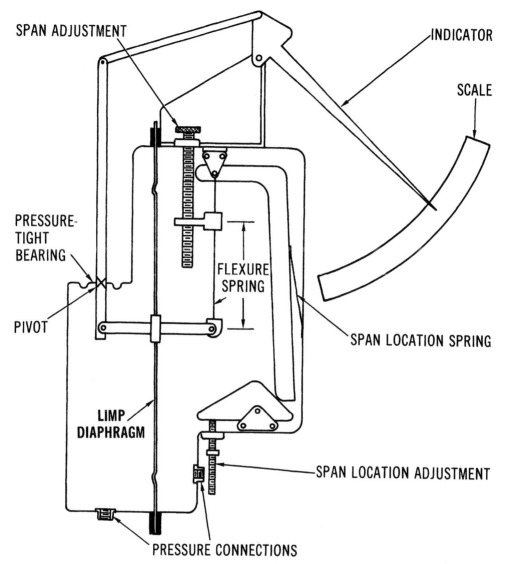

FIGURE 5-1   Limp-diaphragm pressure indicator.

increases, the bellows extends, forcing the spring outward. Expansion stops when the force of the bellows is matched by the spring force. The force of the bellows is equal to the pressure times the effective end area of the bellows.

A second arrangement has the bellows and spring in a "can," as shown in Figure 5-3B. In this device, the measured pressure is applied between the outside of the bellows and the inside of the can. The pressure tends to compress the bellows and extend the spring. The system is balanced when the spring equals the bellows force.

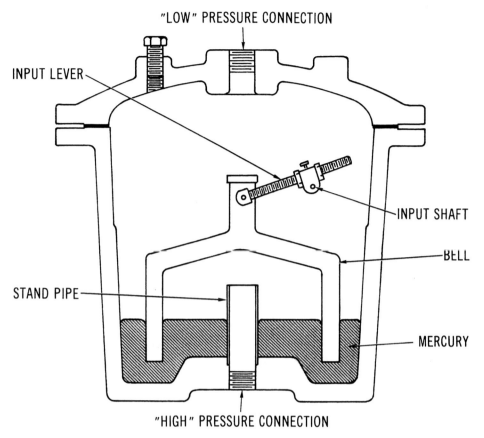

FIGURE 5-2   Bell dispacer manometer.

Construction of the bellows has the output taken from its closed end. The resulting movement and its linkage drives a pen, indicator, or controller. Remember that movement of the bellows is determined by a spring. For each spring, bellows deflection is fixed for a given pressure. As a result, the linkage is designed with a number of adjustments. The adjustments match the bellows to the pen travel. These adjustments are angularity, multiplication, and zeroing.

Two bellows can be installed in one can, as in Figure 5-3C. Pressure applied to this unit causes one bellows to apply force to the other, making it possible to measure differential pressure and absolute pressure. To measure absolute pressure, air from one of the two bellows is expelled.

The output of a bellows must be taken from its closed end. A problem occurs when two bellows units are used in the same unit. How is linkage attached to the closed ends of the two units. The solution is the use of a flexible pressure-tight seal bellows. It is important in this case to recognize the function of each bellows. The pressure sensing bellows must have the output attached to it. The seal bellows serves only as a connecting link.

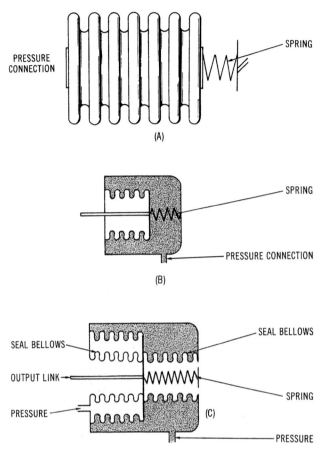

FIGURE 5-3   Bellows pressure devices. (A) Simple bellows; (B) Bellows in can; (C) Bellows in can with seal bellows.

Diaphragm Pressure Device

A flat metal diaphragm clamped between two flanges is an important pressure measuring device. Pressure is applied directly to the diaphragm, which causes the diaphragm to bow out. The diaphragm pressure device responds as the bellows, but with one important difference. The diaphragm will only move a few ten-thousands of an inch. As a result, this limited movement cannot directly drive a mechanism. Additional parts are needed to make the device functional. Diaphragm movement is connected through a motion bar to a baffle-nozzle unit. This assembly controls the amount of air pressure applied to a bellows. The motion bar is forced back by pressure against the diaphragm. The bellows force then balances its pressure against the diaphragm force. The baffle-nozzle unit will be discussed in more detail in chapter 10.

   An important use of the diaphragm device is in differential pressure measurement. This application has pressure applied to both sides of the diaphragm. The resulting net

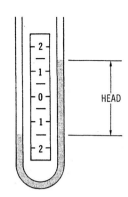

FIGURE 5-4   U-tube manometer.

force is the difference in pressure times the area of the diaphragm. Differential pressure measurement is probably the most important application of this device today

Manometer

The manometer may be described as a liquid balance. It is similar in operation to a mechanical laboratory balance. A laboratory balance is used to weigh an unknown mass by putting the unknown mass on a scale pan and balancing the unknown mass by placing known weights on another pan.

An unknown pressure is applied to a manometer. This pressure forces a change in the level of liquid or its weight. When the column weight equals the force exerted by the applied pressure, the manometer is balanced. The height of the column of fluid multiplied by its density is equal to the pressure at the bottom of the column. This pressure equals the pressure being measured.

There are several varieties of the manometer. Perhaps the simplest is the glass U-shaped manometer shown in Figure 5-4. Pressure is connected to one side and pressure upsets the liquid level until the difference between the two sides balances the applied pressure. An indication of zero on a scale identifies the balance point of the applied pressure values.

A handy roll-up version of the U-tube manometer is shown in Figure 5-5. This instrument is called a slack-tube manometer by its manufacturer. It is used by simply unrolling the plastic tubes and placing it so that the results can be viewed. The device is equipped with magnetic clamps so that it can be clamped to a steel surface. It provides accuracy that is equivalent to a laboratory U-tube device. This instrument can be easily rolled up for storage. Manometers of this type are available in operating ranges from 4-0-4 to 60-0-60 inches of mercury or the metric equivalent.

A modified version of the U-tube manometer, called a well manometer, is shown in Figure 5-6. When pressure is applied to the well, liquid is forced into the column. The liquid continues to rise until the liquid head balances the applied pressure. The capacity of the well is large compared to the tube. Liquid forced up the tube is only a small portion

FIGURE 5-5   Slack-tube manometer. (*Courtesy of Dwyer Instruments, Inc.*)

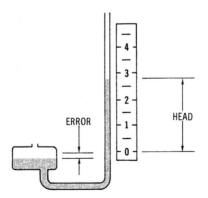

FIGURE 5-6   Well-type manometer.

of the total liquid in the well. Since only a small part of the total liquid moves, the well level stays fairly constant. The head balancing the unknown pressure is equal to the height of liquid in the glass.

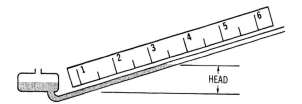

FIGURE 5-7　Inclined manometer.

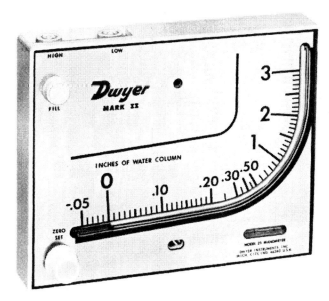

FIGURE 5-8　Inclined-vertical manometer with logarithmic scale. (*Courtesy of Dwyer Instruments.*)

Well manometers are easier to read than the U-shaped instrument. Only a single observation is needed to determine pressure with a well device. Two observations are required of the U-shaped device. Both liquid levels are examined and the difference is determined, providing an indication of the applied unknown pressure.

A third type of manometer is shown in Figure 5-7. This instrument is a modified U-shaped device. One side of the U-tube is inclined. Small pressure variations produce movement of liquid along the inclined scale. These variations can be easily observed without increasing the head. Remember that the head is always read vertically.

A low-cost, molded plastic, inclined-vertical manometer is shown in Figure 5-8. The large diameter curved tube of this instrument forms an easily readable display of low values on a logarithmic scale. These instruments are used to measure values above and

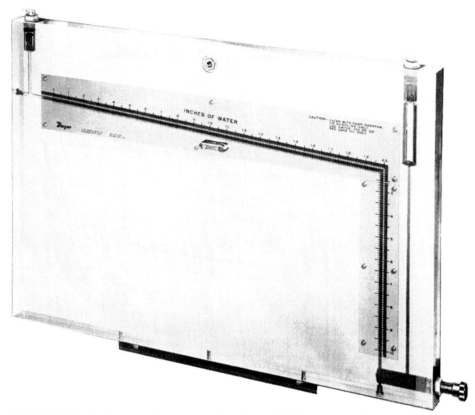

FIGURE 5-9   Inclined-vertical manometer with ± 0.025 accuracy. (*Courtesy of Dwyer Instruments, Inc.*)

below atmospheric pressure, pressure differentials, and air velocities. Range sizes are 0-3 inches and 0-7 inches or their metric equivalents. The tube is filled with oil that has a specific gravity of 0.826 or 1.9, depending on the range of the instrument. The instrument measures pressure in inches of a column of water.

Figure 5-9 shows an inclined-vertical manometer. This instrument has an accuracy of ± 1/4 of 1percent of the scale. Instruments of this type are designed for precision measurement of low differential pressures. The length of the inclined range is 20 inches. The construction provides ample multiplication of indicating liquid movement in the lower parts of the range. The selected oil of this instrument coats the inside surface of the clear tube, providing a well-shaped meniscus that can be easily read.

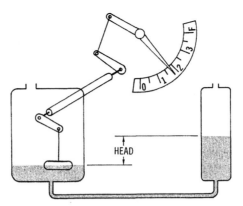

FIGURE 5-10   Mechanical manometer.

Figure 5-10 shows a mechanical manometer. The output of this device is recorded mechanically rather than observing a column of liquid. The output is obtained by placing a float on one of the liquid surfaces. The float rides up and down with changes in level. The level changes with variations in the applied pressure. Attached to the float is a lever that rotates a shaft through a pressure-tight bearing. The indicator and control levers are attached to this shaft. Changes in level are transposed into the pressure being measured causing the shaft to rotate which, in turn, positions the indicator levers.

Discussion of the manometer to this point has focused on measuring single pressure values. Remember that all manometers have two sides. Differential pressure values can be connected to each side of the manometer. The difference in pressure causes a balance between the two values. Through this procedure a manometer can be used to measure differential pressure. The mechanical manometer is primarily used to make differential pressure measurements.

THE BOURDON ELEMENT

A high percentage of all industrial measurements involve the process variables of pressure, temperature, flow, and liquid level. Measurement of this type requires devices that react to these variables. If the variable is pressure, then a device that is sensitive to pressure is required. Such a device is called pressure-sensitive. If the device converts pressure to another medium, it is called a transducer. For example, a transducer is used to change pressure into electricity. One of the most widely used pressure-sensitive devices is the Bourdon element. The remainder of this chapter deals with the Bourdon element.

Construction

A Bourdon element is an oval-shaped tube closed at one end and curved to appear in one of three forms. The form may be a section of a circle, several continuous circles, or a spiral coil. The open end of the tube is attached to a base. The closed end is free to move.

The pressure being measured is connected to the open end of the tube. The closed end is directly attached to an indicator or through a suitable mechanism.

Operation

The Bourdon element is a flat oval shaped tube that is bent to form part of a circle or several complete circles. The cross-sectional area of the tube increases when its volume increases. This expansion tends to straighten out the tube. The total movement of the free end is very small. Typical changes of 3/8-in are normal.

Consider the Bourdon tube diagram of Figure 5-11. The solid lines of the cross-section drawing represent a segment of the Bourdon tube. When pressure is applied, the tube tends to become round. The rounded shape of the tube is represented by dashed lines. Notice the arc lengths of lines a'b' and ab. Since the metal of a Bourdon tube does not compress, lines a'b' and ab must be the same length. They both represent the same physical portion of the tube. Since a'b' and ab are equal, the arc shown as a'b' must fall on a different-radius circle, represented by points o' and o. What was said of arcs a'b' and ab is also true of x'y' and xy. The intersection of x'o' and y'b' extended is the center of the circle that represents the expanded Bourdon tube.

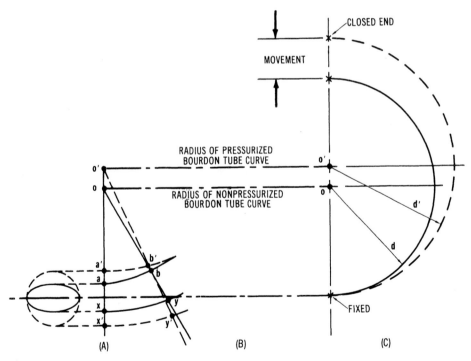

FIGURE 5-11   Theory of operation of Bourdon tube. (A) Cross-sectional distortion; (B) Determination of radius change of Bourdon tube curve; (C) Movement of center line of Bourdon tube.

Bourdon Element Types

The Bourdon tube may be manufactured in several different shapes. The most common is the semicircular, or C-shaped tube shown in Figure 5-12A. This particular shape is widely used as a pressure gauge.

In Bourdon tube operation, a short piece of tubing tends to become straight when pressure is applied. This movement is usually quite small. When indicating pressure, it is

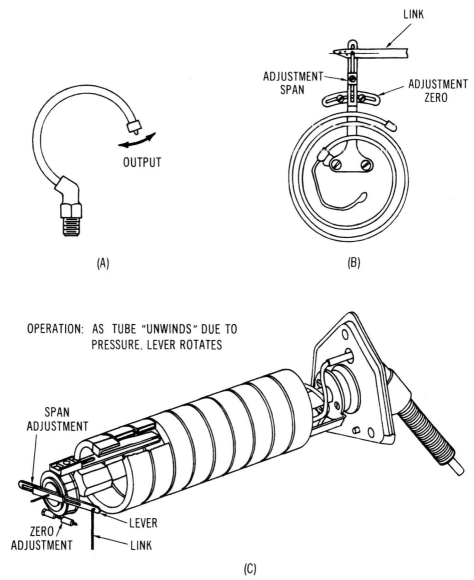

FIGURE 5-12   Types of Bourdon elements. (A) Semicircular or C-shaped; (B) Spiral. (C) Helical.

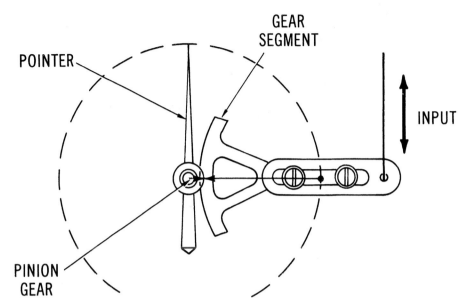

FIGURE 5-13   Motion multiplier mechanism used in Bourdon element pressure gauge.

necessary to multiply the motion of the tube. Multiplication is achieved by the pinion gear/segment of Figure 5-13. A commercial Bourdon element pressure gauge is shown in Figure 5-14.

Other forms of the Bourdon tube are made with built-in multiplication. This type of tube is adequate to drive the indicator over a chart or scale. To obtain built-in multiplication, the tube forms several complete circles. The spiral tube and helical tube of Figure 5-12 are examples of this construction.

Applications

The Bourdon element is a very rugged, dependable piece of pneumatic equipment used in applications that change with variations in pressure. The element may deflect a meter indicating hand or move the pen of a recording instrument. Different range values may be obtained by selecting alternate elements for various applications. A Bourdon pressure element plug-in module of the type used by a recorder is shown in Figure 5-15. The Bourdon tube is a very reliable piece of equipment when used properly. It must be protected from a number of rather harmful side effects, however, including the following:

1. Excessive pressure.
2. Excessive temperature.
3. Excessive vibration.
4. Freezing.
5. Plugging of the tube opening.
6. Corrosion.

FIGURE 5-14   Bourdon element pressure gauge. (*Courtesy of AMETEK, Controls Div.*)

FIGURE 5-15   A Bourdon pressure element plug-in module. (*Courtesy of Bailey Controls Co.*)

FIGURE 5-16   Three methods of mounting a Bourdon instrument to protect against excessive temperature.

Excessive pressure will permanently deform the tube. The manufacturer of a Bourdon tube will set the maximum limit of its pressure. Protection against overpressure requires installing pressure relief valves on the pipelines. The valves insure that the pressure being measured does not exceed the capacity of the instrument.

A Bourdon element should be able to tolerate temperature of the location where it is being used. However, it may be damaged by the high temperatures of fluid or gas pressure being measured. To protect against excessive temperature, the piping must be rearranged. This will allow a pocket of fluid to collect in the line. The line fluid is insulation for the high temperature fluid being measured. A convenient method of obtaining this pocket is called a pig-tail or siphon. Figure 5-16 shows three possible mounting methods.

Vibration may be a problem when a Bourdon tube is mounted directly on pipelines and vessels. If the equipment vibrates, so will the gauge. To prevent vibration, the gauge should be installed away from the vibrating equipment. In some installations vibration is internal, which may be caused by pulsation of the fluid being measured. Pumps are usually the cause of this problem. Selection of a different or newly designed pump will solve this type of problem.

Freezing has the same effect as excessive pressure. It can damage the tube; a rupture may even occur. Protection against freezing may require sealing the gauge with a non-freeze fluid.

Plugging is a serious problem if the gauge is used to measure fluids carrying solids or slurries. The gauge may be purged with a fluid that is clean of all solids. The clean fluid is connected to the line near the gauge to prevent any of the slurry from getting to the gauge.

A second method of preventing plugging is to seal the gauge with a diaphragm as shown in Figure 5-17. The diaphragm prevents solid material from entering the gauge. A clean fluid or gas is placed above the diaphragm. A change in pressure on the lower side causes a change in pressure on the upper side of the diaphragm. Pressure applied to the lower side of the diaphragm causes a corresponding change in upper side pressure. This type of attachment will permit the gauge to indicate an actual pressure.

Corrosion results from a chemical attack on the Bourdon tube. Corrosion can be prevented by fabricating the tube from a material that is immune to corrosion. It can also be prevented by keeping the tube from direct contact with the material.

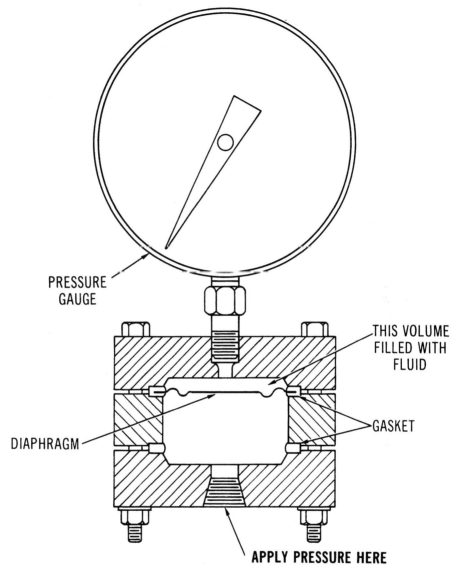

FIGURE 5-17   A sealed pressure gauge.

Measuring Temperature with a Bourdon Element

The measurement of temperature is second only to pressure, the most important process measurement. Measuring temperature with a Bourdon element is done electrically or mechanically. The electrical method uses a temperature-sensitive device that produces a change in resistance or voltage. The mechanical method uses a temperature-sensitive device that reacts to temperature by changing a mechanical property, such as the force of a lever, the deformation of a bellows, diaphragm, or Bourdon element.

The element of a Bourdon tube is a pressure sensitive device. Components added to the element can change temperature to an equivalent pressure. In gas pressure there is a definite relationship between temperature and pressure. This is true only when its volume remains constant. A filled system converts temperature to pressure. The Bourdon element converts pressure into mechanical movement.

Components

The components of a filled system for measuring temperature are:

1. A bulb.
2. A pressure-sensitive component (usually a Bourdon tube, a bellows, or a diaphragm).
3. A capillary tube to connect the bulb and pressure-sensitive component together.
4. A material filling the closed system.

The pressure-sensitive components of a filled system are used to drive a mechanism. These components are used to move the position of an indicator or controller mechanism. The operation of a bellows or diaphragm is similar to the Bourdon element. All three devices are pressure-sensitive.

TYPES OF FILLED SYSTEMS

The principle of operation of the filled system depends on what material runs through the system. There are three types of filling material: liquid, gas, or a mixture of both. Each system will be described separately. The operation is similar for each type of system.

Liquid-Filled System

The liquid-filled system is usually filled with a hydrocarbon or mercury. Hydrocarbons are organic compounds containing hydrogen and carbon atoms. A mercury-filled system is an important division of liquid-filled systems. A filled system with associated links and levers is shown in Figure 5-18.

In Bourdon elements, operation depends on how a change in pressure affects the cross-sectional area. As the pressure increases, the oval tends to round out to a circle. This larger cross sectional area, multiplied by the length of the tube, results in an increase in volume within the tube, causing it to straighten. This occurrence is used for a temperature measurement and shows that a fluid will expand when heated. The amount of expansion depends on the temperature. As the expanded fluid takes up more space, the increase must go somewhere. In the Bourdon element, the only place available is in the filled system. The increase in volume of the fluid causes the Bourdon element to change its volume. This results in the element straightening a small amount.

Movement of the Bourdon element depends on the volume of fluid. This volume of fluid, in turn, depends on the temperature. The movement of the Bourdon is a function

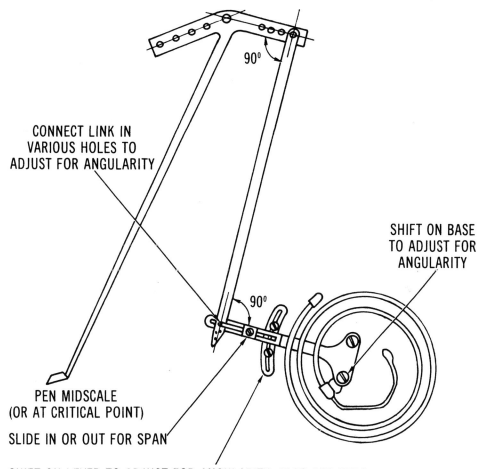

CONNECT LINK IN
VARIOUS HOLES TO
ADJUST FOR ANGULARITY

90°

SHIFT ON BASE
TO ADJUST FOR
ANGULARITY

90°

PEN MIDSCALE
(OR AT CRITICAL POINT)

SLIDE IN OR OUT FOR SPAN

SHIFT ON LEVER TO ADJUST FOR ANGULARITY; ALSO FOR ZERO

FIGURE 5-18   A filled system.

of temperature. The movement drives a linkage that is identical to the linkage of a Bourdon pressure instrument.

Gas-Filled System

In a liquid-filled system of a Bourdon, the relationship between pressure and temperature are interdependent. The same holds true in a gas-filled system of a Bourdon.

Vapor Pressure System

A body of liquid, given enough time, will vaporize into a gas. If the liquid is in a closed container and the gas cannot escape, a pressure will build up that is called vapor pressure. The amount of pressure that builds up depends on the particular fluid and the temperature

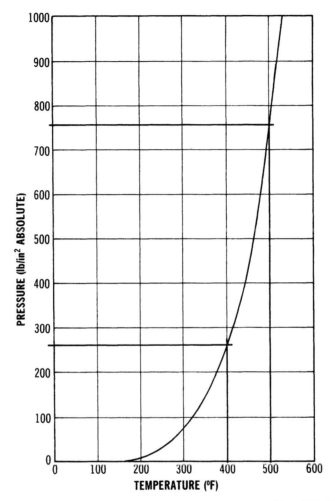

THE ABOVE CURVE SHOWS THE TEMPERATURE/PRESSURE CURVE FOR WATER. OTHER LI-
QUIDS HAVE SIMILAR APPEARING CURVES. NOTE THAT A CHANGE IN TEMPERATURE
FROM 200° TO 300°F RESULTS IN A CHANGE OF PRESSURE OF ABOUT 50 lb/in^2,
WHEREAS A CHANGE IN TEMPERATURE FROM 400 TO 500°F RESULTS IN A CHANGE OF
PRESSURE OF 500 lb/in^2.

FIGURE 5-19   Pressure/temperature curve for a filled system.

of the fluid. If this vapor pressure is connected to a Bourdon element, it is possible to
obtain an output that is determined by temperature.

The relationship between pressure and temperature do not run parallel to one another
as Figure 5-19 shows. For each level of pressure there is a corresponding level of tem-

perature. The practical consequence of this is that the chart or scale of a vapor pressure instrument is nonlinear.

COMPENSATION ▄▄▄▄▄▄▄▄▄▄▄▄▄▄▄▄▄▄▄▄▄▄▄▄▄▄▄

In all the systems discussed, it has been shown that pressure depends on the temperature of the fluid or gas. This introduces a problem: there is fluid or gas in the capillary and Bourdon element. In addition, there is fluid or gas in the bulb. This capillary and Bourdon fluid is at ambient temperature. Ambient temperature is the temperature of the atmosphere surrounding the capillary tube and the element. This temperature is different from that of the bulb, hence an error is introduced into the assembly. Steps are taken to minimize this error or to eliminate it. These methods are called temperature compensation.

A second problem arises when the bulb of the tube is located above or below the recorder, which causes the fluid or gas to exert a pressure that is either positive or negative. This introduces a head error. The methods to eliminate the head errors are called head compensation.

A third possibility for error arises because of ambient pressure. With the use of a Bourdon element, these errors are extremely minor. Where a diaphragm or bellows is used, there must be some way of compensating for changes in atmospheric pressure.

TEMPERATURE COMPENSATION ▄▄▄▄▄▄▄▄▄▄▄▄▄▄▄▄▄▄▄▄

Liquid Systems

Temperature errors are minimized by increasing the volume of fluid or gas within the bulb. There is a limit to how far the capillary volume can be reduced and the bulb volume increased. Doing this slows down the response of the instrument. The practical way of reducing capillary volume is to simply reduce the length of the tube. Some compensation is required if the capillary tube exceeds 10-ft.

There are methods for compensating for Bourdon errors called case compensation. To compensate for the Bourdon and capillary, a second method is required. This method is referred to as complete compensation.

There are two methods for case compensation for temperature errors. The use of a bimetallic strip and a second Bourdon make up these two error reducing methods. The bimetallic strip is placed in the linkage. As the temperature within the instrument case changes, the linkage is repositioned.

The second Bourdon element operates approximately in the same way as the bimetallic strip. This element moves as the temperature changes. The movement is added to or subtracted from the measuring Bourdon movement, thus correcting for case temperature variations.

Complete compensation is achieved by adding a second Bourdon element and a capillary tube, as shown in Figure 5-20. The compensation capillary runs parallel with the measuring capillary. The compensation Bourdon is mounted with the measuring Bourdon so that the resulting output depends on the difference between the two capillary systems.

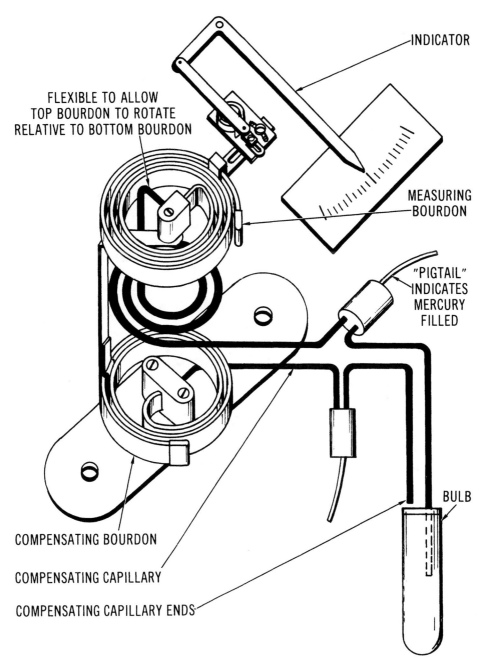

INDICATOR

FLEXIBLE TO ALLOW
TOP BOURDON TO ROTATE
RELATIVE TO BOTTOM BOURDON

MEASURING
BOURDON

"PIGTAIL"
INDICATES
MERCURY
FILLED

BULB

COMPENSATING BOURDON

COMPENSATING CAPILLARY

COMPENSATING CAPILLARY ENDS

FIGURE 5-20   Complete temperature compensation in a filled system.

Gas Systems

Compensation in gas-filled systems is usually necessary if there is a large bulb volume. The ratio of bulb volume to Bourdon and capillary volume should not be less than 8 to 1.

Vapor Pressure Systems

No temperature compensation is required for vapor pressure-filled systems. The pressure within the system depends on the temperature of the surface of the liquid. When no compensation is given, a situation is introduced which is called the cross-ambient problem.

   A simple vapor pressure system will not do a satisfactory job for temperatures that go above or below ambient temperature. If this system is used, the liquid surface will flow to the coolest part of the system which can result in an erratic indicator operation. A specially constructed vapor pressure system will eliminate this problem.

HEAD COMPENSATION ▬▬▬▬▬▬▬▬▬▬▬▬▬▬▬▬▬▬▬▬

When gravity acts on the contents of a filled system, there is a need for head compensation. The amount of fluid exerts a pressure to which the Bourdon responds. Since the Bourdon should respond to pressure due only to temperature, the head pressure presents an error.

   Head pressure is not a problem with gas-filled systems since gas has very little weight. The problem of head errors with vapor pressure systems depends on whether the capillary is filled with gas or fluid. The capillary will be filled with fluid if the capillary is cooler than the bulb. Suppose that the temperature being measured is one that could occur around the capillary and the case. If the capillary becomes cooler than the bulb, liquid will collect in the capillary and will cause the system to have a fluid head pressure. If the temperature around the capillary becomes greater than at the bulb, the liquid in the capillary will boil out. When this happens liquid head is eliminated. The indicator will respond to these heads and will act erratically when the temperature of the bulb and the capillary cross. This crossing of temperature is referred to as a cross-ambient problem. If the capillary remains filled, the zero of the indicator shifts towards zero and subtracts the pressure head.

   The head pressure of liquids is a problem in a filled system. Liquids are relatively heavy, and the head pressure would also be substantial. This is solved by forcing the liquid into the system so that the system would be under an initial pressure. In the case of the Bourdon, the element would exert a force on the liquid equal to the force with which the liquid was put into the system. If the Bourdon force is greater than the force due to head pressure, the Bourdon will not expand. Head pressures that do not exceed the initial pressure of the system have no effect. Approximately 30 ft of bulb elevation is about the maximum for liquid-filled systems. Keep in mind that liquid-filled systems operate on a change in volume of the internal liquid due to temperature. These volumetric changes exert a tremendous pressure on the Bourdon element. All of this takes place when the volume of the Bourdon increases. When used in liquid filled systems the Bourdon element is not pressure-sensitive, but is volume-sensitive.

CALIBRATION AND SERVICE

The linkage considerations for filled temperature instruments are identical with those for pressure instruments. Adjustments are the same. The calibration standards for temperature instruments are atmospheres of a known temperature. For example, the bulb may be immersed in boiling water for the high temperature. Ice water is used for the low temperature standard. Other temperature ranges listed on the instrument may be handled. This is done by inserting the bulb into known temperatures. The indicator must be adjusted to agree with these temperatures in order to properly move.

Service problems are those relating to mechanical difficulties of a filled system. Filled systems must be absolutely pressure-tight or else the temperature indication will not be accurate.

SUMMARY

Several different methods are used to measure pressure today. For pressure values less than 5-psi, limp diaphragms and bell manometers are used. For medium pressure, up to 25-psi, bellows manometers, diaphragms, or Bourdon elements are used. For pressures in excess of 25-psi, Bourdon elements are almost exclusively used.

The Bourdon element is one of the most widely used pressure-sensitive devices in pneumatic instrumentation. The element is an oval-shaped tube which is closed at one end and curved to form either a section of one circle, a spiral, or a helix. The open end of the tube is attached to a base and a closed end is connected to a pen or an indicator. When pressure is applied to the element the indicator will move.

Excessive temperature, excessive pressure, vibration, plugging, and corrosion may impede the performance of a Bourdon element. Proper precautions are outlined to prevent this from occurring.

In order to measure temperature with a Bourdon element, a fluid-filled system is formed. As the pressure increases, the oval tends to round out into a circle, resulting in increased volume within the tube and causing the element to straighten. In normal operation, the Bourdon element expands when heated due to a change in pressure.

Filled-systems need some sort of compensation in order to reduce the possibility of errors that may arise when they are being used. The two methods are head and temperature compensation.

ACTIVITIES

MANOMETERS ▬▬▬▬▬▬▬▬▬▬▬

1. Obtain a "U" tube manometer and place it on a vertical stand or hold it in an upright position.
2. Fill the tube with water until a center balance is reached.
3. Place a scale in the center of the manometer and position it to the center balance line.

4. Connect a 0 to 5-psi air line to one side of the manometer. Carefully adjust the air pressure to 1-psi. Notice the height difference of the balance line.
5. Measure the height difference of the water line.
6. Increase the input pressure to a value of 1.5 psi. Measure and record the height difference of the water line.
7. Repeat this procedure for 2, 2.5, 3 and 3.5 psi.
8. Empty the water from the manometer and refill it with alcohol.
9. Repeat the above steps while recording the height difference of the alcohol line.
10. Empty the alcohol from the manometer and refill it with a light oil.
11. Repeat the procedure while recording the height difference of the liquid line.
12. Empty the manometer and clean the tube.

BOURDON ELEMENT

1. Refer to the operation manual of the instrument being used in this activity.
2. Disconnect it from the system or have the system turned off.
3. Remove the housing or outside cover of the instrument.
4. Locate the element of the instrument. Identify the type of element used in the instrument.
5. Make a sketch of the element.
6. Explain how the element responds to a change in pressure.
7. Does the element have any adjustments?
8. What does the element drive or manipulate in the operation of the instrument?
9. If possible, apply a small pressure value to the instrument. Notice the response of the element. Does it change according to your explanation?
10. Reassemble the instrument.

QUESTIONS

1. What effect does atmospheric pressure have on a limp diaphragm?
2. How is a bellows pressure device constructed?
3. What is meant by a slack-tube manometer?
4. What are some representative pressure values that can be measured with a limp diaphragm?
5. What type of pressure measuring instrument is used to determine pressure in excess of 25 psi?
6. What are some common types of Bourdon elements?
7. What are the harmful side effects of which a Bourdon tube must be protected?
8. What is the primary function of a Bourdon element?
9. What are the basic components of a filled system?
10. What is meant by the term compensation?
11. Describe the operation of a liquid-filled Bourdon element system.
12. Why is head compensation a problem in a filled system?

6
Liquid-Level Measurement

OBJECTIVES

Upon completion of this chapter, you will be able to:

- Explain what is meant by the term level.
- Define the term displacer.
- List some of the techniques used to determine level.
- Explain the operation of the purged pipe level measuring system.
- Identify the components of a purged pipe level measuring system.

KEY TERMS

Archimedes' principle.   A reaction that occurs when an object is immersed in a liquid. First, the object displaces some of the liquid. Second, the object is buoyed up by the liquid and weighs less than it does in air.

Buoyancy.   The tendency of a body to float or rise when submerged in a fluid.

Buoyant force.   The power of a fluid to exert an upward force on a body placed in it.

Diaphragm box.   A hat-shaped flexible diaphragm mounted in a closed compartment that is used to determine pressure of a container.

Displacer.   A level sensor that operates on buoyant force.

False head.   A pressure that occurs in a purged-pipe system not due to fluid depth.

Friction head.   The loss of pressure that occurs when gas or liquid flows through a pipe.

Level. The height of material in a container.

Pressure repeater. A force-balance pneumatic mechanism that duplicates an applied pressure.

Purge. To clear, clean, or evacuate something from a container or housing by replacing it with an acceptable material.

Purge pipe. A container that has air evacuated from it in the operation of a level measuring procedure.

Standpipe. A water tower or vertical pipe used to secure a uniform pressure.

Static head. Air or gas pressure that is present on the surface of liquid in a tank.

Torque. The product of a force about a specific axis and the distance from the force to the axis. Another way of describing the term moment.

Torsion spring. A spring that reacts to a twisting or turning force.

INTRODUCTION

The measurement of liquid and solid material levels in a tank or vessel is an important industrial process for nearly all industry regardless of the product being manufactured. The level of material in a tank, bin, hopper, or other container indicates the amount of material available for accomplishing a process. The level also is used to determine the length of time that a particular production run can be achieved. It may also be used to determine the amount of material consumed during a manufacturing operation.

Equipment used in level control is divided into liquid level and solid level instruments. There is some overlapping of the basic operating principles of this equipment. We will direct our attention to liquid level instruments as these instruments are used in a large number manufacturing operations.

LEVEL MEASURING TECHNIQUES

The first function of level determination is measuring. Level must be measured and evaluated before any control can be achieved. Level is either measured directly or it can be inferred. The direct method responds to some physical change of the material. A change in the amount of liquid in a container is a direct indication of level. An inferrential method uses some outside variable to indicate a condition change. The pressure of liquid at the bottom of a tank can be used to identify level. Level can be measured directly or be inferred by some other variable.

DISPLACERS

Buoyant-force level sensors respond to the Archimedes' principle, which was discovered around 250 B.C. A body placed in water is buoyed up by a force that is equal to the weight of the water it displaces. The body in this case can be called a float. In effect, the body is actually a displacer. A specific amount of liquid is displaced by a body when it is placed in liquid. The weight of the body determines the amount of liquid it displaces. A displacer is actually a liquid level to mechanical-motion transducer. A large number of liquid level instruments use displacers in their operation.

The components of a displacer level instrument are:

1.   A displacer mechanism.
2.   A mechanism for suspending the displacer in the fluid.
3.   A spring to ''weigh'' the displacer.
4.   A mechanism to detect the weight of the displacer.
5.   A pressure-tight connection to bring out the weight of the displacer.

A displacer is suspended in fluid and is attached to a spring. A change in force on the spring is passed through a pressure-tight bearing. It is then detected by a pneumatic mechanism. The pneumatic mechanism converts the spring deformation to an equivalent air pressure. This air pressure is read from a pressure gauge that is calibrated in inches, feet, centimeters, or meters.

Principle of Operation

A displacer must be heavy enough that it will sink if not supported by a spring. Suppose that a displacer is weighed on an ordinary spring scale and found to be 10-lbs. The displacer and scale are then hung in a vessel that is gradually filled with fluid. The fluid will rise until it reaches the displacer. The displacer will begin to ''feel'' the fluid rising. It will be buoyed up by a force equal to the weight of the fluid displaced. If the weight of the displaced fluid is 1-lb, then the scale will indicate 9-lbs. As the fluid continues to rise, more of the displacer will be submerged. Its effective weight will continue to be reduced until the displacer is completely submerged.

To calibrate a displacer, it becomes necessary to change scale weight to values of length. A scale indicating weight in pounds would be changed to inches and feet. This can be done if the following are known:

1.   The volume of the displacer.
2.   The weight of the fluid.

If these two quantities are known, then the weight of the displacer depends on the level of the fluid. Figure 6-1 shows how weight is converted to level.

DISPLACER INSTRUMENT ▰▰▰▰▰▰▰▰▰▰▰▰▰▰▰▰▰▰

A rather simple level mechanism was used to explain the operation of a displacer. An actual displacer is somewhat more complicated. In principle, its operation is identical to the spring scale/displacer. Complications arise when instruments need to measure fluid

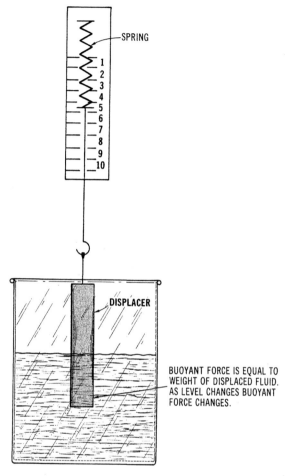

SPRING

1
2
3
4
5
6
7
8
9
10

DISPLACER

BUOYANT FORCE IS EQUAL TO
WEIGHT OF DISPLACED FLUID.
AS LEVEL CHANGES BUOYANT
FORCE CHANGES.

FIGURE 6-1   Principle of the displacer.

levels in vessels that are under pressure. Provisions must be made for getting the weight of the displacer out of the high-pressure atmosphere of the vessel. This is achieved by using a pressure-tight bearing on the mechanism arm.

The solution to measuring level in a pressurized vessel is to use a torsion spring. This is a spring that reacts to a twisting force. Such springs usually are thin-walled pipes called torque tubes. Attached to the movable end is a lever. The displacer is attached to the lever, causing the tube to twist a small amount. The effective weight of the displacer is then reduced by the submerging fluid, which causes the spring to unwind. Its twisting motion changes to an even smaller amount. This twisting is measured by placing a solid shaft inside the torque tube. The shaft is then fastened to the lever end of the tube. As the torque tube rotates, motion is passed down the solid shaft. This twisting or angular

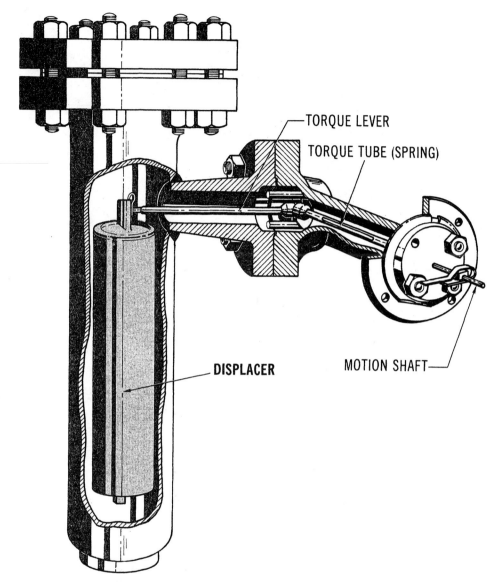

FIGURE 6-2   Displacer level instrument.

rotation is detected by a pneumatic mechanism. It converts rotation to equivalent units of air pressure. See Figures 6-2 and 6-3.

PURGED PIPE LEVEL MEASUREMENT

Liquid level measurement is one of the four major types of industrial processing and is followed by the measurement of temperature, pressure and flow. There is a variety of

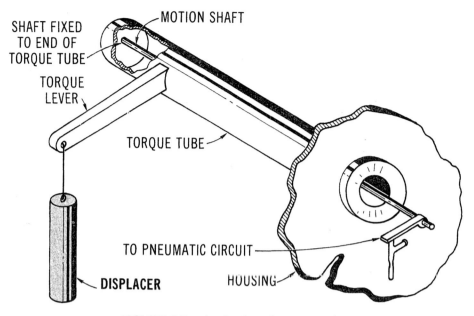

SHAFT FIXED
TO END OF
TORQUE TUBE

MOTION SHAFT

TORQUE
LEVER

TORQUE TUBE

TO PNEUMATIC CIRCUIT

DISPLACER

HOUSING

FIGURE 6-3   Application of a torque tube.

level-measuring systems. One of the more widely used is the purged-pipe method, also known as the bubble-pipe or the standpipe method.

In a purged-pipe system the head of a fluid is converted into an equivalent pressure value. Through this method, level can be equated to pressure. Head has been described as another way of indicating pressure. One or more pipelines are used to connect the measured head to the level instrument. Air or some inert gas flows through these lines.

Components

The purged-pipe system is shown in Figure 6-4. This particular system consists of:

1.   An indicator or recorder.
2.   A purged-pipe.
3.   Two connecting lines.
4.   A source of air or gas.
5.   A sight-flow indicator.
6.   An optional constant-flow pressure regulator.

Arrangement

The purged-pipe is immersed vertically in the fluid being measured. A pipeline is connected to the top of the vertical pipe and to the indicator. The indicator of the system is a pressure-sensitive device. A source of air or gas is connected to the line joining the vertical pipe and the indicator.

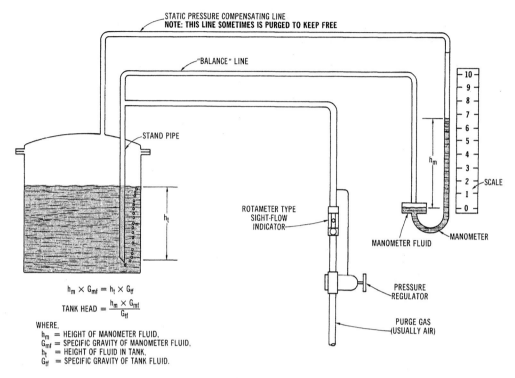

FIGURE 6-4   Purged-pipe system for liquid level measurement.

Through this type of connection, the indicator and the standpipe are linked together by common gas pressure. Gas then flows through the sight-flow indicator. This makes it possible to see how much gas is flowing.

Principle of Operation

Assume that the gas flow into the purged-pipe system is shut off. The level within the standpipe will be the same as in the vessel. Allow a small amount of gas to flow into the system. The gas pressure builds up and forces the fluid down the standpipe. The increase in pressure balances the pressure differences between the fluid and gas.

As the gas continues to flow into the system, pressure continues to build. A point is finally reached where the pressure balances. What is balanced this time is a difference in level that is equal to the length of standpipe covered by the fluid. No further increase will be obtained even though gas continues flowing because any excess gas escapes from the bottom of the standpipe. The build-up in gas pressure is equal to the head of the fluid.

Remember that the fluid head depends on density and height. If head is in inches of water, then density will be expressed as specific gravity. Or,

$$\text{head} = \text{height} \times \text{density}$$

Fluid level can be determined from the measurement of gas pressure. This gas pressure must then be expressed as head. Conversion changes pressure to an equal amount of water in inches or feet. This head must then be divided by the specific gravity of the fluid.

The indicator of a purge system is usually a glass-tube manometer. The system gas pressure is converted to an equal head at the indicator. This means that the fluid column in the indicator balances the gas pressure. The gas pressure will then equal the head times the density of that fluid. Gas pressure also equals manometer fluid height times the specific gravity of the fluid. The gas pressure equals the tank height times the specific gravity of the fluid. This can again be shown as:

$$h_m \times G_{mf} = h_t \times G_{tf}$$

where,

h_m is the height of the manometer fluid,

G_{mf} is the specific gravity of the manometer fluid,

h_t is the height of the fluid in the tank

G_{tf} is the specific gravity of the tank fluid

The fluid height in a manometer is multiplied by the specific gravity of that fluid. This will balance the head of fluid multiplied by its own specific gravity. The link between the entire vessel and the indicator is gas pressure. This link forms the equal sign in the mathematical expression.

COMPENSATION ■■■■■■■■■■■■■■■■■■■■■■■■■■■■■■

Level measurement in a purged-pipe system is a pressure measurement. The indicator is also a pressure-sensitive device. Any factors that cause a change in pressure will affect the device. This in turn will cause an error in measurement. Two important sources of error are:

1. Static head on the liquid being measured.
2. Friction head of the purging gas.

Compensation for Static Head

To compensate for a static head, a second line is brought from the vessel to the indicator. This line is connected to the top of the vessel. The connection senses the static pressure of the gas on the liquid. The other end is connected to the alternate side of the manometer. Gas static pressure acts on the fluid, causing a false high head. This false head is connected to the manometer, where it is forced down an equal amount. This force causes the fluid measurement to increase.

Compensating for Friction Head

The friction head depends on three factors: the amount of gas flowing, the size of the pipe, and the length of the connecting pipe. To minimize friction head, the length of pipe should be kept short. The size of the connecting line should be large and gas flow kept to a minimum. Because of equipment location, it is not always possible to keep the line length short. A slow rate of gas flow results, making the measurement unable to keep up with rapidly increasing levels. This is especially true if the pipeline is large. A short, large-diameter line with small air flows offers only a limited solution to the problem.

A better method of compensating for friction head is to subtract the friction head from the indicated instrument reading. This would work well if the friction head remained the same. The friction head would remain the same if the gas flow remained constant. Gas flow would remain the same if the level remained constant. This, of course, does not occur normally. However, constant-flow controllers are available that will keep gas flowing constantly regardless of level variations. A better solution for friction head compensation is to feed the purge gas through a constant-flow regulator.

A third method of compensation consists of reducing the length of connecting pipeline to almost zero. The purge gas is then admitted into the system at the top of the standpipe. Connecting-line friction is reduced because there is no flow of gas in the line. This results in two lines to the tank in addition to the static-pressure line. Employing this method also increases the cost. Despite increased cost, the two lines are needed if the sight-flow device is located at the instrument. To facilitate service, the sight-flow indicator should be located in the instrument.

Compensating for Standpipe Length

The standpipe end should always be positioned slightly above the bottom of the tank. This positioning is used to clear the pipe of any accumulated solids. Compensating for clearance is achieved by shifting the zero of the indicator scale by the amount that the standpipe clears the bottom of the tank. Another way of doing this adds the amount of clearance to the indicator reading.

SERVICE PROBLEMS ▬▬▬▬▬▬▬▬▬▬▬▬▬▬▬▬▬▬

Service problems are related to the kind of instruments used. Additional problems arise because:

1. Measurement is a pressure value where level is converted into an equivalent pressure.
2. Purge air must be continuously furnished to the standpipe.

To avoid the first item, no pressure should be allowed to leak out of the system. The system should be leak-free. No false pressures should be permitted to enter the system. False pressures could occur if the standpipe or a line is not permitted to flow freely. Leaks, plugs, or partial plugs could result in friction heads.

In order to avoid the second item, air must be continuously furnished to the stand-pipe. If a leak is large enough, the gas may never get to the end of the standpipe. The supply pressure must be greater than the greatest head to be measured. If it is not, then the gas will not flow into the system at the high heads. The sight-flow indicator is used to see gas flow.

OTHER LEVEL-MEASURING INSTRUMENTS ▬▬▬▬▬▬

In addition to the purged-pipe system and the displacer method of level measurement, there are others. These include devices that use a float, other differential-pressure meth-ods, the diaphragm box, and the pressure repeater. The pressure repeater is a force-balance pneumatic mechanism and will be discussed in Chapter 14. Except for the displacer and float devices, all of these instruments respond to pressure.

Float Devices

There are many instruments that use a float as a sensing element. These are all based on the fact that the float will follow the surface of the liquid. There is then a problem of designing a mechanism to follow the float. There are two methods to solve this problem. The first is to arrange to have the float position a lever. This lever motion is then used to drive linkages or pneumatic mechanisms. The second method follows the float with a cable which is usually a wire or flat metallic tape. The cable then serves as the input to pneumatic, electrical, or mechanical mechanisms.

The lever method has the advantage of simplicity. Its disadvantage is that the range of lever measurement is sharply limited because the lever must be kept fairly short. The cable method does not have this disadvantage. The range is limited only by its cable lengths. The cable method requires a fairly formidable take-up mechanism because its measurement is to be displayed on recording devices.

Both float following mechanisms share the problems of a pressure-tight bearing. This occurs when measurements are made on liquid in a pressurized vessel. For the lever device, the bearing is usually a packing in a stuffing-box. The lever shaft must rotate through the packing material. Stuffing-box friction is usually not a problem. The float can be made large enough to furnish the driving power. A pressure-tight bearing for a float or cable device presents an extremely difficult problem. The only practical solution is to use a gas-tight liquid seal through which the cable passes.

One acceptable method for avoiding a pressure-tight bearing is to follow the float with a magnet. The magnet must drive the cable. The cable and the magnet are inside of a closed pipe immersed in the liquid. The float is guided by the pipe, moving up or down with changes in the level, and the magnet follows the float.

Differential-Pressure Devices

The differential-pressure devices discussed thus far can be used directly to measure liquid level. The device that is chosen depends on the nature of the fluid being measured, static pressures, and the type of display desired. The fluid of the manometer differential-pressure devices must not combine with the fluid being measured. Most fluids being measured do not combine with mercury. Because of this problem, mercury manometers are quite popular.

A second problem is the static pressure of a level measurement within a pressure vessel. If these pressures are 100-psi or greater, glass-tube manometers probably should not be used. However, there are some special manometers for high-pressure service which include the mechanical manometer, the bellows, and the diaphragm.

The third consideration is the type of display desired. The glass-tube manometer is used where a visual display at the location is desired. If this is preferred, the mechanical manometer, the bellows, or the diaphragm must be used. Figure 6-5 shows how differential-pressure devices can be used.

Pressure Devices

Pressure gauges, in general, are not sensitive enough to be satisfactory for level measurements. In view of this, differential-pressure devices are used with one connection. The other connection is common to atmospheric pressure.

Diaphragm Box

The diaphragm box is a closed system composed of a flexible hat-shaped diaphragm. The diaphragm is then mounted in a box which forms a closed compartment. The compartment is connected to a low-range recorder with a capillary tube. The box is lowered into the liquid being measured. This liquid forces the diaphragm, squeezing the air in the closed system. The squeezing builds up the pressure, and the pressure is then registered on the recorder. The pressure depends on the height of the fluid multiplied by its density. If the density is known, the pressure gauge can be calibrated in feet and inches. This type of system is shown in Figure 6-6.

The diaphragm box must be pressure-tight. The slightest leak will soon bleed out all of the air in this system. As a rule, this type of system needs periodic maintenance to keep it operational.

SUMMARY ■■■■■■■■■■■■■■■■■■■■■■■■■■■■■■■■■■■■■

Archimedes' principle applies to many liquid-level measuring instruments. For example, a displacer is buoyed up by a force that is equal to the fluid displaced. Float/lever sensors respond to this principle by following variations in liquid-level surface changes. Lever/linkage mechanisms or cables are attached to the float to produce an indication of level.

One common application of the term head is in a purged-pipe liquid-level measurement. In this system, equipment is used which changes level to pressure automatically, or converts one head to an equal one.

Pressure is also used to determine level by producing variations in a diaphragm box. Level measurement of this type is considered indirect.

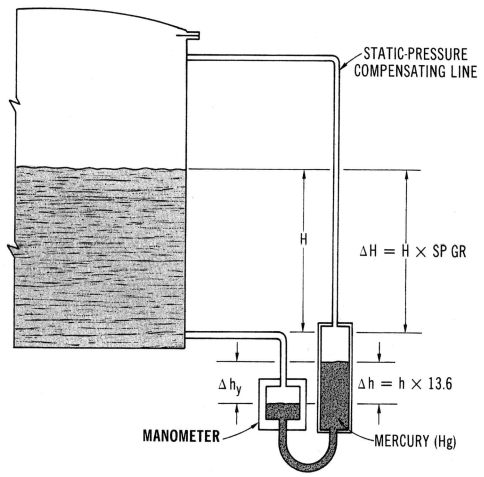

FLUID HEAD BEING MEASURED = FLUID HEAD OF INSTRUMENT
FLUID HEAD BEING MEASURED = H × SP GR OF MEASURED FLUID
FLUID HEAD OF INSTRUMENT = h × 13.6 (SP GR OF Hg)
H × SP GR OF MEASURED FLUID = h × 13.6

$$\text{HEAD} = \frac{h \times 13.6}{\text{SP GR OF MEASURED FLUID}}$$

**NOTE: THIS EXPRESSION DOES NOT TAKE INTO ACCOUNT
THE "FALSE" HEAD Δhy.**

FIGURE 6-5   Manometer measurement of liquid level.

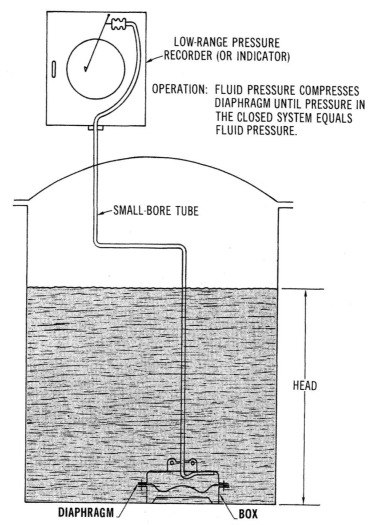

FIGURE 6-6   Level measurement with a diaphragm box.

ACTIVITIES

DISPLACERS

1. Select an object that is to be used for the displacer element. This object needs to be suspended by a string or flexible cord.
2. Attach the displacer assembly to the measuring hook of a spring balance scale.
3. Measure and record the weight of the displacer assembly in air.
4. Place the displacer assembly in a container of water.

5. Measure and record the weight of the displacer assembly when it is fully submersed by water.
6. Measure and record the weight of the displacer assembly when it is partially submersed.
7. How does water level affect the weight of the displacer assembly?

PURGED-PIPE LEVEL MEASUREMENT ▬▬▬▬▬▬▬▬

1. Construct the purged-pipe level measuring assembly of Figure 6-4.
2. Fill the container to its upper level.
3. Adjust the air pressure regulator to produce bubbles from the standpipe.
4. Record the pressure of the manometer.
5. Remove some water from the container and readjust the air pressure to produce bubbles from the standpipe. Record the new air pressure of the manometer.
6. Repeat this procedure for two or three other water levels.
7. How does the air pressure of this assembly relate to water level?

QUESTIONS

1. What accounts for the buoyant force that occurs when an object is placed in liquid?
2. Why does a displacer weigh less when submerged in liquid?
3. Explain Archimedes' principle.
4. What is a purged-pipe system?
5. What are the components of a purged-pipe level measuring system?
6. How is a diaphragm box used to measure liquid level?
7. What does a level measuring system actually determine?

7
Differential-Pressure Instruments

OBJECTIVES

Upon completion of this chapter, you will be able to:

- Define the term differential-pressure.
- Explain how the manometer/float mechanism operates.
- Identify the displays of differential-pressure instruments.
- Explain how the bellows differential-pressure mechanism operates.
- Explain how the diaphragm differential-pressure mechanism operates.
- Explain the calibration procedure of a differential-pressure instrument.

KEY TERMS

Axis of rotation. A straight line about which a body or geometric figure turns or positions itself.

Bellows seal. A tight and perfect closure that prevents the passage of air from the chamber of a bellows but permits movement of the indicator mechanism.

Blowout. To release water or mercury from an instrument when it is subjected to an excessive surge of pressure.

Cock. A faucet or valve device for regulating the flow of liquid or gas.

Dampen. To deaden or diminish the activity or vibration of an object.

Dry calibration. An adjustment procedure that does not incorporate water or liquid in the instrument.

Dry meter. An instrument that does not incorporate liquid in its operation.

Needle valve. A type of valve that has a sharp point plug and a small seat orifice for low-flow metering.

Orifice. An aperture or small opening in a plate used to determine flow.

Run dry. An instrument condition in which no liquid is used in its operation.

Run wet. An instrument condition in which liquid is used in its operation.

Surging. A sudden rise or fall of temperature or pressure that produces a rapid fluctuation in a value.

Tapered displacer. An instrument having a variable diameter cross-section float that is used as a displacer.

Torque tube. A spring mechanism consisting of a thin-walled tube that produces a twisting or turning motion in its operation.

Valve. A device used to control the flow of gas or liquid.

Wet calibration. An adjustment procedure for an instrument that includes liquid in the process.

Wet meter. An instrument that includes liquid in its operation.

MANOMETER/FLOAT TYPES

We have discussed manometer/float instruments as if they were identical to glass-tube manometers. Important differences between the mechanical meter and the glass-tube manometer will be identified. The construction of such instruments will be more thoroughly discussed. Various methods of bringing out the float-position level will also be reviewed.

Manometer/float instruments are frequently used to measure flow. No matter what the final instrument may display, these devices are differential-pressure sensitive. In order to measure flow, the primary instrument must be modified.

Components

A manometer/float differential-pressure instrument consists of:

1. A float chamber.
2. A range chamber.
3. A float.
4. Fluid.
5. A float position mechanism.
6. A display mechanism.

Arrangement

The float and range chambers of a manometer/float instrument are joined at the bottom, forming a U-tube manometer. The fluid used is normally mercury. The mercury is poured into the joined chambers to a predetermined level. A float rides on the mercury level. Float movement is transferred through a suitable mechanism. The output of the mechanism drives a display device. One pressure signal is connected to the float chamber; the second pressure signal is connected to the range chamber.

Principle of Operation

A differential-pressure signal consists of two components: the static pressure component and the part remaining after this component is removed. The resulting pressure difference is the differential-pressure signal. The developed differential pressure signal is of particular interest in the operation of this instrument.

The developed differential-pressure signal is the result of two applied signals. The static-pressure component of each signal is cancelled. In this case, they are both commonly applied to the same manometer. The resulting mercury level of the manometer is due to the difference in applied pressure values. Because of this, the static component is nulled or balanced from the output.

The differential-pressure component of a manometer is represented as a change in mercury level. This change continues until the mercury head balances the pressure component. A change in mercury level causes a positional change in the float. This positional change is used to drive a display mechanism. Movement of the mechanism depends on mercury level. This level is determined by the developed differential-pressure of the instrument. With the two applied pressure inputs the resulting difference is displayed. The magnitude of the differential-pressure will be displayed if the mechanism is properly calibrated.

THE DISPLAY

The elements of the manometer/float differential-pressure instrument are shown in Figure 7-1. Construction of the element is determined by the display. For example, the display might be activated by an air signal. The pressure of the signal is determined by a differential-pressure. In this case, a diaphragm differential-pressure instrument is better suited than a mechanical mechanism. If the display is activated by an electrical signal, then so would the other components of the instrument. If the display has a pen/chart mechanism, other parts are required.

Construction is also determined by the need to rid the meter body of the differential-pressure signal out of high pressure areas. This problem is due to the static pressure component. If measurement is under high pressure, the assembly must be sealed. Float position must be transmitted from a high-pressure region into one of atmospheric pressure. To a large extent, this problem influences the type of display used. If a pneumatic or electrical display is used, there is no problem with high static pressure.

An all-mechanical differential-pressure instrument drives a pen over a chart through direct linkages. The lever is brought out by a device that rotates a shaft through a bearing.

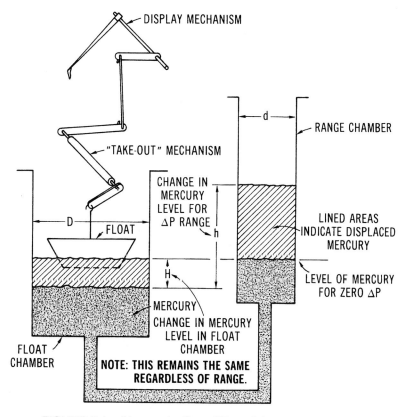

FIGURE 7-1   Manometer/float differential-pressure instrument.

The float drives a lever. The pen linkage is attached to the other end of the shaft. The problem of bringing out the float position is solved by using a pressure-tight bearing. It is important that this bearing be free of friction. Some of the float travel will be lost due to friction in the bearing. Another requirement in using a bearing is that it holds the pressure. A pressure-tight bearing is shown in Figure 7-2. Bearings of this type are an important part of the mechanical differential-pressure instruments.

Shaft rotation in a pressure-tight bearing is caused by the input to a link/lever mechanism. This mechanism is similar to the link/lever assembly previously discussed.

A second type of manometer/float differential-pressure instrument uses magnetism as a means of getting the float position out of the manometer body. Instruments using magnetic coupling between the float and the pen do not need a pressure-tight bearing. Magnetic coupling is normally used with an electric transmission circuit. There are many variations of this type of instrument. They are usually called electric flowmeters. Even though these instruments are called flowmeters, they are still sensitive to differential-pressure. Their operation is similar to that of the all-mechanical type. The difference in pressure upsets the mercury level. This occurs until the mercury head pressure balances the differential pressure. The change in mercury level moves the float. The float serves

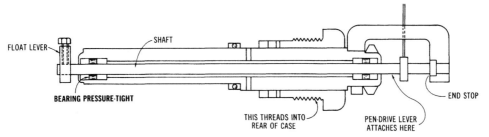

FIGURE 7-2   Pressure-tight bearing.

as an armature with a magnet attached to it. The position of the armature is felt by the external circuit. Due to the nature of magnetism, it is possible for the armature to be inside of a completely closed tube. Many stainless steels meet this requirement. The need for pressure-tight bearings is eliminated. Instruments using magnetic coupling will be discussed in more detail later.

Manometer/float devices are relatively easy to blow. This means that with a large enough pressure, the mercury can be forced out of the instrument which can happen if the instrument is overranged. Overranging occurs when differential pressure is greater than the maximum mercury head that can be developed. The instrument may also be blown if static pressure is applied to only one side. This is a real possibility with manometer/float instruments when they are taken off or put back on the line.

To protect against blowout, check-valves are installed at the exits of the two chambers. These valves are normally float-operated. They close when the mercury level might blow out the instrument, preventing a further transfer of mercury from one chamber to another.

Bypass lines are used to protect an instrument from blowout when taking it off or putting it on the line. This line connects the low-pressure line to the high-pressure line. A valve is placed in this line. A valve is also placed in each of the two pressure lines of the manometer. These three valves must be opened and closed in the proper sequence. The valves must be operated so that the total line pressure is not connected to any side of the manometer. To avoid this pressure, the bypass line valve is opened when one of the pressure lines is opened. The bypass valve is then closed and the remaining valve opened. Only two valves are open at any one time. To take the meter off of the line, the procedure is reversed. Shut one of the pressure lines, open the bypass, and close the second pressure line. In some cases, either the high or the low pressure line be opened first.

A person also needs to protect the meter from surging or rapid fluctuations. In order to do this, a damping restriction is furnished. This is done with a needle valve. As a rule, cock valves are often used. The valve is inserted into the assembly between the range chamber and the float chamber. The valve can be operated by hand or with a tool. If damping is required, the valve needle is moved toward the seat.

RANGE CHANGES

There is a difference between the manometer/float instrument and the glass-tube manometer. For example, the float of the manometer/float instrument must move a fixed amount.

This takes into account the amount of differential pressure involved. This arises out of the need for constructing instruments out of standard parts. A given instrument, with a limited number of component changes, can handle a wide range of applied signals. In a manometer/float device, this means that certain items are the same for all ranges. Specifically, the display mechanism, the float, and the take-out mechanism. If this is done, float travel must also be the same for all ranges. The mercury displaced in the float chamber must be the same regardless of the range. The range, in turn, depends on the difference between the levels in each chamber.

These specifications can be accommodated by having different diameter range tubes. For example, suppose that float travel is one inch and the diameter of the float chamber is four inches. The range desired is 100 inches of water.

1. Convert 100 inches to equivalent inches of mercury as follows:

$$100/13.6 = 7.35$$

This is the head of mercury required to balance 100 inches of water head.

2. The volume of mercury being transferred is taken from one chamber and put into the other chamber, as shown in Figure 7-1. So,

$$V = v$$

where,

V is the volume of mercury associated with the float chamber,

v is the volume of mercury associated with the range chamber:

$$\begin{aligned}
\text{volume} &= \text{area} \times \text{height} \\
&= \pi/4 \times \text{dia}^2 \times \text{height} \\
&= 0.7854 \times \text{dia}^2 \times \text{height} \\
V &= 0.7854 D^2\, H \\
v &= 0.7854\, (d^2\, (h\text{-}H)/4)
\end{aligned}$$

where,

D is the diameter of the float chamber,

H is the amount of the float movement,

d is the diameter of the range chamber,

h is the total mercury head.

In other words, the diameter of the range chamber is to the diameter of the float chamber as the mercury head desired is to the float travel.

BELLOWS AND DIAPHRAGM DIFFERENTIAL-PRESSURE INSTRUMENTS ▬▬▬▬▬▬▬▬▬

The mercury/float differential-pressure instrument is an important industrial mechanism. It should continue to be important in the future. However, there are certain disadvantages to this instrument. First, it is difficult to avoid mercury losses. Second, the mercury meter has fairly high inertia characteristics, which results in the meter not being able to follow a flow that changes rapidly. Because of this, the mercury meter is ruled out of many control applications. To overcome these disadvantages, a number of flow measurements are being made with diaphragm/bellows instruments.

A diaphragm differential-pressure instrument is equivalent to the bellows type of instrument. A diaphragm is simply a very flat bellows. This part of the mechanism imposes a restriction on the design of the instrument. The restriction is due to diaphragm movement. Diaphragm movement is very limited. Values in the order of 0.001-in are normal. As a result, display mechanisms are different from those of the bellows type.

BELLOWS DIFFERENTIAL-PRESSURE INSTRUMENTS ▬▬▬▬▬▬▬

The bellows differential-pressure element is widely used in a number of pneumatic instruments today. Operation of the bellows instrument is not limited by small movements. Substantial movements are taken for the full range of measured differential pressures. The bellows movements are on the order of tenths of an inch.

Components

The components of a bellows instrument are:

1. A high-pressure compartment.
2. A low-pressure compartment.
3. A movable wall between these two compartments.
4. A take-out mechanism that is driven by the movable wall.
5. A display mechanism.
6. A balancing spring.

Arrangement

The two pressure compartments are arranged as in Figure 7-3. This is done so that the end of the two compartments is formed by a movable wall. The wall is movable because it is the closed end of a bellows. Attached to the movable wall is a lever that drives the display mechanism.

Principle of Operation

In a differential-pressure system, signals are connected to the pressure compartments. Differential pressure exerts a force on the movable wall, causing it to move away from

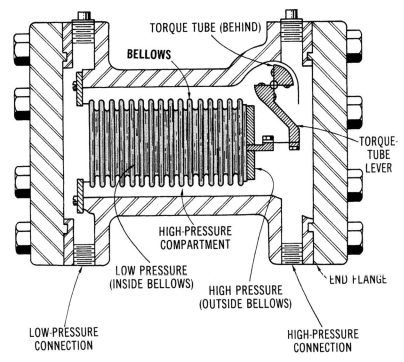

FIGURE 7-3   Single-bellows differential-pressure instrument.

the greater pressure. This movement is opposed by a spring. The difference in pressure balances the movement force and the differential pressure.

Mechanism movement passes through a pressure-tight connection. This connection has several forms. Two common types are the torque tube and the sealed bellows.

A cutaway view of a bellows differential-pressure instrument is shown in Figure 7-4. When pressure is applied, it goes to the two enclosed bellows elements. These elements are enclosed in separate pressure-tight chambers and are connected together by a common center shaft. The difference in pressure applied to each bellows will cause the assembly to move. This is determined by the bellows elasticity, range spring, and torque tube.

Once in operation, the two bellows are filled with a liquid such as ethylene-glycol. When the differential pressure between the bellows elements increases, liquid is forced from one element to another. This then causes the connected torque tube to move. If, for some reason, the difference in pressure exceeds the operating range, movement will continue until an O-ring on the shaft seats against the center plate. The liquid will then be trapped in one bellows element. Instrument damage is prevented because the liquid cannot be compressed. Additional movement of the common shaft will cease.

Mechanical movement of the connecting torque tube can be used to drive a display mechanism. The torque tube drives a pen with a link/lever unit. Torque tube mechanisms are found in many instruments today.

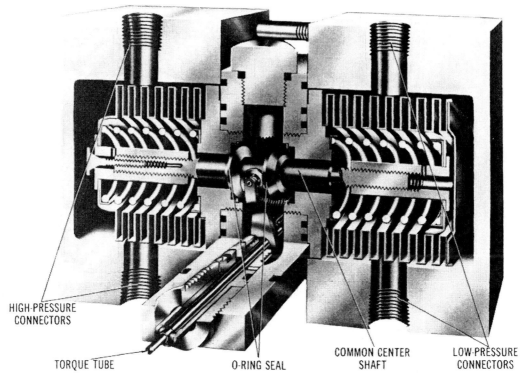

HIGH-PRESSURE
CONNECTORS

TORQUE TUBE   O-RING SEAL

COMMON CENTER   LOW-PRESSURE
SHAFT   CONNECTORS

FIGURE 7-4   Cutaway view of a two-bellows differential-pressure instrument. (*Courtesy of Moore Products Co.*)

Torque Tube

A torque tube combines two functions. First, it serves as the pressure-tight connection device, needed to get movement between a high-pressure area and a low-pressure one. Second, it acts as the spring force that opposes the measured signal.

The structure of a torque tube is shown in Figure 7-5. It consists of a thin-walled tube. One end is fixed to the meter body and the other end is closed. The entire tube is under the high static pressure of the measurement. Attached to the closed end of the tube is a lever. This lever is rotated by the bellows movement. The lever rotates, with the tube acting as its axis of rotation. The lever movement twists the closed end of the tube. Angular rotation of the fixed end of the shaft is passed down a solid shaft. The solid shaft is mounted inside the torque tube. This shaft rotates as the closed end of the tube rotates. Attached to the other end of the solid shaft is the lever that drives the display mechanism.

The operating range of a bellows instrument that uses a torque tube is somewhat limited. To change this range, the tube itself must be changed. The tube also serves as a spring that opposes bellows travel.

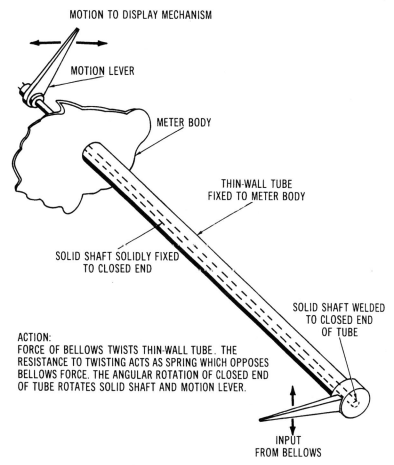

MOTION TO DISPLAY MECHANISM

MOTION LEVER

METER BODY

THIN-WALL TUBE
FIXED TO METER BODY

SOLID SHAFT SOLIDLY FIXED
TO CLOSED END

SOLID SHAFT WELDED
TO CLOSED END
OF TUBE

ACTION:
FORCE OF BELLOWS TWISTS THIN-WALL TUBE. THE
RESISTANCE TO TWISTING ACTS AS SPRING WHICH OPPOSES
BELLOWS FORCE. THE ANGULAR ROTATION OF CLOSED END
OF TUBE ROTATES SOLID SHAFT AND MOTION LEVER.

INPUT
FROM BELLOWS

FIGURE 7-5   Torque tube.

Sealing Bellows

The second important method of taking out the bellows position is through a sealing bellows, or flexure. This flexure is also called a torque tube, possibly because both have similar constructions. Devices that use torque tubes do not use range springs. Where the bellows seal or flexure seal is used, a spring is also used. The seal must be designed to offer as little resistance to the bellows movement as possible. When the sealing bellows is used, the motion of the bellows is opposed by a spring. To change the span of a bellows instrument using a seal, the spring must be changed. This is achieved by changing to a unit with a different spring constant.

Damping

The bellows instrument just discussed was a single-bellows device. The high-pressure compartment was formed of the volume outside the single bellows. The low-pressure compartment was formed of the volume inside the bellows. This arrangement has the advantage of simplicity. A problem arises because it cannot be damped down and is subject to a large amount of pulsation and flutter.

To solve this problem, instruments are supplied with two bellows. This construction is shown in Figure 7-4. The inside volume of each bellows is connected through an adjustable restriction. This volume is filled with a fluid. Differential pressure applied to the outside of each bellows tends to compress them. The fluid inside of the bellows is also compressed, which stops the bellows from contracting. The fluid is squeezed out of one of the bellows through the restriction and into the second. The rate at which the bellows will contract or expand is regulated by the amount of restriction. This is a damping action. The bellows drive a shaft connecting both bellows. The shaft drives a lever through a sealing tube. The spring loading on the bellows system establishes the range of the instrument.

Compensation

Use of the filling fluid introduces a temperature compensation problem. This problem occurs because the fluid volume changes with the temperature. A variable-volume compartment is provided at one end of the bellows system. The increase in fluid, due to temperature, expands into this variable-volume compartment. This is not a compensation for temperature; the compensation is for changes in the volume of the filling fluid to prevent damage to the bellows system.

DIAPHRAGM DIFFERENTIAL-PRESSURE INSTRUMENTS ▰▰▰▰

Diaphragm differential-pressure instruments will now be considered. This discussion will focus on the diaphragm and its take-out mechanism. The display mechanism, which is controlled by a pneumatic or electrical signal, will not be discussed in any detail.

One important characteristic of the diaphragm instrument is its limited movement before becoming taut. This deflection is on the order of 0.001-in. The instrument must be designed so that its diaphragm does not move with the impact of differential-pressure. This is done by using a force that the differential pressure causes when it is applied to the diaphragm. The force is then counterbalanced by outside air pressure.

Components

The components of a diaphragm instrument are:

1. A high-pressure component.
2. A low-pressure component.
3. A movable wall between the two compartments.

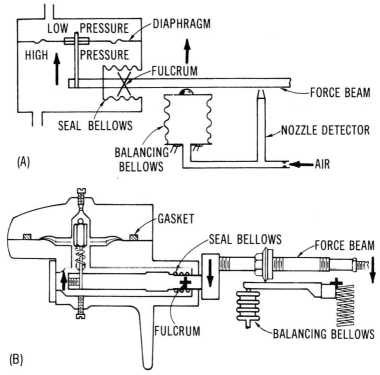

FIGURE 7-6   Diaphragm differential-pressure instrument. (A) Schematic diagram;
(B) Simplified drawing.

4. A force beam.
5. A pressure-tight connection.
6. A display mechanism.
7. A balancing force.

Arrangements

The diaphragm is bolted between two hollowed-out flanges, which forms the high- and low-pressure compartments. The diaphragm is fastened to one end of the force beam. The other end is loaded with the force of a pneumatic bellows. This is the balancing force. The position of the beam is detected by a flapper/nozzle assembly as shown in Figure 7-6.

Operation

The differential pressure across the diaphragm exerts a force back on the diaphragm. This force is applied to the force beam. Because of this, it rotates a very small amount. The axis of the rotation is at the seal bellows. The extremely small movement is detected by a flapper/nozzle mechanism. An air pressure then builds in the feedback bellows. The bellows expands, exerting a force on the force beam. When the differential-pressure force is balanced by the bellows force, the force beam comes to rest. There is air pressure in the bellows, and the additional pressure measures the differential pressure being applied to the instrument. This air pressure is connected to an element that can be calibrated in flow, liquid level, etc.

BELL DISPLACER MANOMETERS ████████████████████████

Another type of differential-pressure instrument, the displacer element, uses buoyancy to measure a differential pressure. Instruments of this type are used to measure low differential pressures produced by flow.

Components

1. A high-pressure compartment.
2. A low-pressure compartment.
3. A sealing fluid.
4. A bell-shaped displacer.
5. A take-out mechanism.
6. A display mechanism.

Arrangements

The bell manometer is used as a level determining element. Refer to the bell manometer of Figure 5-2. The bell-shaped displacer is immersed upside down in a pool of mercury. The open end of the bell sinks in the mercury, forming two compartments. One compartment is inside the bell and is connected to the outside of the meter. The connection is through a standpipe long enough that its end is above the surface of the mercury. The second compartment is the space outside the bell and inside the meter body. Mercury seals separate these two compartments.

Principle of Operation

A bell displacer normally sinks in a pool of mercury. Sinking continues until the weight of displaced mercury equals displacer weight. This action should be recognized as Archimedes' principle. The differential pressure is applied across the displacer. The displacer will lift out of the mercury or push farther into the liquid, depending on whether high-pressure is inside the bell or the outside compartment. The force of differential pressure will displace the bell. Displacement will continue until the buoyant force balances the differential-pressure force.

The bell movement must be extracted from the meter body. In general, this procedure is similar to that of the float manometer.

Suppose the range of the displacer manometer is 10 inches of water. Suppose further that the displacer travel required to stroke a pen is one inch. Assume that high-pressure is applied under the bell. Now suppose that 10 inches of pressure is applied to the instrument. This pressure tends to lift the bell out of the mercury. The force involved will be equal to 10 inches of water pressure times by the area of the bell. As the bell rises, it displaces a smaller amount of mercury. The buoyant force of the mercury is reduced because only part of the bell is lifted out of the mercury pool. Less mercury has been displaced. Buoyant force is reduced by an amount equal to the volume of the bell lifted out of the liquid. For low-range instruments, equal volumes of bell position can be equated to increments of applied differential pressure. This occurs because wall thickness is the same throughout. As a result, a linear relationship between pressure input and bell movement occurs.

THE CHARACTERIZED DISPLACER

Suppose that the bell wall is tapered rather than being the same thickness. This would make a section through the wall triangular in shape, as shown in Figure 7-7. Different measurements of differential pressure would be required to raise the bell equal amounts. If the thick part of the displacer were down, then a small differential would raise it at the low end of the range.

Tapered bell thickness is used in a characterized displacer. This displacer will obtain different incremental lifts for the same increments of applied differential pressure. It is possible to design the shape of a displacer to accomplish different math functions. Suppose that for 50 percent of the head range the displacer would move 70.7 percent of its travel. The square root of 0.50 is 0.707. The displacer, in this case, has be characterized.

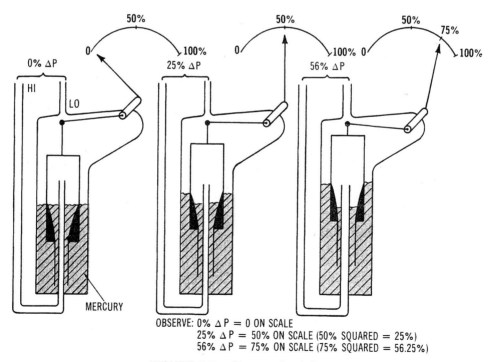

OBSERVE: 0% Δ P = 0 ON SCALE
25% Δ P = 50% ON SCALE (50% SQUARED = 25%)
56% Δ P = 75% ON SCALE (75% SQUARED = 56.25%)

FIGURE 7-7   Characterized displacer.

It causes a display to equal the square root of the applied differential pressure. The value of a characterized displacer will be better appreciated when orifice flow measurements are studied.

CALIBRATING DIFFERENTIAL-PRESSURE INSTRUMENTS ▬▬▬

The calibration of differential-pressure instruments represents a big part of the work in instrument service. When these devices are used to measure flow, a number of problems arise. A flowmeter may employ an orifice in its construction. The differential pressure which develops is measured upstream and downstream of the orifice. The resulting output is the square-root of these pressure values. A second problem arises out of the false head developed by mercury instruments when these devices run wet. A third problem arises out of the terminology used to describe types of meters and how they are applied. The fact that differential-pressure instruments are expected to respond to fractions of inches of water or millimeters of mercury further compounds all of these problems. We will devote some time to these problems. In addition, calibration procedures will be developed that will apply to all differential-pressure instruments.

TERMINOLOGY ▬▬▬

In differential-pressure instrument calibration, a terminology problem is caused by the meaning of the terms wet and dry. It is also the result of how these terms are used in an instrument. For example, a dry meter may be run wet and calibrated wet. But it may also be run dry and calibrated wet. The terms needing definition are:

wet meter	dry meter
run wet	run dry
wet calibration	dry calibration

A wet meter is a mercury meter. A dry meter is a nonmercury instrument. A bellows or diaphragm is a dry instrument. Run wet means that the instrument has a fluid in the meter body, usually water. It may also be a transfer fluid, gasoline, oil, or alcohol. Run dry means that the instrument has air, carbon dioxide, or other gas in the meter body.

The terms wet and dry calibration are not as easily defined. The most common definition of a wet calibration is one made with water in the meter body. A dry calibration is one with air in the meter body. The confusion of wet versus dry calibration is that there is no difference in the magnitude of the input signal. The problem is simply one of technique and convenience. There is a well-established view held that a meter that runs wet must be calibrated wet; however, this is not the case.

THE "FALSE HEAD" PROBLEM ▬▬▬

A false head problem arises whenever a wet meter is used or calibrated with a liquid. Mercury moves up or down when differential pressure is applied to the meter. The water head of the meter also moves because the space vacated by the mercury is replaced with water. The addition of a water head contributes to the differential pressure being applied. As a result, the total head consists of the applied differential pressure plus the differential water head. These occur because the mercury level in the float chamber differs from the mercury level in the range chamber. A meter could be easily calibrated with water in the meter body. The differential pressure would be less than it would be if there were no water at all. The differential pressure is reduced by a factor of 0.93. Fluids of other densities require a different factor. The density of the transfer fluid establishes this factor. Figure 7-8 shows how transfer fluids alter the false head problem.

WET VERSUS DRY CALIBRATION

A meter can be calibrated with or without water in the meter body. In either case, the total differential pressure on the meter is the same. In wet calibration, it is necessary to apply only part of the total head. The remaining part is automatically made up by the water filling the space vacated by the mercury. The part to be applied is 93 percent of the range value. For example, calibrate with water in the meter body a mercury meter in the range of 0 to 50 inches. Get calibrating values from column B of Table 7-1. The values in column B are 93 percent of the values in column A.

Observe that the calibration of a meter, either wet or dry, results in exactly the same procedure. The end use of the meter is not a consideration when calibration is to be made. The determination of whether to calibrate wet or dry is one of convenience. Assume that a field check is to be made and the meter has water in its body. It is more convenient to calibrate the meter wet than to dismantle the meter and remove the water.

In meters that do not use mercury, the problem of a water head is added. As a differential pressure is being applied, it is no longer a consideration. There is no displacement of mercury in a dry meter. Therefore, the applied differential pressure always equals the total differential pressure. No correction factor needs to be used when making a wet calibration.

The replacement of a wet meter with a dry meter of the same range will result in an error in flow calculation. Instrument calibrations remain the same and are correct. The error lies in assuming that for a 100 percent chart reading for a meter running wet, a differential pressure of 100 percent is being applied. It is not. The applied differential pressure is the differential pressure being delivered by the orifice. To drive the pen to 100 percent of the chart, only 93 inches of water need be delivered by the orifice.

TABLE 7.1.   Values for Wet Calibration

Percent of Scale	(A) Ranges Values	(B) Calibrating Values
90	45 inches	41.85 inches
50	25 inches	23.25 inches
10	5 inches	4.65 inches

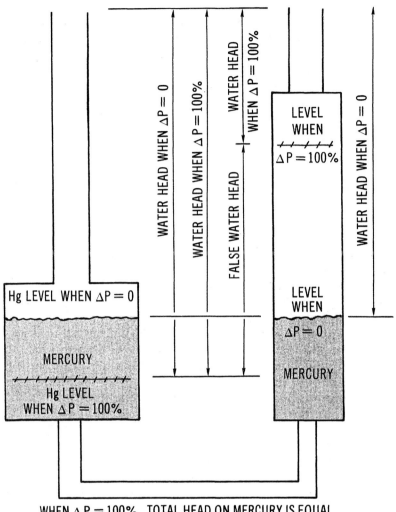

WHEN $\Delta P = 100\%$, TOTAL HEAD ON MERCURY IS EQUAL
TO APPLIED HEAD PLUS "FALSE HEAD"

FIGURE 7-8   "False head" problem.

CALIBRATION EQUIPMENT

The usual calibration standard is a column of fluid in a glass-tube manometer. The fluid may be water or some of the "red oils" of a known specific gravity. A common heavy red oil has a specific gravity of 2.94. The use of this oil cuts the length of the manometer by a factor of 1/2.94. For all calibration, except wet calibration, the following equipment is required:

1.   A calibration standard gauge.

2. An input signal pressure source.
3. A visual display.
4. Tubing and fittings to assemble the equipment.

Item 1 is a precision gauge of sufficient accuracy. Item 2 is usually compressed air that passes through a regulator. Item 3 is the meter or recorder. If the instrument is a transmitter, the display is a calibrated device that is sensitive to the transmitter output. If the system has an air pressure that varies from 3- to 15-lb/in^2, the display would be a calibrated gauge or manometer of a suitable range. A four-foot manometer using mercury and calibrated in terms of inches of water forms an excellent display. Item 4 is the necessary fittings and piping needed to assemble the equipment. Figure 7-9 shows equipment setup for a dry calibration.

WET-CALIBRATION

A wet-calibration unit has the same basic components as a dry calibration unit. There are a few differences, which can be seen in Figure 7-10. The pressure signal is obtained by raising and lowering a can of water. This action changes the head on the meter. The bottom of the can is connected to the high-pressure side of the instrument being calibrated. The calibration standard is a water-filled manometer. In order to reduce meniscus errors, the glass tube of the manometer should be no smaller than 1/2-inch in diameter. All connections should be 1/2-inch pipe or larger. A general consideration is the need to get all the air out of the connections, tubing, and instrument. Larger pipe sizes make it easier to eliminate the air.

CALIBRATION PROCEDURE

Mechanical Stops

1. Are the mechanisms free to travel their range without hitting stops?
   a. With no mercury in the body, the pen should go slightly below zero or above 100 percent depending on the meter. If it does not, bring the pen to slightly above 100 percent or below zero by adjusting the pen micrometer.
   b. Add mercury until the pen goes slightly below zero or above 100 percent. Gently tap the meter body while adding the mercury. Some time is required for mercury levels to equalize. Both check valves must be unsealed during this procedure.

Accuracy

1. Is the mechanism free of friction, dead space hysteresis and sufficiently sensitive? Check the link/lever mechanism and pressure-tight bearing for friction.
   a. Feel for end play. There should be some end play, and the bearing should freely move in and out.

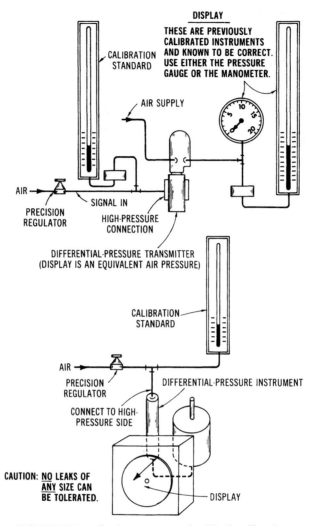

FIGURE 7-9　Equipment setup for "dry" calibration.

b.　Observe the position of the pen.

c.　Gently push the float into the mercury. Use a stick through the vent connection to move the float 3/8-inch.

d.　Very slowly lift the stick, allowing the float to return to its starting position.

e.　Observe the position of the pen. It should return to within 1/32-inch of the original pen position. If it does not, recheck the link/lever mechanism. This procedure was discussed in chapter 3. If this does not correct the condition, overhaul the pressure-tight bearing.

2.　Check the pressure-tight connections mentioned in items a through d.

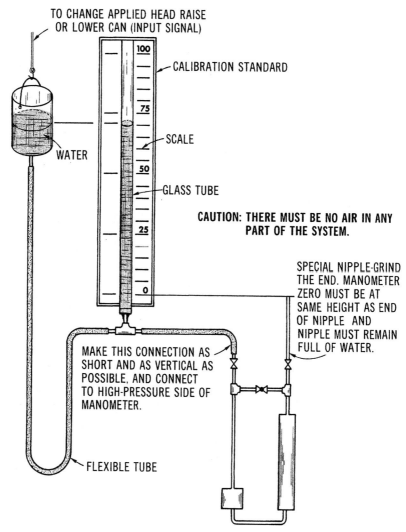

TO CHANGE APPLIED HEAD RAISE
OR LOWER CAN (INPUT SIGNAL)

CALIBRATION STANDARD

SCALE

WATER

GLASS TUBE

**CAUTION: THERE MUST BE NO AIR IN ANY
PART OF THE SYSTEM.**

SPECIAL NIPPLE-GRIND
THE END. MANOMETER
ZERO MUST BE AT
SAME HEIGHT AS END
OF NIPPLE AND
NIPPLE MUST REMAIN
FULL OF WATER.

MAKE THIS CONNECTION AS
SHORT AND AS VERTICAL AS
POSSIBLE, AND CONNECT
TO HIGH-PRESSURE SIDE OF
MANOMETER.

FLEXIBLE TUBE

FIGURE 7-10   Equipment setup for "wet" calibration.

Calibrate

1. Does the instrument reproduce the input signal as indicated by the calibration standard? The methods of adjusting span, angularity, and zero are the same as those of a link/lever mechanism. These adjustments were discussed in Chapter 3.

   a. Apply a 50 percent-of-range head to the instrument. In flowmeters, 75 percent of the chart is used for this critical point adjustment. Make this adjustment to minimize angularity errors.

b. Check the span by feeding in a 10 percent head, then a 90 percent head. Adjust as for any link/lever instrument.

c. Check zero by feeding in a head of any value between 10 percent and 90 percent. Adjust as for any link/lever mechanism.

d. Recheck as in item a by putting in a 10 percent, 50 percent, and 90 percent head. If the pen does not fall on all three points, readjust angularity as for any link/lever mechanism.

e. Repeat steps a through d as required.

Wet Calibration Procedure

The procedure just discussed is applicable for a wet calibration. There are two precautions that must be considered. First, the instrument must be free of all air bubbles. Vent connections are provided in the meter for freeing air pockets. A fair amount of tapping is required to shake the instrument free of air bubbles. It is equally important that the tubing of the test rig also be freed of air. Some tapping of the tubing is required to do this. To free the tubing of air, locate the manometer relative to the instrument. Tubes and connections should be as short and vertical as possible.

Second, the special nipple of the low-pressure compartment must be filled with water at all times. The manometer zero should be level with the top of this nipple.

SUMMARY

Manometer/float differential-pressure instruments are similar in many ways to the glass-tube manometer. Instruments like this are used as flowmeters and used to make liquid-level measurements. The instrument must be sensitive to differential pressures.

A differential-pressure system must have two parts: a static-pressure and the difference components. These two signals are each connected to a separate chamber. The static component is cancelled by the alternative signal. The differential signal is the resulting output.

The elements of these instruments are used to develop either an electrical or pen-on-chart display. In both cases, these display instruments are easily blown. This means that a large applied pressure will force the mercury out of the instrument.

There are several types of differential-pressure instruments. A diaphragm differential-pressure instrument is one. This instrument produces a deflection on the order of 0.001-in. A bellows differential-pressure instrument generates movement of tenths of an inch. The bell displacer manometer is a buoyancy instrument. This mechanism produces the greatest amount of display.

Calibration of this class of instruments represents a problem in maintenance. A known input will produce an equal output if calibration procedures are followed.

In the calibration of a differential-pressure instrument, false head problems arise. This can be coupled with wet/dry problems and square root problems. A precision gauge can be useful in overcoming these problems. Also, an input pressure, a display, and assembly components are helpful.

Calibration involves adjustment of mechanical stops, accuracy adjustment, and evaluation of the output. Air must be cleared from the instrument and tubing to reduce errors.

ACTIVITIES

DIFFERENTIAL-PRESSURE INSTRUMENTS

1. Refer to the manufacturer's operational manual for the instrument used in this activity.
2. Identify the function of the instrument.
3. Remove the housing or cover from the instrument.
4. Locate the differential-pressure element of the instrument.
5. Identify the type of differential-pressure element used in the instrument.
6. Indicate where the inputs and output of the mechanism are located?
7. Does the element have adjustments?
8. Locate the adjustments and describe the function of each.
9. Reassemble the instrument.

INSTRUMENT CALIBRATION

1. Refer to the manufacturer's operational manual for the instrument used in this activity.
2. Examine the calibration procedure for the instrument.
3. Connect a calibration standard instrument to the system.
4. Apply an input air pressure source to the system.
5. Is the uncalibrated instrument mechanism free to travel its range without hitting stops?
6. Does the uncalibrated instrument reproduce the input signal as indicated by the calibration standard?
7. Apply a head pressure of 50 percent of range head pressure to the system.
8. Compare the value of the standard instrument and the uncalibrated instrument.
9. Adjust the uncalibrated instrument so that it compares with the standard instrument reading.
10. Check the span by feeding in a 10 percent head, then a 90 percent head. Adjust the link/lever mechanism if necessary.
11. Turn off the pressure source and remove the standard instrument from the system.

QUESTIONS

1. What are the basic components of a manometer/float differential-pressure instrument?
2. What is differential-pressure?
3. What are the basic components of a bellows differential-pressure instrument?

4. Explain the operation of a bellows differential-pressure instrument.
5. What are the basic components of a diaphragm differential-pressure instrument?
6. Explain the operation of a diaphragm differential-pressure instrument.
7. What are the basic components of a bell displacer differential-pressure instrument?
8. What is meant by the term calibration?

8
Flow Measuring Instruments

OBJECTIVES

Upon completion of this chapter, you will be able to:

- Define the term flow.
- Identify different types of orifice flow instruments.
- Explain how differential pressure is used to determine flow.
- Explain how an orifice is used to measure flow.
- Identify flow meters that use a rotameter in their operation.
- Distinguish between direct and inferred flowmeters.
- Identify instruments that use the velocity principle in their operation.
- Explain how the volumetric principle is used to determine flow.

KEY TERMS

Anemometer.   An instrument for measuring the rate or velocity of airflow.

Annular space.   An area formed by a ring around the inside or outside of an object.

Batching.   A process that involves a quantity of material prepared or required for one operation.

Beta ratio.   A ratio that compares the inside pipe diameter to the orifice diameter. This is used to determine the flow velocity and differential pressure of a system.

Centrifugal.   Acting in a direction away from a center or axis.

Concentric.   A device that has a common center or axis in its construction.

Constant.   Something that is invariable or unchanging such as a number that has a fixed value in a given situation.

Cross bred (cross).   A hybrid produced by combining two or more varieties of devices into a single unit.

Eccentric orifice.   A circle or hole that is off center or not having an exact center as used in the construction of an orifice plate.

Empirical.   Factual information obtained as the result of experimentation or observation, without reliance on scientific theory.

Flare.   An enlargement at the end of a piece of flexible tubing or solid material.

Inferential.   The act of relating one condition or value to another condition or value that is believed to follow the original.

Kinetic energy.   Energy related to the motion or movement of a body or material.

Nutate.   A rocking motion or wobble that occurs when a disk rotates.

Pipe taps.   Upstream and downstream differential pressure measurement points in the pipe of an orifice installation.

Plummet.   A solid piece of material that is placed in the tapered tube of a rotameter and responds to the flow of liquid or gas.

Rotameter.   A variable area flowmeter consisting of a tapered metering tube and a plummet that responds to the flow of gas or liquid.

Segmental orifice.   A hole in a metal plate that forms a sector or part of a circle in the construction of an orifice.

Stroked.   A sudden action or process that occurs when an instrument moves through an operational cycle.

Tachometer.   A device for indicating the speed of rotation.

Ultrasonic.   A device that responds to alternating current. Electricity that has a frequency higher than the human range of hearing. Typical frequencies are in the range of 40 kHz, or kilo Hertz.

Velocity.   The speed of a moving object or material.

Vena contracta.   A point downstream of an orifice where the velocity is maximum and the pressure is minimum.

Venturi tube.   A short tube of varying cross sectional area that causes a pressure drop in the smallest section due to the velocity of flow.

Viscosity.   A measure of the thickness of fluid or the resistance to its flow.

INTRODUCTION ■■■■■■■■■■■■■■■■■■■■■■■■■■■■■■■■■■

Nearly all manufactured products are related in some way to the flow of materials such as gas, liquids and solids. But this also includes the control of large amounts of oil or gas passing through a pipe-line. The measurement of flow is a very important process control application.

Flow processing is somewhat simplified when compared with other instruments. This process is primarily used to perform a measurement function. It may then be used to achieve some type of control. The complexity of this process is in the measuring technique. The sensing elements of a flowmeter are usually of the inferred type. This element can decide on one value by observing a change in another value that is believed to follow the original. For example, temperature is measured by observing the length of a column of mercury. Flow rate can be determined by observing a change in differential pressure. The output of the sensing element may be either mechanical or electrical, depending on the type of sensor being used. This discussion will concentrate on mechanical flow principles that relate to pneumatic instrument applications.

ORIFICE FLOW ■■■■■■■■■■■■■■■■■■■■■■■■■■■■■■■■■■

Differential pressure may be used to measure a process variable. In this discussion, primary devices are used to convert flow into an equivalent differential-pressure signal. The most common flow to differential-pressure converter is an orifice plate. Investigation of the orifice plate is used to show some of the problems and limitations of orifice flow measurement.

Orifice is an abbreviation of the term orifice plate. An orifice plate is a disk of metal about 1/16 to 1/4-in thick. A hole, or orifice, is cut from the center, or near the center, of the disk. The outside diameter (OD) of the orifice plate fits within the bolts of a standard pipe flange. An important consideration in selecting an orifice plate is the ratio of its opening (d) to the inside diameter of the pipe (D). This is called the beta ratio. If the d/D ratio is too small, the pressure is too great. If the beta ratio is too great, the loss of pressure is very small. Small pressure values are unstable and difficult to detect. Beta ratios of 0.2 to 0.8 provide the best accuracy.

The construction and dimensions of an orifice plate are shown in Figure 8-1. The complete name for this piece of equipment is the orifice in a thin plate. A thin plate is defined as a piece of material not thicker than 1/16-in. Orifice plates are generally thicker than 1/16-in. In order to meet the 1/16-in requirement, the plate is thinned down near the center. This involves beveling the orifice to 1/16-in around the hole. A second requirement is that the leading edge of the orifice be a square edge. The leading edge is the side that faces the flow. To square the plate means that the edge be free of sharp edges. It also must be free of extended wire or burrs.

These requirements are necessary because the performance of an orifice is determined only by actual testing. These tests are made with orifices meeting these requirements. The bulk of information is on pipe sizes less than 12-in using air, water, or steam as a fluid. The significance of these observations is that the calculation of the flow/differential-pressure relationship is questionable for uncommon fluids. For example,

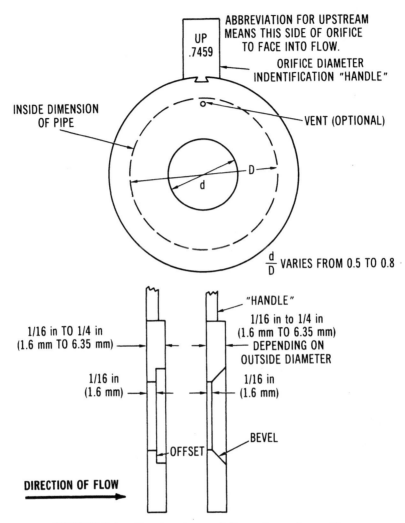

FIGURE 8-1   Construction and dimensions of an orifice.

sulphur, sodium nitrate, or hydrocarbons are questionable if their physical properties are not known. Remember that the kind of fluid and the larger pipe sizes may make it more difficult to predict flow values. The major source of problems with orifice differential-pressure installations is the piping between the orifice and the differential-pressure instrument.

PRINCIPLE OF OPERATION

One of the basic laws of physical science states that energy can neither be created nor destroyed. The orifice utilizes this law in its role as a flow-to-differential-pressure converter. Figure 8-2 shows how this law is applied.

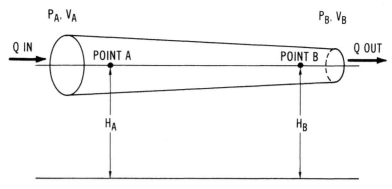

FIGURE 8-2   Basic flow laws.

Assume that a flow of Q enters the pipe at point A. Q is measured in gallons per minute, or gpm. It can be said that if Q gpm enter the pipe, then Q gpm must also leave the pipe. Observe that the pipe diameter at B is less than the diameter at A. If the same amount of fluid is to pass by B as passed by A, the speed or velocity of the fluid must be greater. Examine point A:

1. The fluid has internal pressure which will be called P_A and is the pressure head at A.
2. The fluid has kinetic energy which will be called V_A and is the velocity head at A.
3. The fluid has elevation above a predetermined height. This elevation is called H_A.

At point B, the same kinds of heads are present as seen at point A—namely, P_B, V_B, and H_B. The law of conservation of energy states that the total energy at A must equal B. Mathematically, this is stated as:

$$P_A + V_A + H = P_B + V_B + H_B$$

To simplify this relationship, suppose points A and B are at the same height or $H_A = H_B$. The relationship then becomes:

$$P_A + V_A = P_B + V_B$$

Due to the difference in diameter at points A and B, the velocity of the fluid at B is greater than at A. Examine the last equation. If V_B is greater than V_A, then P_B must be less than P_A if they are to be equal.

In other words, the restriction of the smaller-diameter pipe at B causes the velocity to increase at B. To balance this increase in velocity, the pressure must drop. There is a difference in pressure between P_B and P_A. This is a differential pressure whose magnitude depends on the velocity. The velocity for a given system is a measure of the quantity for flow through a pipe. It is in this way that flow is converted to differential pressure. Suppose this pipe is squeezed together so that it is about 3-in long. This is the equivalent of an orifice in a short section of pipe. This same principle also applies to an orifice plate placed in a pipe.

Basic Flow Law

The relationship between flow and differential pressure depends on another ratio. This is the beta ratio of the orifice and the physical properties of the fluid. The actual relationship is stated as inches of water pressure for a given distance. The basic expression is:

$$Q = cA\sqrt{2gh}$$

where,

Q is the flowing rate (volume per unit of time),

c is a number which takes into account the physical properties of the fluid being measured,

2g is due to the force of gravity (32.4 ft/sec/sec) × 2, or 64.8,

h is the differential pressure in inches of water,

A is the area of the orifice.

An important point to note here is that flow is related to the square root of the differential pressure. This fact introduces a square-root problem that will be discussed in Chapter 9.

Orifice Variations

Discussion of the orifice thus far assumes that it is a hole in a thin plate. There are several variations of the orifice plate. The theory of the orifice in a thin plate applies to these other devices.

The Venturi is the most accurate and the most efficient primary device. This device is shown in Figure 8-3A. The efficiency of an orifice is measured by the recovery it has from a pressure drop. For example, a common differential pressure for the orifice at its peak flow is 100-in of water pressure. On most thin-plate orifices, 30 percent to 50 percent of the head is recovered, depending on the ratio of the diameters between the orifice and the pipe. The Venturi pressure recovery may be as high as 90 percent.

The flow nozzle is a flared piece usually designed to be installed between flanges. Nozzles have a higher throughput capacity as shown in Figure 8-3B.

The Venturi nozzle is a combination of a nozzle with a tapered downstream recovery system. The nozzle is shown in Figure 8-3C.

The Dall flow tube is a variation of the Venturi. It is designed to give pressure differentials with a high recovery factor.

An orifice has several different variations. They are:

1. Concentric.
2. Eccentric.

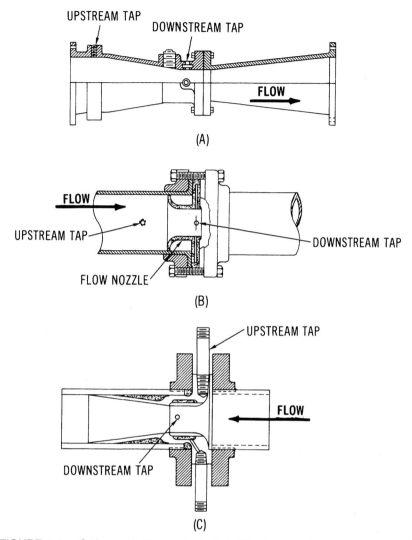

FIGURE 8-3 Orifice variations. (A) Venturi; (B) Flow nozzle; (C) Venturi nozzle.

3. Segmental.

These differences are shown in Figure 8-4.

The orifices may be provided with a small vent hole which falls just inside the pipe diameter. The vent hole allows gases in a liquid stream to flow by the orifice. This hole also allows condensed liquids in a gas stream. If the vent is for gases in a liquid stream, the vent is at the top. If the vent is for passing liquid in a gas stream, the vent is near the bottom.

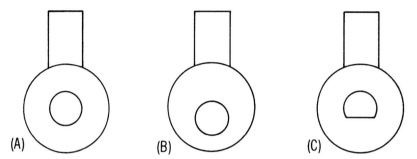

FIGURE 8-4   Types of orifices. (A) Concentric; (B) Eccentric; (C) Segmental.

ORIFICE TAPS

An orifice installation may have the actual pressure connection located in several positions relative to the orifice. Three examples are shown in Figure 8-5.

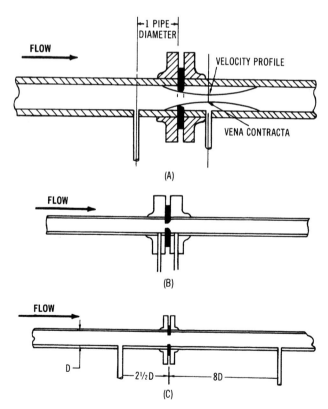

FIGURE 8-5   Tap locations. (A) Vena contract connections; (B) Flange taps;
(C) Pipe taps.

The flange taps are made through the flange body. The orifice flanges are the same as standard flanges except that the flange is thicker. A person may drill through the flange into the pipeline with this thickness. These openings are then tapped. Orifice flanges may be screwed or welded onto the pipe. The weld-neck and the slip-on welded orifice flanges are shown in Figure 8-6.

Pipe Taps

Pipe taps are made into the pipeline at a fixed distance from the orifice. Typical locations are 2-1/2 pipe diameters upstream and 8 diameters down stream. Differential pressure measured at the taps is the permanent pressure drop due to the orifice.

(A)

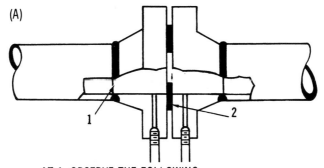

AT 1, OBSERVE THE FOLLOWING:
INSIDE OF WELD MUST BE CLEAN.
FLANGES MUST BE CONCENTRIC ON PIPE.

AT 2, OBSERVE THE FOLLOWING:
ORIFICE CONCENTRIC WITH FLANGE.
GASKET ID MUST BE NO SMALLER THAN PIPE ID.

(B)

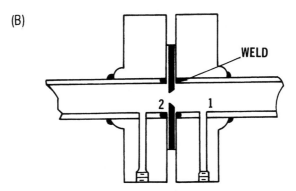

1: HOLE THROUGH PIPE WALL MUST BE
FREE OF BURRS OR WIRE EDGES.

2: BURN-IN WELD AND FILE SMOOTH.

FIGURE 8-6   Weld-neck and slip-on orifice flanges. (A) Weld-neck; (B) Slip-on.

Vena Contracta Taps

One connection is made about one pipe diameter upstream from the orifice. The second is located at the vena contracta. The vena contracta is the point downstream from the orifice where the velocity is maximum but the pressure is minimum. The maximum velocity is not at the orifice. It is where the stream "pinches down" for a short distance after passing through the orifice, which is due to the kinetic properties of the stream when it becomes narrow and is only for a short distance after passing through the orifice. The location of the vena contracta depends on the beta ratio. This is orifice diameter (d) to inside pipe diameter (D). Location of this tap represents the maximum pressure drop measuring point.

Flange Taps

A large number of new orifice installations use flange taps. The reason is one of convenience. The installation of a connection in a pipe wall is a critical job. The entrance must be precisely at right angles to the pipe wall and at a proper distance from the orifice. The inside of the pipe must be smooth and the hole free of burrs. Under field construction conditions, it is difficult to consistently get good fabricated connections. The weld-neck orifice flange eliminates most of the problems of making an entrance into the pipe. This device works easily because all the connections are pre-machined in the flange.

In flange installations there is still a problem of the butt weld, located upstream from the orifice. The flange must be concentric with the pipe. No welding ring should be used. The inside of the pipe must be clear of weld build-ups.

Slip-on and screwed flanges do not eliminate the problem of entering the pipe. After the flanges are put on, the pipe wall must be drilled through. The hole through the wall must be freed of burrs and wire edges and must be smooth. Since these connections are near the flange, they are accessible. This is not the case with pipe tap connections. The slip-on flange should be located with the face of the flange and the end of the pipe almost flush. Setting the conventional slip-on flange one pipe wall thickness back will not work here. The welder must burn in the weld bead. Also, a lot of filling is needed to smooth the face of the flange.

The vena contracta, flange tap, and pipe tap are three standard connections. In addition, there are several special connections. For pipe sizes less than two inches and over twelve inches, these special arrangements are needed. These variations do not represent any service problems. However, a technician should be alert to these variations. It is important to see that the connections agree with the flow data.

METER TYPES

Three types of differential-pressure instruments have been discussed. These are the float/manometer, the bellows, and the diaphragm. Piping for each instrument is determined by its internal volume. The float/manometer and bellows instruments displace a relatively large quantity of measured fluid. Assume now that a manometer/float instrument is stroked. This could cause the float to move one inch in a four inch diameter chamber.

12-1/2 in^3 of fluid must move in and out of the chamber during this operation. A diaphragm element requires a negligible quantity of fluid to stroke the instrument. Something in the order of hundredths of cubic inches is normal. A diaphragm element moves very little fluid in and out of its piping. The float/manometer or bellows, by comparison, move large quantities of fluid through the connecting pipes.

MEASURED FLUID PROPERTIES

The properties of fluids that determine piping arrangements are:

1. The state of the measured material at flowing temperatures and at the ambient temperature of the differential-pressure instrument.
2. Chemical properties such as corrosiveness of the material.
3. Mechanical properties such as viscosity and solids content.

The term state of the material is a rather general way of describing the conditon of a fluid. In this regard, fluids may be gaseous, liquid, or solid. The state of a material is largely determined by its operating temperature. Pressure also has some influence on the state of a material. All parts of an installation operate at the same pressure. Pressure is important only in determining such things as the pressure rating of meters, flanges, valves, and piping.

In flow installations, a material may be elevated in temperature as it passes through the pipes. This is reduced to ambient temperature when it passes through the instrument. This difference in temperature may result in a material changing state. For example, steam in a pipeline is in a gaseous state. However, steam condenses into water at the ambient temperature of the instrument. The state of material in the instrument is liquid. This is an example of gas changing to a liquid. Suppose that carbon dioxide is flowing through a system at a reduced temperature, which causes carbon dioxide to be in a liquid state. When it reaches the ambient temperature of the instrument, it may change into a gas. This is an example of a liquid-to-gas change of state.

A fair number of chemicals are liquids at elevated temperatures and solids at ambient temperatures, meaning that a number of chemicals will become solid in the meter body. Solutions to the change-of-state problem are:

1. Prevent the change in temperature by raising or lowering the meter body.
2. Allow the change in state.
3. Introduce a second fluid as a transfer liquid.

It is necessary to prevent a change in state by methods 1 or 3. In method 1, it is possible to raise the temperature of the meter only a limited amount. A diaphragm instrument can be raised to approximately 150° Celcius. The bellows instrument uses air pressure in its display. By using pressure, it may be raised to approximately 100° Celcius, providing that it is not a double-bellows device with a damping liquid. Nonmercury differential-pressure instruments with an electrical display are limited to temperatures that will not break down the insulation. Electrical transmitters do not allow large temperature increases in the meter body.

The second alternative is to allow the change of state. This is acceptable to many fluids if the change is from a gas to liquid. An example would be to allow steam to condense into water without it freezing. In no way can the material change its state to a solid.

The third alternative is the introduction of a tranfer fluid. This method introduces a second fluid into the system that fills the meter body. The differential pressure of the orifice is transferred through the fluid to the instrument. This method is limited by the small number of transfer fluids available. A fluid, to be suitable must:

1. Be liquid at the meter body temperature. It is possible to raise the temperature of the meter body to some extent.
2. Remain liquid at the elevated temperatures of the material being measured.
3. Not mix the material with another, which would include absorbing and giving off the measured fluid when it becomes a gas.
4. Have a density greater than the material being measured.
5. Have chemical stability over long periods of time.
6. Not react with the material being measured.

The availability of materials meeting these requirements is limited.

Chemical Properties

The chemical properties of a measured fluid are used to determine the piping of a system. If the material is corrosive, then the piping must be made of materials immune to corrosion. Special alloys and nonmercury instruments are used to meter corrosive liquids. If corrosion-resisting alloys are not available, a transfer fluid may be used. The properties of the transfer fluid used for chemical protection are the same as those of a state change. In general, there is no upper temperature limit in chemical protection.

Mechanical Properties

The mechanical property of material passing through a system is an important consideration. Properties include viscosity, the percentage of solids in a liquid stream, and the liquid and/or solids in gas. Viscosity of the stream seriously alters the sensitivity of instruments.

The percentage of solids in a liquid stream tend to separate in the meter and can plug the piping and collect in the meter body. Solids are also likely to collect in front of the orifice, which introduce errors in the sensing element.

As solids increase, the stream becomes more difficult to measure. If the solid content is high, special steps are required to obtain a satisfactory primary flow element. A special primary device, such as a flow nozzle or a Venturi, must be used. The solution to the solids problem may be:

1. The introduction of a purge fluid, either continuously or intermittently.
2. Sealed orifice connections.

The cost of the purge fluid may be important. The possibility of contaminating the measured stream is a possibility. Cost factors permit the use of only very cheap fluids. Materials such as air, water, and carbon dioxide are cmmonly used. Perhaps liquid from a separator might be fed back into the system. The dilution effect of the purging stream must also be considered since the purge materials enter the process. All effects of the purge stream on the process stream must be considered.

Sealed orifice connections are similar in construction to seals used to keep process fluids out of pressure gauges. These seals are made to the pipeline directly without use of the pipe. Such seals are used together with purging fluids entering the seal.

Liquids or solids in gas streams result from gases condensing into liquid droplets. This also results from dusts that have not been separated from the gas stream. Powdered products are sometimes mixed with air to speed transporting.

Liquid in a flow stream can be collected with drip pots located in the piping. This is similar to catching small amounts of solids in the stream. Drip pots are not needed if the liquid content is high. Depending on the kind of liquid, it is possible to let the connecting lines fill with liquid, which is similar to allowing the change in state from a gas to a liquid. Purging, sealing, and introducing a tranfer fluid are also possible solutions, depending on the particular gas and liquid involved.

THE ROTAMETER

The rotameter is an instrument used to measure flow. This instrument employs a tapered metering tube. The tube contains a suspended float that moves up and down. The metering tube is mounted vertically with the small end located at the bottom. Fluid or air enters the tube at the bottom and passes upward around the float. The flow then exits at the top of the tube. Any position change of the float is used to indicate flow.

The following equation is the basis of rotameter operation:

$$Q = cA\sqrt{2gh}$$

This formula shows that a flowing quantity varies with the area (A) times the square root of the head. Thus far, the only comments on rotameter area were that the value depended on the ratio of the orifice diameter and ID of the pipe. Notice that the basic flow equation has two variable elements. Flow varies with the square root of the head if the area stays the same. If the head is constant, then the square root of the head is also a constant. The other possibility is the basis for a large group of flow-measuring instruments. These instruments are called variable area meters. The rotameter is one of the more important variable area meters.

Components

The components for a rotameter are:

1. A tapered tube.
2. A plummet or float.

3. A take-out mechanism.

4. A display.

NOTE: One rotatmeter uses a tube made of glass. The display of the rotameter is the position of the plummet within the glass tube. As a result, there is no take-out mechanism. Some people feel that the design of the tube of glass constitutes a take-out mechanism.

Arrangements

The tube ends of a variable area flowmenter are installed vertically in a pipe. End fittings are part of the instrument. The fittings are inserted into the instrument and packing is placed around the tube, thus making a pressure-tight connection. The end fittings are joined, forming a rigid member. If hazardous fluids are being measured, safety glass is put around the tubes. This prevents the hazardous fluid from entering if the glass tube breaks. An example of a glass-tube rotameter is shown in Figure 8-7.

If the tube is made of metal, the end fittings are welded to the tube. Also, the connecting pipe is screwed or flanged to the end fittings.

The plummet is put into the tube and is free to move. The walls usually guide the plummet, but some are connected to a rod. The plummet is shaped so that the flow tends to center it within the tube. The tube end fittings are equipped with stops. Stops are placed there to contain the plummet within the tube.

The plummet itself may be made of various materials. These include stainless steel or ceramic carbon, to name just two. In most cases, stainless steel is used. Other materials are used for flows that would corrode steel.

The shape of the plummet varies from a simple ball to fairly elaborate ones. The differing shapes are shown in Figure 8-8. The shape of the plummet is not of any consequence, except in the glass-tube rotameter. In these instruments, it is important that the top of the plummet be inserted in the tube with the top side up. The correct end of the plummet must be compared to the scale. Some plummets are designed so that their bottoms are compared to the scale. Others use the top. Still others use a flare to compare the scale. This is somewhere between the bottom and the top. In nonglass rotameters there is little chance of installing plummets backwards because they have extensions.

A molded-plastic rotameter is shown in Figure 8-9. This instrument uses a ribbed-bore tapered tube that guides a ball float. In this construction, the float is held in the center of the flow area. It reduces turbulence and provides a smooth flow through the tapered area. The window area provides a quick and easy display of the float position. Scale sizes vary from 10-in to 2-in. The longest scale provides an accuracy of ± 2 percent.

Principle of Operation

Suppose that with a vertical pipeline, the line has a transparent section with two different diameters. The diameter for the upper section is 2-in and the lower section is 1-1/2-in. A top-shaped plummet that weighs 1/2-lb is dropped into the transparent section. The water is turned on and flows into the pipe. If the water rate is low, the plummet will not move.

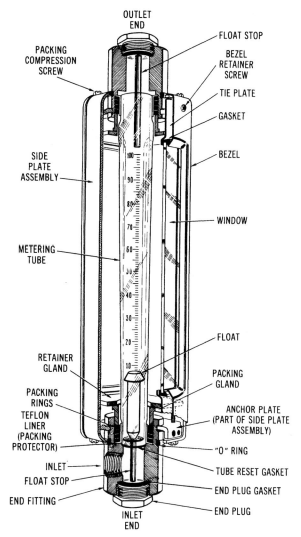

OUTLET
END

FLOAT STOP

PACKING
COMPRESSION
SCREW

BEZEL
RETAINER
SCREW

TIE PLATE

GASKET

BEZEL

SIDE
PLATE
ASSEMBLY

WINDOW

METERING
TUBE

FLOAT

RETAINER
GLAND

PACKING
GLAND

PACKING
RINGS

ANCHOR PLATE
(PART OF SIDE PLATE
ASSEMBLY)

TEFLON
LINER
(PACKING
PROTECTOR)

"O" RING

INLET

TUBE RESET GASKET

FLOAT STOP

END PLUG GASKET

END FITTING

END PLUG

INLET
END

FIGURE 8-7   A glass-tube rotameter.

If the rate is increased, the plummet will rise out of the 1-1/2-in diameter section. But this action will stop when the plummet moves into the 2-in diameter section. If flow is increased further, the plummet might lift out of the section altogether. The reason is that the stream of water and the plummet cannot occupy the same space. As long as the flow rate is small, there is room for both. In this case, the plummet is held down only by its weight. So, the water is able to lift it. When the plummet is lifted into the 2-in section, it cannot rise because there is room for both the plummet and the water, but only until the flow is

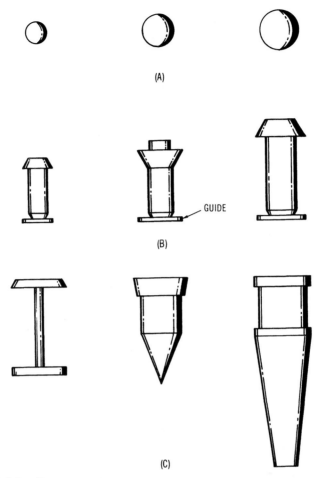

FIGURE 8-8   Rotameter plummets. (A) Ball; (B) Nail head; (C) Special types.

increased. Overall, the position of the plummet within the pipe is determined by the relation between the cross-sectional area occupied by the plummet.

The area available to the water is the ring-like or annular space between the pipe wall and the cylinder. This is shown mathematically as:

$$D^2/4 - d^2/4 = \text{annular area}$$

where,

D is the ID of the pipe,

d is the OD of the plummet.

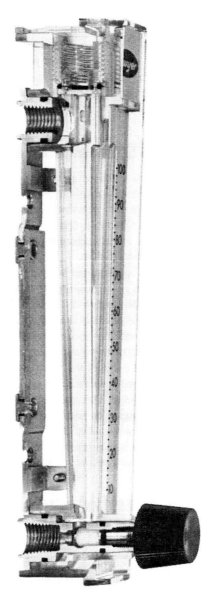

FIGURE 8-9   A molded-plastic rotameter. (*Courtesy of Dwyer Instruments, Inc.*)

   Suppose the transparent pipe is tapered so that the outlet is a larger diameter than the inlet. The plummet, in this case, occupies nearly the entire tube area when at the smaller end. As fluid is allowed into the tube the plummet will be pushed up, allowing water to pass between the outside of the plummet and the pipe wall. It will be pushed up until the pressure across the plummet results in a force. The force will be equal to the weight of the plummet. The effective weight is equal to the weight of the plummet minus the buoyant force of the fluid. For the fluid and a plummet, the effective weight stays the same. As the flow changes, the annular area must change.

Since the variable is the area, the plummet position within the tube can be used as a measure of the area. The position can be used to determine flow. Observe that there is a direct relation between flow and plummet position. So, there is no square-root problem when measuring flow with a rotameter.

TAKE-OUT AND DISPLAY MECHANISM

The take-out mechanism is an important part of the rotameter. For recording or transmitting instruments, the take-out mechanism is almost always a magnetic coupling. The magnetic couplings used with rotameters are similar to the couplings in differential-pressure take-out mechanisms. There is an important difference, however. The plummet or displacer position may vary a small amount in the differential-pressure instrument. In the rotameter it may vary several inches.

For an electrical display, the plummet is used to change circuit impedance. Impedance is a form of resistance for an alternating current circuit. A bridge circuit is used to detect impedance changes in the instrument.

For a recorded display or a pneumatic display, a magnetic following mechanism is used. This method for taking out the plummet position is accomplished by putting an extension on the displacer. A permanent magnet is placed inside the extension at its upper end. The extension runs inside a closed tube mounted on the upper end fitting. The extension is nonmagnetic. A second permanent magnet is located outside the tube. The second magnet is mounted on a lever. This magnet is free to follow the movement of the magnet in the plummet extension. The second magnet may be replaced by an armature mounted on a lever. A link is connected to this lever and then to an indicator lever. If the display is pneumatic, a second link is taken from the lever to drive a pneumatic mechanism. This pneumatic operation is shown in Figure 8-10.

OTHER FLOW-MEASURING INSTRUMENTS

The measurement of flow with the orifice-differential method and the rotameter account for a rather high percentage of all flow measurements. These two methods of measuring flow, however, are only two of many. Some of the other methods will be discussed later in this chapter. A relationship between different types of flowmeters has been developed. Figure 8-11 shows a family tree of flowmeters.

BASIC MEASUREMENT

When talking about flow measurement the response might be to say it is a measurement of moving material. In this regard, what is meant by the term moving material? Obviously it could be a solid, liquid, gas, or a mixture of these materials. The definition raises two further points. One refers to the nature of the material moving through the system. The second deals with how a moving quantity is measured.

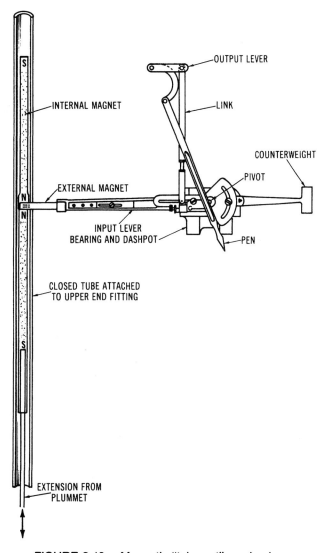

FIGURE 8-10   Magnetic "take-out" mechanism.

There are two ways to measure a moving quantity. One is called measurement on the fly. This method measures the quantity of material as it moves. The second technique interrupts the flow so that measurement counts a series of uniform batches.

There are two ways to measure a uniform batch of material. Each batch can be weighed and the total counted. The volume of each batch must also be determined and totalized. These measuring procedures are generally called totalized flow.

In order to weigh a material, a scale is used. The force of gravity on the material is determined with the scale. To measure material volume, its dimensions must be

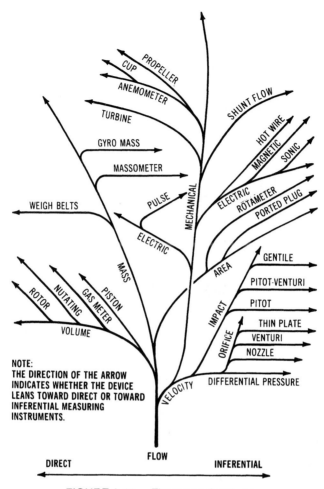

FIGURE 8-11   Flowmeter family tree.

determined. If the material is a fluid, then a number of small containers might be filled. Each container is then counted to determine the total. Either of these methods are considered to be a direct measurement procedure.

There is another way to measure besides the direct method of weighing or measuring flow volume. By evaluating some form of disturbance in the flowing material, the amount of disturbance can then measured. This type of measurement is an inferential process.

Direct and Inferential

The majority of flow measuring instruments fall directly under the categories direct or inferential. Normally, there is a gray area, however. Meters that measure velocity fall into this gray area. One viewpoint on velocity-measuring holds that they are direct. The reasoning is that the volume of material is measured since the length is measured and the diameter is known. That is, a measurement is made of the number of feet of material that flows by a certain point. An analogy might be made using a tube of toothpaste. The quantity of toothpaste used might be determined by squeezing the paste out in a straight line and measuring its length. The length multiplied by the area of the opening in the tube equals the volume of paste used. The second viewpoint would classify the velocity measurement as inferential because flow is determined by measuring a velocity. The velocity is not a direct property of a material like volume and mass.

The approach in this discussion will borrow from both viewpoints. A velocity measurement which is obtained by subjecting the flowing material to a disturbance shall be classified as inferential. Velocity measurement made without a disturbance shall be considered direct.

INFERENTIAL FLOW MEASURMENTS

Inferential flow-measuring device operation is based on the formula:

$$\text{flow} = cA\sqrt{2gh}$$

The rotameter and the orifice-differential instruments fall under the inferential classification. The rotameter is a variable-area device. Variable area instruments are a branch of inferential meters. Differential-pressure instruments are a second branch of meters. The differential-pressure branch includes measurements made with orifices and their variations.

Velocity Instruments

The basis of this grouping is the relationship that quantity equals area times velocity. The differential-pressure method converts the velocity of a flowing stream to a corresponding head. The head is then measured. There is a group of instruments that are designed to measure this velocity more directly.

Inferential Velocity Flowmeters

The Pitot tube is a device that balances the velocity head of a flow with a liquid head. An open-ended tube is placed in a pipe facing the stream. The stream fluid pushes into the tube. The height of fluid driven up the tube is the velocity head. The value of the head is a differential-pressure measurement, made with a differential-pressure instrument. The bent tube is the Pitot.

The Pitot tube and associated devices are referred to as impact devices. They are primary devices in the same sense that orifices are primary. Both groups need a differential-pressure-sensitive instrument in the total installation. Some other devices similar in operation to the Pitot are the Pitot-venturi and the Gentile flow nozzle. These are shown in the family tree of Figure 8-11. Methods of using the Pitot-venturi are shown in Figure 8-12.

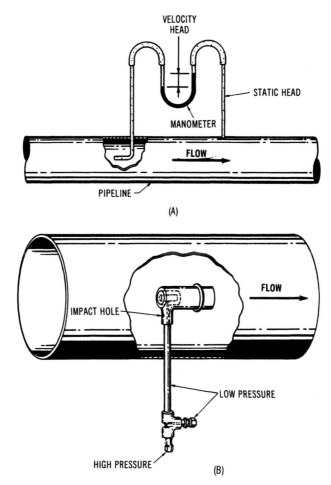

FIGURE 8-12   Flow measurement with Pitot tube. (A) Pitot tube; (B) Pitot-venturi.

Electrical Velocity Flowmeters

Some interesting devices for flow measurement are the hot-wire anemometer, the magnetic flowmeter, and ultrasonic instruments. All of these devices disturb the flowing stream and measure the effect of the disturbance. The hot-wire anemometer is arranged to feed a known quantity of heat into the stream. The increase in temperature of the flowing stream depends on the flow rate. The temperature is measured and calibrated in terms of flow. Ultrasonic instruments subject the stream to high-speed vibrations. These enter the stream at an angle and are reflected. The time relationship of the input vibration and reflected vibration is a measure of flow velocity. This measuring procedure is similar to that of a radar speed detecting unit.

Magnetic flowmeters disturb flow by subjecting it to an alternating current magnetic field. The stream cuts the force lines of the magnetic field. A voltage is generated that is proportional to the rate at which the lines of force are cut. The velocity of a flowing stream is then measured as a value of generated voltage. This measuring procedure is similar to the operation of an electric generator.

Velocity Flowmeters-Propeller Type

It is possible to measure the velocity of a stream by placing a propeller in the stream. The speed of rotation of a propeller depends on the velocity of the flow. A measure of the speed is also a measure of the velocity. This measurement is made by electrical and mechanical means. These devices fall in an area between direct and inferential devices. They lean toward the direct category since the stream is not disturbed by an outside impulse.

Velocity Flowmeters-Electrical Output

An example of a device falling under this heading is a pulse-generated turbine. The turbine rotor contains a magnet. As the rotor turns, the magnetic field produced by the magnet cuts through a coil of wire. When it cuts the coil, a pulse of current flows through the coil. These pulses are then counted electronically. The output of these counters can be displayed as a series of numbers or converted to a signal for an electronic recorder.

Velocity Flowmeters-Mechanical Output

There are several different types of propeller velocity instruments. These are arranged so that the propeller drives the counter through a series of gears.

The shunt flowmeter is a common flowmeter. It is a cross between an orifice and a propeller velocity instrument. The segmental orifice causes a fixed portion of the total stream to be directed through a shunt. A shunt is a mechanical switching device that is used as a bypass around the orifice. The propeller is located in the shunt. The propeller rotates according to the velocity of the shunted flow. It then drives a magnet through a reducing gear train. The reduction is such that the rotation of the magnet is about one revolution per minute (rpm). Attached to the shaft between the propeller and the gear train is a blade. The blade provides damping action for the propeller. This blade and the gear train run in a fluid. Water causes a drag on the propeller.

Magnet rotation is driven by the propeller. This is followed by the rotation of a second magnet that drives a counting mechanism. Between the two magnets is a diaphragm. The diaphragm forms a pressure-tight connection eliminating packed bearing problems. The construction of the shunt flowmeter is diagrammed in Figure 8-13A.

Anemometers

Anemometers generally imply velocity measurements of a gas. These could be propeller operated or something similar. The cup anemometer used to measure wind velocities is an example. The hot-wire anemometer is a different type of instrument. It does not respond to the rotary principle. Hot-wire anemometers add heat to the flowing stream and determine velocity by a change in temperature.

DIRECT-FLOW MEASUREMENT ━━━━━━━━━━━━━━━━━━━━━━━━

Direct-flow measuring devices are designed to respond to volume or weight. Those measuring volume are described as volumetric meters. Those measuring weight will be called mass meters.

It is difficult to categorize flow instruments under the heading of direct of inferential. For this discussion it is important to develop a scheme that makes it easier to remember what devices use what principles. Some direct flowmeters are partially inferential. Yet others are classified as direct, but at the same time lean towards inferential. They take on both aspects because the stream is disturbed by an outside force. The effect of this disturbance is measured; the amount depends on the mass or weight of the flowing material.

Volumetric Meters

Volumetric meters are designed to sample a specific amount of value being measured. These volumes are then counted. The size of each batch is known. The number of batches is then multipled by the volume of each batch. The result is a measure of the quantity of fluid passing through the meter.

Some of the more important volumetric meter types are:

1. Piston.
2. Rotor.
3. Nutating disk.
4. Liquid-seal gas meter.

Meters of this kind are frequently described as being positive displacement instruments. Of the four types mentioned, the piston and the nutating disk are perhaps the most common in chemical plant operation.

Nutating Disk

The nutating disk meter of Figure 8-13B is used to measure domestic water flow. The action of this meter defies an adequate description, but at the same time its operation is extremely simple. The moving part of the meter is a flat, thin, circular plate. Through the center of the plate and at right angles to it is the vertical axis. The bottom of the axis rests in a socket. The upper end is allowed to move and then fall over. The plate forms an angle of about 15 with the vertical axis of the meter housing.

The flat plate is slotted from the outside of the disk to the axis. The disk is installed in the meter housing so that a fixed vertical partition lies within the slot. This arrangement slows down disk motion and causes it to move in a see-saw pattern. One end of the diameter lifts while the other end drops.

Inside of the meter body a section is formed so that when the disk is in place, two compartments are obtained. One compartment lies under the disk while the second one is above the disk.

Water entering the meter body causes the disk to nutate, or wobble. With each nutation, a known volume of water is passed through the meter. The free end of the axis of

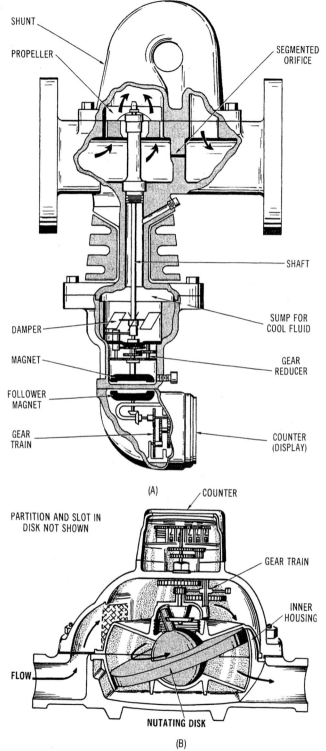

SHUNT

PROPELLER

SEGMENTED
ORIFICE

SHAFT

DAMPER

SUMP FOR
COOL FLUID

MAGNET

GEAR
REDUCER

FOLLOWER
MAGNET

GEAR
TRAIN

COUNTER
(DISPLAY)

(A)

COUNTER

PARTITION AND SLOT IN
DISK NOT SHOWN

GEAR TRAIN

INNER
HOUSING

FLOW

NUTATING DISK

(B)

FIGURE 8-13   Flowmeters. (A) Shunt flow; (B) Nutating disk.

the disk moves in a circular path as the disk nutates. The circular motion of the axis drives a gear train and a counter.

Piston Meter

The piston meter is a piston water pump in reverse. The water enters the meter through a valve, forcing the piston up. At the upper end of the piston's travel a second port opens. This action allows the fluid to leave the cylinder. In order to get a self-starting meter, a number of pistons are used. These cylinders and pistons are arranged in a circle. While one piston is filling, the opposite piston is discharging. The pistons are attached by piston rods to a circular crankshaft that is a nutating gear. The action of the pistons causes the crankshaft to nutate. The counting mechanism is similar to a nutating-disk meter.

Mass Meters

The simplest mass meter consists of a conveyor section on a scale. As the material passes onto the conveyor, it is weighed by a scale, as shown in Figure 8-14. The conveyor is driven at a known speed. The length of the belt is also known. The information determines the amount of flowing material. A conveyor weigh meter is one of the ways to measure the flow of solid materials.

VARIATIONS

A problem that has not been discussed is the cross-breed flow-measuring device. Cross-breed means that the flow-measuring principles of two devices are combined. For example, an area meter may be used in series with a turbine meter or a rotameter and an orifice. The shunt flowmeter is a cross between an orifice and a turbine meter. In certain respects, it is similar to the orifice-rotameter cross.

Volumetric meters are different from all other types mentioned because they are batching devices. The stream is cut into batches before the batches can be counted. Such an action is an integrating action. The display of these meters indicates the number of batches that have passed through the meter in a period of time. In a sense these devices are not flowmeters as much as they are integrators. The display can be used to get a reading other than the count. That is, a mechanism will react to the angular velocity of the disk, turbine, or rotor that is being used. A common arrangement is to attach a voltage generating tachometer to the disk or turbine. The tachometer converts angular velocity to a proportional voltage. This voltage is measured by a voltage-sensitive instrument that is calibrated in flow terms.

SUMMARY

Flow is a very important part of industrial processing. Flow is defined as the movement of liquid, gas, solids, or mixtures through a structure. Most applications of flow deal with the movement of material through pipes. Flow can be measured directly or by inference.

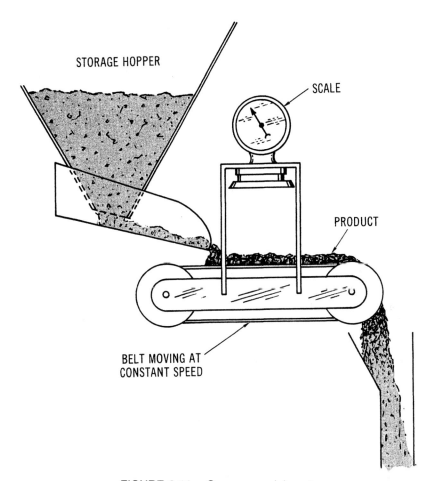

FIGURE 8-14   Conveyor weigh meter.

Direct flow measurement is made by weight and volume. Inferred flowmeters decide on a value by observing a change in another value. Flow rate can be measured by observing a change in differential pressure.

An orifice plate is a thin metal disk with a hole cut from the center. When placed in a pipe an orifice plate obstructs flow, causing a differential pressure to appear across the orifice. Differential pressure can then be equated to the velocity of a flowing stream.

A rotameter is a variable area flowmeter. It has a plummet that changes its position according to the rate of flow through a variable area. Rotameters are mounted vertically. Flow is measured by observing the position of the plummet on a graduated scale.

Direct-flow measurements are made with a piston, rotor, nutating disk, or liquid-seal gas meter. These instruments are volumetric flowmeters. The mass of material flowing on a conveyor can be weighed to determine its value.

ACTIVITIES

ORIFICE FLOW INSTRUMENTS

1. Locate an orifice installation in an operating system.
2. Note the construction of the flange assembly.
3. Describe how the orifice is installed.
4. Locate the orifice taps.
5. What type of taps are used in this installation?
6. Locate the pressure measuring element.
7. Identify the type of element used in the installation.
8. If possible, turn off the flow.
9. Remove the orifice from the flange.
10. Identify the type of orifice used in the installation.
11. Describe the construction of the orifice.
12. Return the orifice to the assembly.

ROTAMETERS

1. Locate a rotameter in an operating flow installation.
2. Note the location of the inlet and outlet of the instrument.
3. Locate the plummet of the instrument.
4. What type of plummet is used?
5. Describe the construction of the plummet.
6. Locate the metering tube of the instrument.
7. Describe the construction of the metering tube.
8. Measure and record the flow indicated by the instrument.
9. Explain how flow is measured with this instrument.

QUESTIONS

1. How is an orifice installation used to measure flow?
2. What are three different types of orifices?
3. How does an orifice change flow into differential pressure?
4. How does a venturi differ from an orifice?
5. What are the properties of fluids that influence piping?
6. What are the essential parts of a rotameter?
7. Explain the operating principle of a rotameter.
8. What flow instruments are inferrential?
9. What are some direct-flow instruments?

9
Square-Root and Integration Problems

OBJECTIVES

Upon completion of this chapter, you will be able to:

- Discuss the square-root relationship between a measured differential pressure and its corresponding flow rate passing through an orifice.
- Outline the methods used to compensate for square-root.
- Discuss the importance of the square-root displacer.
- Describe the importance of the integrator.

KEY TERMS

Cam.   A rotating or sliding piece that gives motion to a roller moving against its edge.

Compensate.   To neutralize or counteract the effects of a known error.

Helix.   A spiral formed from wire or tubing that is used in various instruments.

Inferential instrument.   A device producing a value in one form that is believed to follow the original value in a different form. For example, temperature is measured by the height of a column of mercury.

Integrator.   A device which continually adds up the value of a quanity during a particular period of time.

Linear output.   An output that is a constant multiple of the input.

Square-root displacers.   Instruments that respond to the Archimedes principle.

Square-root mechanism.   Any mechanism that translates the square-root of an input signal to a linear output.

Square-root problem.   The problem of converting a differential-pressure measurement into a display of flow with a linear output.

Synchronous motor.   An alternating current motor that operates at a constant speed regardless of the load applied.

THE SQUARE-ROOT PROBLEM

In orifice theory, it was shown that flow varies as the square-root of the differential pressure across the orifice. The differential pressure is measured by a differential-pressure-sensitive device, which include the float/manometer, diaphragm, or bellows instruments. These devices convert a measurement signal to a display. Since the instrument senses the differential-pressure, the display will also be differential pressure. If flow is to be determined, the square-root of the pressure should be displayed. Converting a differential-pressure measurement into a display of flow has a square-root problem that exists whenever flow through an orifice is to be displayed.

The differential-pressure sensitive instrument remains pressure sensitive whether it is tied to an orifice or not. The instrument becomes part of the square-root problem only when the response is altered. This, then, displays the square-root of the pressure it senses. The altered instrument is one solution to the square-root problem. Several other methods of solving this problem will also be considered.

COMPENSATING FOR SQUARE-ROOT

Methods that compensate for square-root are:

1.   Paper and pencil.
2.   Square-root charts.
3.   Square-root displacers.
4.   Square-root mechanisms.

Paper and Pencil Compensation

The paper and pencil method is rarely used on industrial recorders or indicators; yet, technicians should be quite comfortable with this method. In principle, the square-root problem is rather simple. Under field conditions, the question of what device should be doing at a specific time can get rather involved.

Remember, the basic flow equation is:

$$\text{flow} = cA\sqrt{2gh}$$

This equation states that flow varies as the square-root of the differential head.

First, a differential-pressure instrument is calibrated. Next, the indicator signal is matched to the pressure being applied to the instrument. Since flow varies as the square-root of the differential pressure, the indicator does not show flow. Flow can be determined by taking the square-root of the indicator reading. For example, the indicator reading equals 50 percent of the chart. Find the flow that causes the indicator to show 50 percent of the chart. To determine this flow, find the square root of 50 percent. This equals 70.7 percent, meaning that 70.7 of the total is flowing when the indicator is on 50 percent of the chart.

This calculation can be repeated for all indicator positions between 0 and 100 percent. While this procedure makes it possible to determine the actual flow for every display reading, it is tedious and impractical for a continuously operating flow measurement. For calibration and servicing of an instrument, the paper and pencil method is very practical. In some instances, it may be the only way. It is important, therefore, to learn how to convert from head to flow or vice versa.

Square-Root Chart

The easiest way to handle this problem on a continuing basis is to use a square-root chart. If the chart or scale is used, a differential-pressure sensitive instrument is needed. This device must be calibrated by applying a specific amount of differential pressure. Basically, the chart simply alters the reading. When 50 percent of the pressure range is applied to the instrument, the indicator will display 50 percent. For this 50 percent indicator position, the chart will be marked 70.7 percent of the flow. Or, if 81 percent of the range is applied to the meter, the indicator will go to 81 percent on the chart. Yet, the square root of 50 percent is 70.7 percent and the square root of 81 percent is 90 percent. This chart is nonlinear. It is similar to the charts used for measuring temperature in a vapor filled system. In Figure 9-1, the square-root chart appears crowded at the smaller values. This is necessary since 70.7 percent of the divisions are recorded below 50 percent on the chart.

An interesting point arises when considering the head of the instrument when the indicator displays 70.7 percent of the flow. This would be similar to having a given flow value on a square-root chart and asking for the corresponding head. Assume that the instrument range is 100-in of water. In this case, the flow value would be squared as follows:

$$0.707^2 = 0.499 = 50 \text{ percent}$$

and since the range in the original problem is 100-in of water, the head is 50 percent of 100-in, or 50-in.

Square-Root Displacer

The square-root displacer, or the characterized displacer, was introduced earlier. It is mentioned here again because it is a solution to the square-root problem. The displacer makes possible the ability to record, indicate, and transmit flow in equal units. The charts

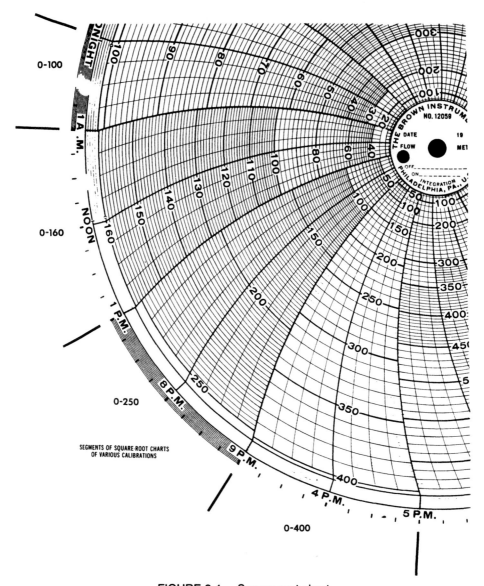

FIGURE 9-1   Square-root chart.

used on these recording instruments have equal divisions. Instruments that use this displacer are called square-root compensated.

Square-Root Mechanism

The square-root mechanism is any mechanism that changes the square-root of an input signal to a linear output. Such devices are used frequently in flow measurement. They are

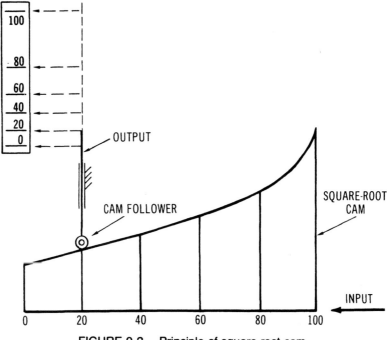

FIGURE 9-2   Principle of square-root cam.

used in the measurement section of the instrument or the display. One display indicates the accumulated total units that have passed through the orifice. Devices that accomplish this are called integrators. Integrators, used on nonbalancing instruments, need a square-root mechanism in their design.

The square-root mechanism is usually a cam. It is designed so that the radius at any point on the cam varies as the square-root of the number of degrees to that point. When the cam is rotated to 50°, the radius will be $0.1 \sqrt{50}$, or 0.7-in. The 0.1 is a constant of the cam. When the cam rotates 80°, the radius will be $0.1 \sqrt{80}$, or approximately 0.9-in. When the cam rotates to 20°, the radius will equal $0.1 \sqrt{20}$, or, 0.45-in approximately. Notice that the radius changes 0.2-in for a 30° angular rotation in the region from 50° to 80°. But for the same 30° angular rotation in the region from 20° to 50°, the radius change is 0.25-in.

Figure 9-2 shows the principle of the square-root cam. With input movement applied to the slider, the cam follower is the output. The cam is similar in principle and is more practical for rotary movements. The square-root cam of Figure 9-2 is used on many integrators.

A square-root mechanism used to convert nonlinear pneumatic-flow signals into linear output signals is shown in Fig 9-3. This unit is used for signals applied to indicating recorders or control functions.

THE NEED FOR INTEGRATORS ▰▰▰▰▰▰▰▰▰▰

Occasionally there is a need to know the total flow over a specific time period. For instance, the company furnishing water to a home wants to know how much is consumed in a month. The company is not interested in day-to-day amounts, but in total amounts. In order to determine the total amount, it is necessary to know how much is flowing at all times. This rate of flow is multiplied by the time that the total amount is flowing. The result would be the total flow for that time period. For example, the rate of flow is 21-gallons-per-minute, or gpm. This rate was maintained for 5 minutes:

$$21 \times 5 = 105 \text{ gallons}$$

Suppose for the next 5 minutes the rate of flow was 23-gpm. Find the total flow at the end of 10 minutes:

$$
\begin{array}{ll}
\text{flow for first 5 minutes} & = \text{105-gal} \\
\text{flow for second 5 minutes} & = \text{115-gal} \\
\hline
\text{total flow} & = \text{220-gal}
\end{array}
$$

In industry, there are many applications for determining total quantities for a time period. Normally this information is used for cost control.

There is a need for knowing total flow without time being specified. For example, there may be a need for 365-ft^3 of material to be added to a mixer. This material is first placed in a batch tank. Its volume is measured. The entire batch is then added to the mixer. The length of time required to add this is of no interest as long as it can be added within a reasonable time period.

A more common method of adding material to a mixer uses instruments. The instruments can be flowmeters of the inferential or direct type.

If a direct instrument is used, the principle is similar to the batch tank operation. Direct instruments are essentially used in batch operations. In this instrument known cup-sized batches are taken from the supply and are dumped into the mixer through appropriate piping. The instrument counts each batch as it is dumped. The total passed is indicated on a counter. It registers the number of cupfuls that have passed through the meter.

INTEGRATORS ▰▰▰▰▰▰▰▰▰▰

Inferential instruments have a problem in determining total flow. This type of instrument is somewhat complicated. Both time and the rate of flow need to be known. The usual solution is to measure flow using conventional meters, then add an integrator that couples the indicator position with a time measurement. There are four things that the integrator must do:

1. Mark off units of time.
2. Know the position of the indicator. The integrator must be arranged to sense the flow rate.

3. Multiply this rate by each unit of time.

4. Add and display each of the answers obtained by the third step.

Components

An integrator is an inferential instrument composed of three parts:

1. A mechanism that couples the indicator and the integrator.

2. A timing mechanism which is usually a small electric motor. This motor performs the timing function. It may also serve to power the mechansims that sense the indicator position.

3. An adding and counting mechanism that indicates the total flow.

Arrangements

The basic structure of an integrator is somewhat unusual. A continously running driver rotates at a known speed. A follower engages this drive for a portion of a fixed cycle. That portion is a function of the indictor position. For example, assume the pen were 80 percent of the chart. Then the follower engages the driver for 80 percent of the fixed cycle. The follower drives a counting mechanism which registers the turns of the follower. The number of turns is determined by the time that the follower engages the driver. Time and pen engagement are establish by the position of the pen. A count is registered which is a function of flow rate. The counter displays the total units counted. These units are received by multiplying flow rate by a period of time.

ALIGNMENT AND ADJUSTMENT ■■■■■■■■■■■■■■■■■■■■■

Alignment and adjustment deals with three things an integrator must do. The coupling mechanism is designed to feel the indicator position. It must not restrain movement. If it were to do so, the integrator measurement would be false. It should be noted that an instrument must be calibrated before any adjustments are made. The integrator is designed to reflect the pen position of the indicator.

Adjustments make the follower engage the driver for a proper portion of the fixed cycle. For each indicator position there is a specific count for a unit of time. This procedure, first, disconnects the input link which frees the indicator. Then, the indicator is fixed at various places over the chart. This is a check to see if the counter is registering the correct amount. If it is not the proper count, an adjustment is made to change the duration that the follower engages the driver. This is similar to the zero and multiplication adjustments of a link/lever mechanism.

There is very little adjustment of the timing mechanism required, which consists of a gear train and a small electric motor. The gears might require replacement and the motor might fail or slow down. A visual or aural inspection will usually locate faults in the gear train and motor. In some cases, it may be advisable to check the driver cam with a stop watch.

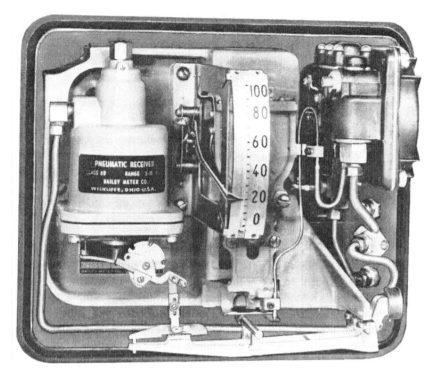

FIGURE 9-3   A square-root mechanism. (*Courtesy of Bailey Controls Co.*)

If the flowmeter is not compensated for square-root, then a square-root mechanism is added to the integrator. This is usually accomplished in the coupling mechanism by using a square-root cam or helix. Figure 9-3 shows a square-root cam mechanism.

In flow transmitters it is common to arrange the integrator to be coupled to flow through the transmitted signal. For example, if the transmitted signal is an air-pressure signal of 3- to 15-psi, the integrator will be coupled to this air pressure.

SUMMARY

The square-root relationship is between a measured differential pressure and the corresponding flow rate. The flow rate passes through an orifice and poses a problem in pneumatic instruments. Square-root correction must be provided in order to display information on a graduated scale.

Several methods used to compensate for square-root problems include paper-and-pencil correction, square-root charts, displacers, and square-root mechanisms.

Integrators are used to determine the total amount of flow. They can also be used to control the quantity of material used over a specific time period. The function of an integrator is to mark off units of time and sense the flow rate. It must also multiply these values and add them together to display the answer.

ACTIVITIES

SQUARE-ROOT COMPENSATION ▬▬▬▬▬▬▬▬▬

1. Using the paper and pencil method of calculation determines the cubic feet per second of flow occuring in an orifice installation that has a 4-in OD, a differential pressure of 100-in of water, a constant of 1.0, with the force of gravity being 32.2 feet-per-second squared.

2. Refer to an instrument that employs a square-root chart in its operation.

3. Locate a value of 70.7 percent on the chart. Calculate the differential-pressure needed to produce this reading.

4. Locate 90 percent on the chart. Calculate the differential-pressure needed to produce this reading.

5. Refer to an instrument that has a square-root displacer. Describe how the displacer compensates for square-root.

6. Examine an instrument with a square-root mechanism. Refer to the operational manual of the instrument.

7. Locate the square-root mechanism. Describe how the mechanism accomplishes the square-root function.

INTEGRATORS ▬▬▬▬▬▬▬▬▬▬▬▬▬▬▬▬

1. Determine the total amount of flow that occurs in 12-min for an instrument that has a flow rate of 25-gpm.

2. Determine the total amount of flow that occurs in 15 minutes at a rate of 15-gpm, for 20 minutes at a rate of 30-gpm, and for 10 minutes at the rate of 40-gpm.

3. Examine an instrument that has an integrator. Refer to the operational manual of the instrument.

4. Locate the integrator mechanism in the instrument.

5. Explain how the timing function is accomplished?

6. Describe the coupling mechanism.

7. Explain how adding and counting is accomplished.

QUESTIONS

1. What is meant by the term square-root problem?

2. Why does someone need to know how to compensate for the square-root problem of flow when there are instruments that are specifically designed to calculate this mathematical problem?

3. What is the importance of the square-root displacer?

4. What is the importance of the square-root mechanism?

5. How is the total quantity of water over a specific period of time measured?

6. Name several things that an integrator must do?

7. Describe how an integrator is maintained?

10
Self-Balancing Instruments

OBJECTIVES

Upon completion of this chapter, you will be able to:

- Determine if a self-balancing mechanism is used in an instrument.
- Explain the operation of a self-balancing instrument.
- List the basic components of a self-balancing instrument.
- Explain the function of an error detector in a pneumatic self-balancing instrument.
- Explain the operation of a flapper-nozzle mechanism.

KEY TERMS

Backpressure compartment.   Consists of that volume between the nozzle and the restriction and includes all tubing control devices.

Baffle.   A flexible metal-plate device that regulates flow from a nozzle.

Baffle/nozzle.   A mechanism which regulates the flow from a nozzle.

Baffle/nozzle detector.   A device that determines the backpressure that occurs in the nozzle/baffle backpressure compartment.

Detector.   A device used to discover or sense the presence or disguised character of data in a signal.

Error-detector mechanism.   A combination baffle/nozzle/pilot mechanism that senses a change in air pressure.

Nozzle.   A duct or tube of changing cross section in which fluid velocity is increased.

Nozzle backpressure.   A pressure that builds up between the nozzle and a restriction in the nozzle backpressure compartment.

Pilot.   A piece of equipment or device that guides or steers an operation.

Recorder.   An instrument designed to tabulate or make a record of information applied to its input.

Restriction.   An obstruction placed in the flow line to allow a predetermined amount of flow.

Self-balancing instrument.   An instrument whose output matches the input.

Transmitter.   An instrument designed to send variable information from a sensor over a long distance to another instrument.

INTRODUCTION ■■■■■■

A class of instruments wherein the output is compared to the input will now be considered. From this comparison, there is an indication of the measure of the input. This kind of instrument has a self-balancing aspect. It is arranged so that the output always balances the input within the range of the detector. The device is a feedback, a self-balancing, or a null-balanced type of instrument. These terms all refer to the same basic type. In this discussion, self-balancing instruments will be described. Later, self-balancing instruments will be divided into motion-balance and force-balance devices.

Components

The essential components of a self-balancing instrument are:

1.   An input mechanism.
2.   An error-detector mechanism.
3.   An output or balance mechanism.

Arrangements

These three mechanisms are arranged on a linkage or beam mechanism so that any difference between the input and output mechanisms is sensed by the error detector. Any difference is an error. The error-detector mechanism operates in such a fashion as to change the output mechanism. Therefore, the output balances the input.

Principle of Operation

The operation of a self-balancing instrument can be likened to a weighing scale. Figure 10-1 shows a laboratory weighing scale. Suppose that an unknown quantity of weight is placed in the left-hand tray of the scale and its value is to be determined. This determination is made by adding weights of a known quantity to the right-hand tray until the

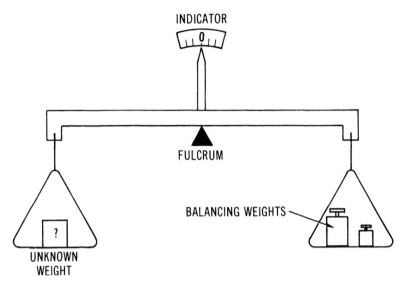

FIGURE 10-1   Balancing principle illustrated by a weighing scale.

indicator lines up with the zero mark, which would show that the unknown quantity in the left-hand tray is equal to the weight of the right-hand tray. To determine the unknown, the unknown weights are added together.

After balancing, the indicator returns to the same position no matter the size of the unknown weight. The indicator displays when the output weight equals the input weight. Determination of the size of the input is made by balancing it with known weight. It is then possible to determine the input by an examination of the output.

This type of scale is a conventional laboratory balance. There is nothing automatic about its operation. Suppose that the scale is then rearranged. When the indicator is away from zero, weights will automatically be added to the right hand tray. The addition of weights will bring the indicator back to zero, which will make the instrument a self-balancing scale.

These concepts can now be applied to an instrument. Suppose the same beam mechanism is used, however, the input tray is replaced by a bellows. This construction is shown in Figure 10-2. The purpose of this instrument is to determine the pressure being applied to the input bellows. An unknown pressure is applied to the input bellows. A bellows instrument is identical to the input bellows. It is positioned the same distance from the fulcrum point as the input bellows. The output bellows is connected to an air supply. In this line are the valves that are shown in Figure 10-2. By opening and closing these two valves, all pressures between zero and the supply pressure can be obtained. It is then possible to position the valves to obtain a pressure that balances. The indicator shows when these pressures are balanced. If the indicator is to the right of zero, it must increase the balancing pressure. If it is to the left of zero, the balancing pressure must decrease. When the indicator is at zero, the pressures are equal.

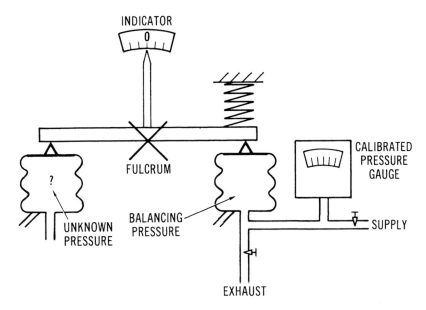

FIGURE 10-2   Balancing principle in a pneumatic instrument.

So far, no work has been accomplished because the pressure in the balancing bellows is unknown. As shown in Figure 10-2, a pressure gauge is connected. This gauge will tell the balancing pressure. It will also tell the unknown pressure if the indicator is at zero. The unknown pressure can then be determined by examining the balancing pressure. So far, this instrument is not self-balancing.

Two valves are then added and arranged to the instrument in such a way that the indicator positions the valves. When the indicator is at the right of zero, the exhaust valve is slightly closed. The supply valve is then opened a small amount. The instrument becomes self-balancing, as shown in Figure 10-3.

Some indicators, called error-detector mechanisms, are designed to drive two valves at the same time. These indicators are available for electrical and pneumatic installations.

ERROR-DETECTORS

The error-detector mechanism can be considered as the heart of the self-balancing instrument. It is the error-detector mechanism that makes possible self-balancing instrumentation. It can be found in every self-balancing controller, transmitter, recorder, or valve positioner. In all of these applications, the function is identical. There are two basic pneumatic error-detector mechanisms and a combination of the two. These are:

1.   The baffle/nozzle mechanism.
2.   The pilot.
3.   A combination baffle/nozzle and pilot mechanism.

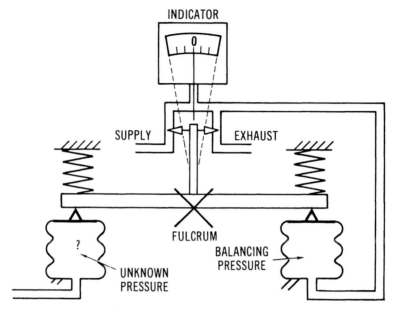

FIGURE 10-3   Self-balancing principle.

The combination unit is by far the most common of this list. Pilots driven by baffle/nozzles are often called relays.

Components

The elements of the baffle/nozzle error-detector mechanism are:

1. The baffle.
2. The nozzle.
3. The restriction.
4. The nozzle backpressure compartment.

It should be noted that the nozzle backpressure compartment is an integral part of the error-detector mechanism.

Arrangements

The baffle is independent of the nozzle. The nozzle and the restriction are located in front of the baffle. The supply of air is connected to the mechanism ahead of the restriction. A backpressure compartment consists of that volume between the nozzle and the restriction. This includes all tubing devices connected between the nozzle and the restriction. An arrangement of the baffle/nozzle mechanism is shown in Figure 10-4. The instrument can be used in a wide variety of forms. These mechanisms are essentially the same and consist of the four elements previously mentioned.

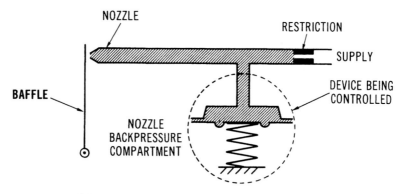

FIGURE 10-4   Typical baffle/nozzle mechanism.

An error-detector mechanism is designed to convert a small mechanical movement to an equal backpressure. The input balances the output. This movement clearance is the variation between the nozzle and the baffle. The backpressure is the pressure built up in the backpressure compartment. Some of the properties of the baffle/nozzle error detector mechanism will be considered later in the chapter.

Principle of Operation

Suppose the position of a baffle is some distance from the front of the nozzle and the air flowing through the nozzle is unrestricted. If the nozzle diameter is larger than the restriction diameter, the pressure in the nozzle backpressure compartment will be near atmospheric.

Now assume that the baffle is brought into contact with the nozzle. Flow through the nozzle will shut it off, that is, except for a small amount of leakage. Leakage occurs between the baffle and the nozzle. In such a case, backpressure will build up to the supply pressure. If the supply pressure is 20-psi, the nozzle backpressure will approach 20-psi. If the supply pressure were 40-psi, then the nozzle backpressure would be 40-psi. The nozzle is a very small distance away from the baffle and causes more air to flow through the nozzle. If the nozzle is moved farther away, the resulting backpressure would be proportional, as shown in Figure 10-5. The baffle/nozzle is essentially a variable restriction. Notice that the pressure between the two restrictions is the output. This will always be the case.

The baffle/nozzle clearance required to change nozzle backpressure from zero to the supply pressure varies according to mechanism manufacturers. Typically, the clearance is 0.002-in. There are numerous cases where the clearance is less and some cases where it is greater. In this discussion, a clearance of 0.002-in is used as a standard. Keep in mind that this number is commonly used, but there are numerous instances where other clearances are used. Assume that nozzle backpressure verses clearance between the baffle and nozzle is plotted. This would show that pressure is not strictly proportional to the distance between baffle and nozzle.

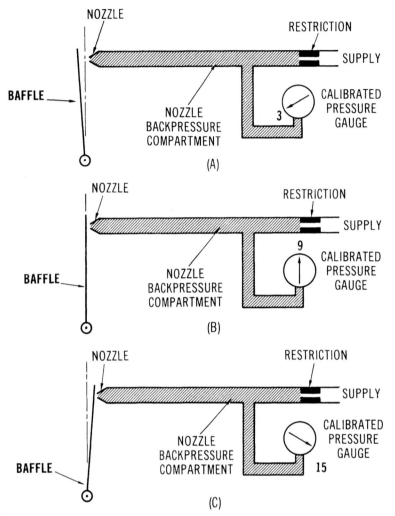

FIGURE 10-5   Effect of various baffle/nozzle positions. (A) Nozzle uncovered;
(B) Midposition of baffle/nozzle; (C) Nozzle covered.

Examine the baffle/nozzle clearance curve of Figure 10-6. Notice that the portion of the curve between 3- and 15-psi is fairly linear, or in a straight line. It is this portion of the curve that is most widely used. Most final elements are designed to operate at signals between 3- and 15-psi. Baffle/nozzle detector operation has been described as a movement of the baffle toward the nozzle. It is important to recognize whether the clearance between the baffle and the nozzle changes. In some cases the baffle is stationary and the nozzle moves, the nozzle is stationary and the baffle moves, or they both move. In all cases, however, it is the clearance that is changed.

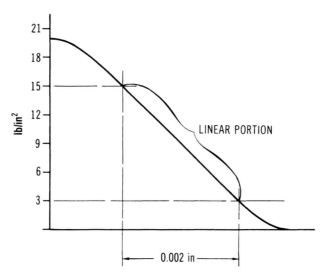

FIGURE 10-6   Baffle/nozzle clearance versus nozzle backpressure.

PROPERTIES OF THE BAFFLE/NOZZLE DETECTOR ▬▬▬▬▬

The baffle/nozzle detector is a rather remarkable device. Signal level gains of 10,000 are typical, and higher gains are possible. If a bellows element is used with the baffle/nozzle detector, power gains of 10,000 are possible. One of the problems of the baffle/nozzle detector mechanism arises out of its high-gain properties. In chapter 21, attention is given to the problem of reducing detector gain.

A second limitation of the baffle/nozzle is the forcing effect of the air blast on the baffle. Power furnished by the input element is required to force the baffle toward the nozzle. The required force is measured in thousandths of ounces or Newtons. Unless the instrument is carefully designed, the baffle/nozzle can introduce an error in measurement. To minimize this error, reduce the force required to position the baffle in front of the nozzle. The force can be reduced either by reducing the size of the nozzle, decreasing the air pressure, or both. If the air consumption of the baffle/nozzle is kept small the forcing effect is lessened. This is frequently done; however, decreasing the nozzle size can be self-defeating. If the nozzle size is reduced, its ability to rapidly change the backpressure is reduced. The time it takes to change the nozzle backpressure is increased. In many applications this longer time can be excessive.

Now consider the problem of the time it takes to change the pressure. The nozzle backpressure compartment consists of instrument tubing volume and volume of the device being operated. If the device is a control valve, a large volume of air is added to the baffle/nozzle system. If the distance between the instrument and the valve is large, the volume of the connecting tubing becomes large. Yet, all the air that is required to change the pressure in that large volume must pass through the restriction. If the nozzle is small, the restriction must also be small. In all cases, the nozzle is bigger than the restriction. The full supply pressure of 20-psi must be reduced to zero by restrictions in the baffle/

nozzle detector. One is the baffle/nozzle clearance and its variable. The other is the fixed restriction. That portion of the 20-psi that is dropped across the baffle/restriction is the output of the baffle/nozzle detector.

The position of the valve must be able to respond to rapid changes. To do this, a large restriction is provided that controls the air flowing into the nozzle backpressure system. If the restriction is large, the forcing effect of the baffle is also large.

With the introduction of a smaller nozzle it is now possible to have a forcing effect plus the ability to rapidly change the pressure of a closed system. These conditions are met by the introduction of a relay. Relays are used in conjunction with the baffle/nozzle detector.

The forcing effect on the baffle is not necessarily a liability. In fact, there are instances where this forcing effect is used to an advantage. For example, in electric-to-air transducers it is the force of the nozzle blast that is used to balance the electrical signal. Some controllers use the blast effect of a nozzle to balance the input.

ALTERNATE ARRANGEMENTS

There are a variety of baffle/nozzle configurations. Some of these are shown in Figure 10-7. Also of interest are some variations in the baffle/nozzle detector mechanism. The physical appearance of the elements can be confusing in understanding the mechanism.

Examine Figure 10-8. This mechanism is the baffle/nozzle detector mechanism used with a differential-pressure transmitter. It is identified as a standard baffle/nozzle detector element. Notice that there is a restriction, a nozzle, a baffle, and a nozzle backpressure compartment. Note also that air flow through the nozzle appears to be backward. A backward air flow could have an adverse effect on its operation.

In Figure 10-4, it was mentioned that the supply of air is connected ahead of the restriction and flows through the restriction first. This is also true of the mechanism in

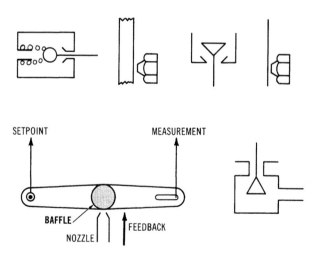

FIGURE 10-7   Various baffle/nozzle configurations.

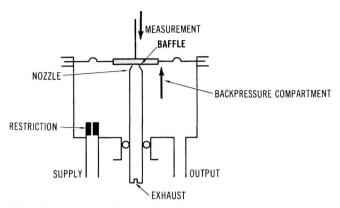

FIGURE 10-8   Baffle/nozzle differential-pressure transmitter.

Figure 10-8. In Figure 10-4, the flow of air out of the backpressure compartment is through the nozzle and out to the atmosphere. This is also the case in the differential-pressure mechanism of Figure 10-8. Note that this backpressure compartment is made up of the air between the restriction and the nozzle. The backpressure compartment has volume between the nozzle and the restriction.

Now consider the baffles of the two mechanisms. In both mechanisms the baffles move as the input signal causes them to move. This movement varies the baffle-to-nozzle clearance. Therefore, the nozzle backpressure changes. Notice again that the output is that pressure between the two restrictions. An examination of two dissimilar-looking mechanisms shows that they are functionally identical.

SUMMARY

A self-balancing mechanism is used in instruments where the output is compared to the input. The resulting output is an indication of the measure of the input.

A self-balancing instrument is made up of an input mechanism, an error-detector, and an output or balance mechanism.

An error-detector mechanism senses the difference between the input and the output. This mechanism will balance out the two. An error-detector is found in any self-balancing instrument. The function is basically the same.

ACTIVITIES

BAFFLE/NOZZLE INSTRUMENTS ▬▬▬▬▬▬▬▬▬▬

1. Refer to an operational manual of the instrument used in this activity.
2. Identify the primary function of the instrument.
3. If the instrument is attached to a functioning system, do not alter its operation.
4. If possible, remove the cover or housing from the instrument.

5. Locate the baffle/nozzle mechanism.

6. Identify the components of the mechanism.

7. Is the baffle/nozzle mechanism used in a balancing operation?

8. Explain the operation of the baffle/nozzle mechanism using the components of the instrument used in this activity.

9. Can the baffle/nozzle mechanism be adjusted?

10. Reassemble the instrument.

QUESTIONS

1. Explain the principle of operation for a self-balancing instrument?

2. What are the essential parts of a self-balancing instrument?

3. What is the importance of an error detector in relation to a self-balancing instrument?

4. In an error-detector mechanism, what is meant by the term nozzle backpressure?

5. Name several properties of the baffle/nozzle detector.

6. Explain the operation of a baffle/nozzle mechanism.

11
Pilots and Relays

OBJECTIVES

Upon completition of this chapter, you will be able to:

- Discuss the function of a pilot valve.
- Describe the makeup and operating principle of a pilot valve.
- Describe the transformation of a pilot valve into a relay-type detector.
- Explain the function of a relay in pneumatic instrumentation.
- Describe the pressure-to-force converter.
- List the forms of a pressure-to-force converter.
- Identify the force-ratio differences between pressure-to-force and backpressure-to-force converters.
- Define the terms nonbleed relay and bleed relay.
- Identify the advantages and disadvantages of bleed and nonbleed relays.

KEY TERMS

Amplification.   A process which draws power from a source other than the input signal. It also produces an output that is an enlarged reproduction of the input.

Amplifying relay.   A device that uses a small flow of air pressure to control the same pressure at a larger flow.

Baffle/nozzle displacement.   Movement caused as a reaction to relay backpressure.

Bellophragm.   A diaphragm assembly that is controlled by the action of a bellows. Also called a rolling diaphragm.

Blast effect.   The response of air applied to the baffle of a baffle/nozzle mechanism.

Bleed relay.   A relay that releases air even when the backpressure is held constant or in a balanced condition.

Bleed valve.   A valve that controls air when the backpressure is held constant or in a balanced condition.

Compound pilots.   A pilot valve with two sets of seating surfaces.

Converter.   A device or instrument that changes energy from one type to another such as pressure-to-force.

Exhaust port.   An opening which expels or releases air pressure from a device or instrument.

Nonbleed relay.   A relay that uses air only when it is in an unbalanced condition.

O-ring.   An elastic ring assembly that is placed in a scribed groove that forms a seal or pressure retaining element.

O-ring piston.   A piston that is enclosed with an O-ring seal that forms a backpressure-to-force converter.

Pilot detector.   A device that is characterized by two variable restrictions.

Pilot valve.   An auxiliary mechanism that actuates or regulates another mechanism.

Relay.   A mechanism that opens or closes supply and exhaust ports in the control of air pressure. Relays can reduce, reverse, amplify, and totalize air pressure values.

Seat.   That portion of a valve against which the valve presses to effect shutoff.

Supply port.   An opening which allows air from the source to enter a system or device.

Valve plug.   A mechanism that impedes or alters the flow through a chamber or housing.

Valve stem.   An element attached to a valve that is used to alter or change its position.

THE PILOT VALVE

The baffle/nozzle detector converts a small displacement to an equivalent backpressure. The detector consists of a fixed restriction, a variable restriction, and a backpressure compartment. A pilot valve is somewhat different. It consists of two variable restrictions. Functionally, these two devices are identical.

Componets

The components of a pilot valve are:

1.   Two variable restrictions.
2.   A mechanism for simultaneously varying two restrictions.
3.   A backpressure compartment.

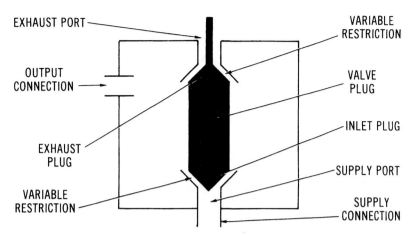

FIGURE 11-1   Pilot valve construction, diamond-shaped plug.

Arrangements

The pilot valve can be likened to a three-way valve, as shown in Figure 11-1. Notice that this valve has supply and output connections and an exhaust port. Mounted between these ports is a two-seated valve plug. The valve stem is used to position the valve plug. The output is taken off between the supply and the exhaust port. This is a common arrangement. However, it is important that one not become confused by the physical appearance. Pilot valves can be found in a wide variety of forms. In all cases, however, the function is identical. The pilot valve has two restrictions that are varied at the same time by input displacement.

Principle of Operation

Assume that the valve plug of Figure 11-1 is placed in its uppermost postion. This causes the exhaust port to close and the supply to be fully open. For this valve plug position, the output pressure equals the supply pressure. Now move the valve plug against the supply port. In this position, the exhaust port is fully open and the supply port is closed. As a result, all of the backpressure is exhausted into the atmosphere; the backpressure is zero. As the plug lifts away from the supply port, it permits the backpressure to increase. If it continues to lift, the output pressure increases further, which shows that the backpressure is proportional to the valve plug position. Therefore, the backpressure is proportional to the displacement of the valve stem. It is in this way that a small displacement is converted to a proportional pressure. As the valve plug moves up, the backpressure increases.

Suppose now that a pilot valve similar to the one in Figure 11-2 is being used. As before, raise the plug to its extreme position. The supply will be closed and the exhaust will be fully opened. Therefore, the backpressure will drop to zero. Assume now that it is moved to its lowermost postion. The exhaust will be closed and the supply will be open. The backpressure will equal the supply pressure. In the center position, the presure is proportional to the valve stem displacement.

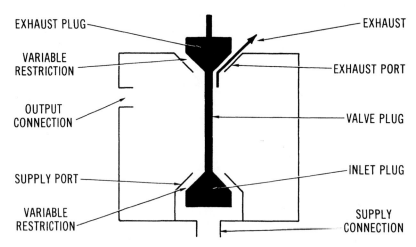

FIGURE 11-2   Pilot valve construction, dumbell-shaped plug.

Notice in Figure 11-2, that the plug is shaped like a dumbbell. In Figure 11-1, the seating surfaces on the plug face away from each other. These two plug forms are characteristic pilot-valve forms. In the study of relays, a third form will be discussed. This relay has seating surfaces facing in the same direction.

Alternate Arrangements

Figure 11-3 is a diagram of a pilot valve mechanism used in a controller. This mechanism is somewhat difficult to classify. Is it a baffle/nozzle detector or a pilot-valve detector? One might be tempted to conclude that this detector is neither. It may be something different because it appears to have two nozzles and one baffle.

Further study of the mechanism will show what kind of detector is being used. Examination of the diagram shows, that there is no fixed restriction. It was previously determined that a fixed restriction is an essential element of the baffle/nozzle detector. Therefore, this mechanism connot be classified as a baffle/nozzle detector. If it is not a baffle/nozzle detector, it must be a pilot detector. Remember that a pilot detector has two variable restrictions. This mechanism has two nozzles with a baffle between them. Still further examination of the mechanism is needed before it can be classified.

Assume now that the baffle of Figure 11-3 is moved to the right. In this position, the supply nozzle is closed and the exhaust nozzle is open. The backpressure exhausts into the atmosphere. Now move the baffle to the left against the exhaust nozzle. The supply nozzle is fully open for this baffle position. The backpressure will equal the supply pressure. Notice that the backpressure is taken off between the supply and exhaust nozzles. When the baffle is moved, it varies two restrictions. This is an essential of the pilot valve. The detector mechanism shown is a pilot valve. What appears to be a baffle is in fact a very flat valve plug. The nozzles are simply supply and exhaust ports. This will serve as a second example of how appearances can be confusing. Keep in mind that the element is the key factor in analyzing and classifying mechanisms.

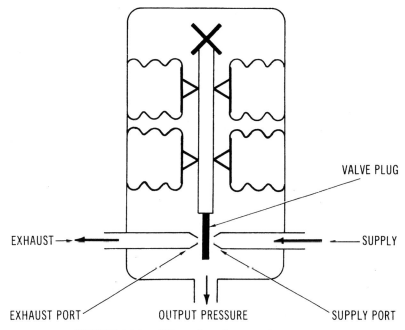

FIGURE 11-3   Pilot valve of unusual construction.

Compound pilots are found in some mechanisms. A compound pilot has two valves with two sets of seating surfaces. Pilots of this kind generally have double-acting pistons. These pilots are mentioned here to avoid confusion when attempting to classify different mechanisms.

RELAYS

By far the most widely used detector consists of a baffle/nozzle mechanism combined with a pilot. A pilot so arranged is called a relay. Remember that the baffle/nozzle mechanism has some serious limintations. If the restrictions were made large enough to rapidly change backpressue, excessive amounts of air would be required. If the restrictions were kept small, the time required to change would become excessively long. The second limitation with the baffle/nozzle is its blast effect; that is, the air blast from the nozzle forces the baffle. This makes it necessary for the input mechanism to drive against this blast force. If the restriction is large, the forcing effect is large. If the restriction can be made small, the forcing effect will remain small. If the supply pressure to the baffle/nozzle mechanism is lowered, the blast effect is lowered for a given restriction. The relay makes it possible to do these things.

Figure 11-4 shows an assortment of multipurpose relays. These pneumatic devices can be used to achieve a number of specific functions. More than sixty different control functions and operations can be achieved by these six units. The outward appearance of

FIGURE 11-4   An assortment of multipurpose relays. (*Courtesy of Moore Products Co.*)

a relay takes on many different shapes and sizes according to its work function. Operation is, however, the same in nearly all respects.

Components

The essential components of the relay are:

1.   A pilot.
2.   A pressure-to-displacement converter.
3.   A nozzle backpressure compartment.

Arrangements

The pressure-to-displacement converter is arranged to drive the pilot stem. The nozzle backpressure is connected to the pressure-to-displacement converter. The converter appears in various forms. It may be a bellows, a diaphragm capsule, or a relay diaphragm. The relay valve plug may be one of three types. The seats on the plug may face each other, as shown in Figure 11-5. The seats may face away from each other, as shown in Figure 11-6. They may also face in the same direction, as in Figure 11-7.

Relays may be of the bleed type or of the nonbleed type. The nonbleed relay will be discussed in detail later in this chapter. Baffle/nozzle backpressure is connected to the nozzle backpressure compartment of the relay valve, as shown in Figure 11-5.

Principle of Operation

If the baffle covers the nozzle, the nozzle backpressure will build up to the supply pressure. Backpressure applied to the pressure-to-force converter causes displacement that is equal to the backpressure. This displacement acts through the stem of the relay. It positions the valve plug so that the exhaust port is closed and the supply port is open. This

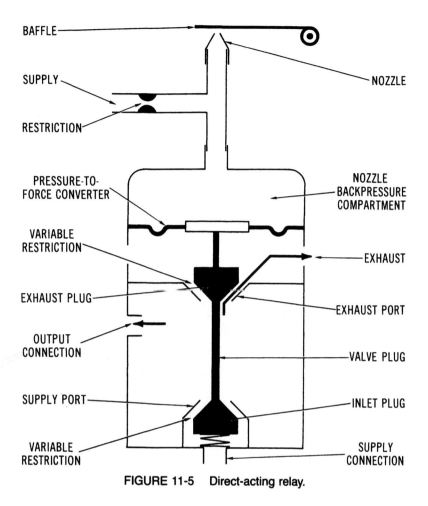

FIGURE 11-5   Direct-acting relay.

causes relay backpressure to equal the supply pressure. Remember that pilot backpressure is equal to stem displacement, which means that stem displacement is equal to nozzle backpressure. Relay backpressure is also equal to baffle/nozzle displacement.

It is important to know whether small baffle/nozzle displacement can be converted to a proportional output pressure. The answer is yes. But in the mechanism of Figure 11-5, conversion takes place in two stages. Baffle/nozzle clearance is first converted to a proportional pressure. This pressure is then converted to a displacement. Proportional displacement is reconverted to a proportional pressure. This takes two stages of amplification. Notice that the output of each stage is the pressure between two restrictions. The output is the pressure between the baffle/nozzle and the fixed restriction. This is the input to the relay. The output of the relay is the pressure between the two restrictions.

This arrangement makes it possible to use a baffle/nozzle with small restrictions. Further, it is possible to reduce the blast effect by changing the nozzle backpressure. Typically, nozzle backpressure will vary from 2- to 3-psi, which will cause full stroking of the

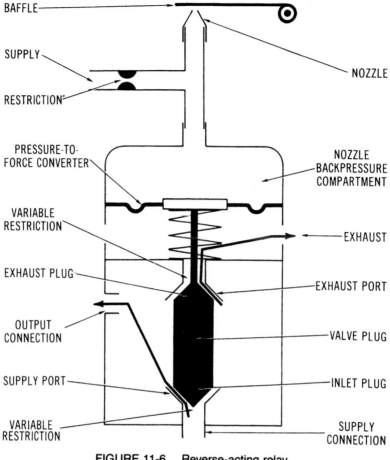

FIGURE 11-6   Reverse-acting relay.

relay valve plug. This is made possible by making the range of the pressure-to-displacement converter 2- to 3-psi rather than 3- to 15-psi. In this way the relay makes a fast response using a small restriction.

The device just described is direct acting. As the nozzle backpressure increases, the output air pressure increases. If a diamond-shaped valve plug is used rather than a dumbbell-shaped plug, the output pressure will decrease as the nozzle backpressure increases, making a reverse-acting detector similar to the one in Figure 11-6.

Alternate Arrangements

A relay valve reduces the size of the restriction used in a device and reduces the blast effect. It does not reduce air consumption. The supply port of the pilot needs to be large if the relay backpressure is to be changed rapidly. Assume now that the nozzle backpressure is in its midposition. The valve plug is positioned so that the exhaust port and the

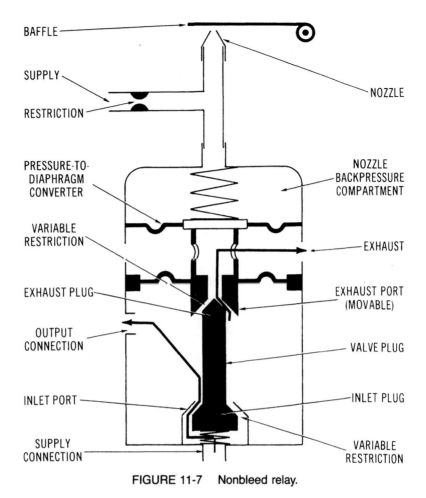

FIGURE 11-7   Nonbleed relay.

supply port are half open. This causes large amounts of air to bleed through the valve, even though the backpressure is held constant.

There is a valve arrangement that maintains fixed backpressure without exhausting air to the atmosphere, as Figure 11-7 shows. Notice the location of the exhaust port. It is mounted so that it can move as the nozzle backpressure changes. As the nozzle backpressure increases, the exhaust port moves down. If this continues, it will seat against the valve plug and close the exhaust port. Notice the point where the exhaust port is closed. This point also closes the supply port. If both ports are closed, the relay backpressure will stay constant. For a balanced condition, there is no air consumption. If the nozzle backpressure continues to increase, the exhaust port will move downward. The valve plug will also be moved downward. This will open the supply port. The exhaust port is still closed and the supply port is open. Consequently, the relay backpressure will increase. If the

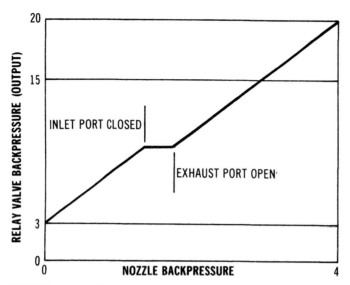

FIGURE 11-8   Nozzle backpressure versus relay backpressure.

nozzle backpressure continues to build up, the supply will fully open, causing the relay backpressure to equal the supply pressure.

Now follow the operation of the relay as the nozzle backpressure decreases. As it decreases, the diaphragm will move upward permitting the exhaust port and the valve plug to move up while closing the supply port. As the nozzle backpressure continues to decrease, the valve plug continues to rise until it closes the supply port. If the nozzle backpressure drops off further, the exhaust port will start to open. The valve plug, seated against the inlet, cannot rise any further. The opened exhaust port will permit the relay backpressure to fall to the atmospheric pressure.

To summarize, notice that the only time the relay uses air is when the supply port is open. The only time the supply port is open is when the exhaust port is closed. Hence, no air exhausts from the relay when it is balanced. When the relay is balanced, the nozzle and relay backpressure are constant.

A relay so constructed is called nonbleed. The key idea in the nonbleed relay is the movable exhaust seat, wherein the valve plug seating surfaces face in the same direction. The movable exhaust seat makes it necessary to use two pressure-to-force converters. One of these converts the nozzle backpressure to a proportional force. The other has a dual function. It converts the relay backpressure to a proportional force. It also seals the exhaust port so that it is possible for the seat to move.

Nonbleed relays are economical in air consumption, but have the disadvantage of a flat spot in their operation as shown in Figure 11-8. This figure shows nozzle backpressure verses relay backpressure. As the nozzle backpressure increases, the relay backpressure increases proportionally. This is normal except at the point where the exhaust port and the inlet port are closed. At this point, there is a flat spot where the nozzle backpressure increases. Though slight, this flat spot can be objectionable on certain control applications. To eliminate this flat spot, many nonbleed relays have a slight air bleed from the relay backpressure compartment. In other words, it is a bleeding nonbleed valve. The presence of this bleed eliminates the flat spot. The nonbleed valve can be identified by observing that the exhaust seat moves. This requires pressure-to-force elements. It is possible to see two diaphragms in the construction of this relay. This permits it to be identified without disassembly. Examination of the valve plug shows that nonbleed valves have two seating surfaces facing the same direction.

Notice that these two pressure-to-force converters are arranged so that the forces oppose each other. If these two forces are equal, the exhaust seat takes a fixed position. If there is a difference in force, the exhaust seat will move up or down depending on which force is greater. Assume now that the upper diaphragm has six times the area of the lower diaphragm. The upper diaphragm will cause a downward force six times as great as the upward force of the lower diaphragm. This means that 6-psi on the lower diaphragm would balance a 1-psi change on the upper diaphragm. A 1-psi change in nozzle backpressure will cause a 6-psi change in relay output, thus changing the ratio of the bellows area makes it possible for a small nozzle backpressure to change a much larger backpressure. Therefore, it is possible to operate with a reduced nozzle backpressure and still obtain a standard 3- to 15-psi output pressure.

RELAY VARIATIONS

There are several common relay variations, however, their operation is functionally identical. Variations in appearance are because there are a variety of pressure-to-force converter elements. In Figure 11-5, the pressure-to-force elements were diaphragms. Remember that in addition to diaphragms, there are several other pressure-to-force elements. These are:

1. The diaphragm.
2. The bellows.
3. The O-ring piston.

A diaphragm can be flat or rolling. Some manufacturers call a rolling diaphragm a bellophragm. The bellows can be a stacked capsule, a single capsule, or one with corrugated convolutions. There are several manufacturers who use different combinations of pressure-to-force converters. Consider the relay shown in Figure 11-9. This is a nonbleed relay used by Honeywell in their Air-O-Line controllers. Notice that this unit uses two bellows elements instead of two diaphragm elements. The inner bellows permits the exhaust port to move. The outer bellows converts the nozzle backpressure to a force. The relay backpressure, acting on the inner bellows, creates a force that is balanced by the nozzle backpressue acting on the outer bellows. Functionally, the two bellows of Figure 11-9 are equivalent to the two diaphragms in Figure 11-7.

Figure 11-10 shows a detector mechanism used by Fisher Controls Company. Notice in this relay that flat diaphragm and O-ring piston converters are used. The diaphragm forms the nozzle backpressure compartment and converts the nozzle backpressure to a

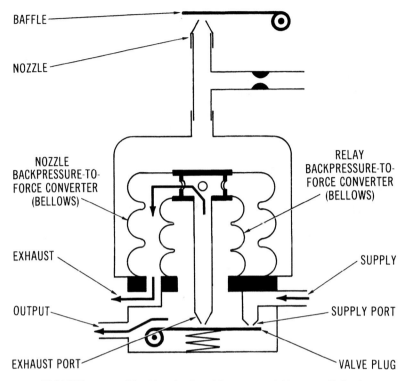

FIGURE 11-9   Nonbleed relay. (*Courtesy of Honeywell, Inc.*)

force. The O-ring and piston form the relay backpressure-to-force converter. These two forces balance each other. Figure 11-11 is a diagram of a Foxboro relay.

Figure 11-12 is a diagram of the detector mechanism used by ABB Kent-Taylor in a transmitter. In this detector, a rolling diaphragm and bellows are used. A rolling diaphragm converts the nozzle backpressure to a proportional force. The bellows converts relay backpressure to a proportional force. These two forces balance each other. Notice that in the three devices discussed, the relay backpressure-to-force converter is smaller in diameter than the nozzle backpressure-to-force converter. It can be concluded that the ratio of nozzle and relay backpressure is not one-to-one. Further, the nozzle backpressure is less than the relay backpressure when the relay is balanced. Essentially, nozzle backpressure times the area of the pressure-to-force converter equals relay backpressure times the area of the relay converter. It can be concluded that output relay backpressure is greater than nozzle backpressure. This permits pressure level amplification to be achieved.

SPECIFIC RELAY FUNCTIONS

A number of specific relay functions are used in the automatic control of pneumatic instruments. Among these are reversing, amplifying, reducing, and totalizing functions.

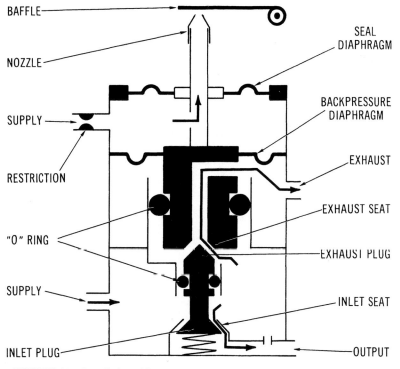

BAFFLE

NOZZLE

SUPPLY

RESTRICTION

"O" RING

SUPPLY

INLET PLUG

SEAL
DIAPHRAGM

BACKPRESSURE
DIAPHRAGM

EXHAUST

EXHAUST SEAT

EXHAUST PLUG

INLET SEAT

OUTPUT

FIGURE 11-10   Relay. (*Courtesy of Fisher Controls International, Inc.*)

These particular functions are singled out to show how a specific relay is used to achieve automatic control.

Reversing Relays

Reversing relays are used to deliver an output that will decrease in direct proportion to an increase in input pressure. Figure 11-13 shows a reversing relay and a cross-sectional view of its internal construction.

In operation, an input pressure acts between the upper and lower diaphragms. The resultant force is exerted in a direction that opposes the force of the spring. With zero or a low input pressure, reversing relays will maintain a set level of output. As a result, the output pressure is balanced according to the action of the spring-loading force.

With a decreasing input signal, the pilot valve will open, which causes the output pressure to increase. An increasing input signal applied to the pilot valve will cause it to close. This, in turn, reduces the output. The spring-loading adjustment can be altered at any time to produce different output-pressure values.

Amplifying and Reducing Relays

Relays are used to amplify or to reduce the output pressure of a pneumatic system. In this application, a 3- to 15-psi control signal can be increased or amplified to 6- to 30-psi to

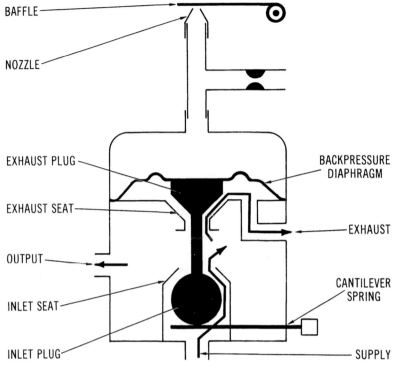

FIGURE 11-11   Diagram of a relay. (*Courtesy of The Foxboro Company*)

operate a control valve. In a similar action, 3- to 15-psi signals can be reduced to operate devices that require lower pressure levels. Figure 11-14 shows an amplifying/reducing relay and a cross-sectional view of its internal construction.

In operation, input pressure acting on the top diaphragm produces a force. This is used to balance a force produced by the output pressure developed across the lower diaphragm. Any imbalance in the forces will move the pilot valve. This, in turn, increases or decreases the air level supply to the output. The amplifying ability of the relay is primarily determined by the input-to-output area ratio of the two diaphragms.

When the input signal increases in value, it forces the pilot valve open, thus increasing the output level from the supply. A decrease in input signal level opens the exhaust port and passes air away from the output.

Figure 11-14 is an amplifying relay. It can be easily changed to a reducing relay by altering the diaphragm assembly. The area of the diaphragm is changed in this modification.

Totalizing Relays

Totalizing is used in applications where multiple input signals are applied to a single relay. In a sense, this relay is capable of achieving arithmetic computing functions such as

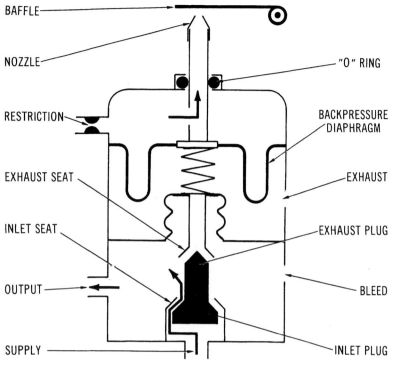

FIGURE 11-12   A relay. (*Courtesy of ABB Kent Taylor*)

adding, subtracting, scaling, or averaging. Figure 11-15 shows a totalizing relay and a cross-sectional view of its internal construction.

Operation of a totalizing relay is based on a combined force applied to inputs A, B, C, and D. Pressure applied to input A enters a chamber that acts downward on a diaphragm with a unit effective area. A signal applied to input B forces an annular or ring diaphragm down. This also has an effective area of unity. Signals applied to inputs C and D similarly act on the diaphragm except in an upward direction. These two diaphragms also have unit-effective diaphragm areas. Any imbalance in the applied inputs moves the diaphragm assembly with its integral nozzle seat. Changes in nozzle seat clearance alter the nozzle backpressure. This results in a corresponding change in output pressure.

SUMMARY

A pilot valve has two variable restrictions in its construction. The terms pilot, pilot valve, relay, and relay valve mean the same thing. A pilot converts a small mechanical displacement to a proportional backpressure by simultaneously varying two restrictions.

A pilot valve can be likened to a three-way valve. It has a supply, an output, and an exhaust port. A two-seated valve plug is mounted between these connections or ports.

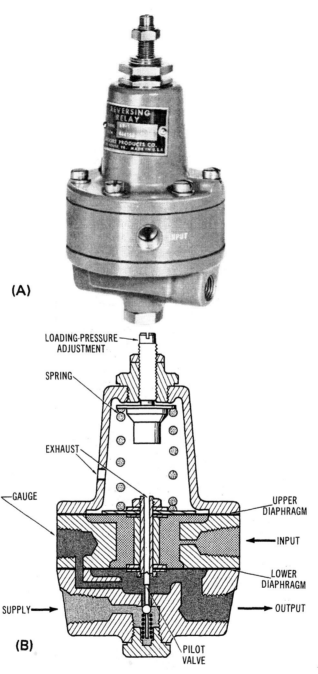

FIGURE 11-13   A reversing relay. (A) Photograph of unit; (B) Cross-sectional drawing. (*Courtesy of Moore Products Co.*)

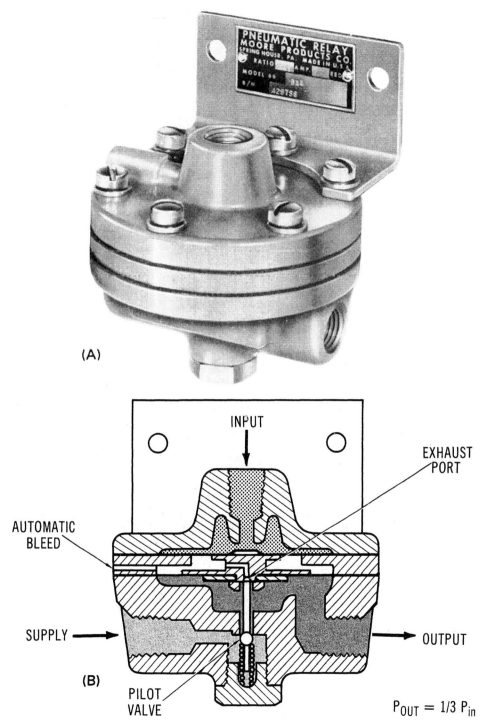

FIGURE 11-14   An amplifying/reducing relay. (A) Photograph of unit; (B) Cross-sectional drawing. (*Courtesy of Moore Products Co.*)

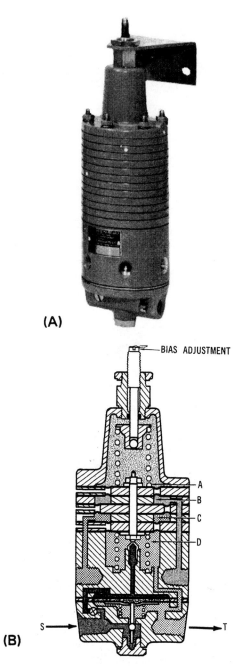

FIGURE 11-15   A totalizing relay. (A) Photograph of unit; (B) Cross-sectional drawing. (*Courtesy of Moore Products Co.*)

A pilot and a baffle/nozzle mechanism are combined to form a relay. When these two mechanisms are combined, a pressure-to-force converter is used to drive the pilot valve. Relays can be direct or reverse-acting, or they may be bleed or nonbleed types. A relay has a pilot valve plug that is operated by baffle/nozzle backpressure.

Pressure-to-force converters can be a bellows, diaphragm, capsule, bellophragm, or any of the other pressure-to-force devices.

Many relays are of the nonbleed type. A nonbleed relay uses air only when it is in an unbalanced condition. It is balanced when nozzle backpressure balances relay backpressure. These forces are directly proportional to the pressures that create them. The ratio of these two pressures is governed by the effective area of the feedback diaphragm to the input diaphragm.

A number of relays are used in automatic control. Among these are reversing, amplifying, reducing, and totalizing. Reversing relays deliver an output that decreases in direct proportion to an increase in input pressure. Amplifying relays increase an applied pressure so that they can operate a control valve. Reducing relays lower an applied pressure so that they can control low pressure devices. Totalizing relays can achieve addition, subtraction, scaling, and averaging.

ACTIVITIES

RELAY FAMILIARIZATION

1. Refer to the opertional manual of the instrument used in this activity.
2. Identify the primary function of the instrument.
3. If the instrument is attached to an operating system do not alter its operation.
4. As a rule, it is not good practice to remove the cover or housing of the instrument unless it is being modified or repaired.
5. Notice how the covering can be removed.
6. In the operational manual find a cross-sectional view or diagram of the instrument.
7. Describe the internal construction of the relay.
8. Identify the components of the relay.
9. Can the operation of the relay be adjusted externally?
10. Explain the operation of the relay.

QUESTIONS

1. What are the fundamental elements of a pilot?
2. What is meant by the terms valve plug, stem, and seat?
3. What are the components of a pneumatic relay?
4. Define the terms bleed and nonbleed as used in relays.
5. How is a relay used to achieve amplification?
6. What is meant by the term direct-acting relay?
7. What is the function of the pressure-to-force element of a relay?
8. How is a relay used to totalize pressure values?

12 Moment-Balance Transmitters

OBJECTIVES

Upon completion of this chapter, you will be able to:

- Explain the role of a transmitter in a pneumatic system.
- Discuss the moment-balance principle as used in pneumatic transmitters.
- Explain how the moment-balance principle applies to differential-pressure transmitters.
- Describe the alignment procedure of a differential-pressure transmitter.
- Identify the component parts of a moment-balance temperature transmitter.
- Explain how a moment-balance pressure transmitter operates.

KEY TERMS

Alignment. To bring into line or to adjust to the correct position with respect to other elements or components.

Beam. A piece of material or bar that is used as part of a balance assembly.

Beam contactor. A connecting element placed between two beams that transfers force between the two beams.

Detector. An indicator or sensor that determines the existence or presence of a signal or some form of energy.

Fulcrum point. A support location around which a lever turns or pivots.

Pivot. A shaft, pin, or point on which something turns or rotates.

Range beam. One of a two-beam mechanisms that makes it possible to make large changes in the ratio of output and input pressure in a transmitter.

Receiver. An instrument or device that converts an incoming signal from a transmitter into some usable function.

Restoring beam. An assembly which consists of a beam and a bellows that returns the force beam to its original position or status.

Seal bellows. A bellows assembly that connects a force beam to the sensing diaphragm through a pressure sealed connection.

Static pressure. A force that is based on weight alone without motion or movement.

PNEUMATIC TRANSMITTERS ▄▄▄▄▄▄▄▄▄▄▄▄▄▄▄▄▄▄▄▄▄▄

Pneumatic transmitters are widely used today in industrial process control systems. These systems are used to transmit air signals from the input sensor element to a remote element. It may be impractical to send small signals over a long distance because errors might be introduced due to measuring lags. In these applications, pneumatic transmitters are employed.

Pneumatic transmitters use a source of compressed air. These systems also use an element that controls the air flow. Output tubing enables the air signal to reach a remote receiving location. The receiver might be a pressure gauge, an automatic controller, a recorder, or a final control element.

A pneumatic transmitter provides equal air output pressure to the amount of the variable being measured. An output pressure of 3- to 15-psi is common and is derived from a supply of 20-psi. Air-signal pressures representing temperature, differential pressure, and other measurements, are controlled by transmitters. A liquid-filled temperature system, for example, could be adjusted so that a low temperature of 30° C equals 3-psi. A high-pressure of 15-psi could equal 100° C. As the temperature variable increases from 30° C, so does the pressure. A linear relationship between temperature and pressure is generally desired.

THE MOMENT-BALANCE PRINCIPLE ▄▄▄▄▄▄▄▄▄▄▄▄▄▄▄▄▄▄▄▄▄

The moment-balance principle was discussed earlier along with basic lever priciples. The term moment was referred to as the turning effect that occurs when an unbalanced force is applied to a lever. The moment of a force is equal to the force multiplied by the distance of force from the fulcrum about which it turns. The term is applicable to the operational principle used by transmitters. A lever, fulcrum, or flexure, and a force are used to control the supply pressure applied to the transmitter output. Figure 12-1 shows an assortment of differential-pressure transmitter assemblies.

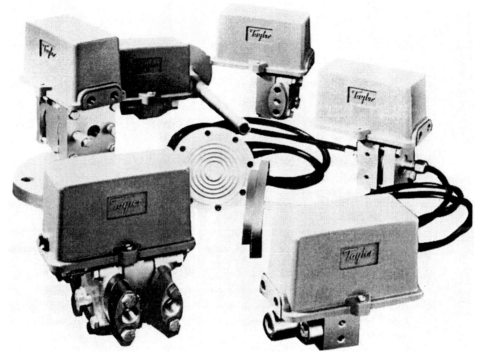

FIGURE 12-1   Differential-pressure assemblies. (*Courtesy of ABB Kent Taylor*)

Components

The essential components of a moment-balance instrument are:

1.   An input element to convert a process variable to force.
2.   A balancing element to convert an air pressure to force.
3.   A beam mechanism on which the two forces are set.
4.   A detector mechanism to detect an unbalance and cause a change in balancing force.

Arrangements

The input element is arranged so that it applies a force to the left-hand side of the beam, as shown in Figure 12-2. The beam is pivoted in the middle. The balancing component is located on the right-hand side of the beam. The balancing component is a pressure-to-force converter such as a bellows, a diaphragm, a bellophragm or any other pressure-to-force converter. A bellows has been shown. The baffle is mounted vertically on the beam. As the baffle moves the beam moves. The nozzle remains fixed. The nozzle back-pressure is connected to the balancing bellows. In the diagram, a simple baffle/nozzle mechanism is shown. The detector mechanism, in some instruments, is a pilot valve. In most instruments, the detector mechanism is the baffle/nozzle relay detector.

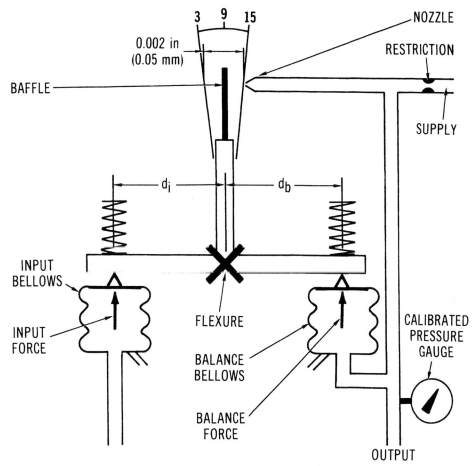

FIGURE 12-2   Moment-balance mechanism, equal lever lengths.

Principle of Operation

The detector converts a small displacement to a proportional pressure. Suppose that the input signal is a pressure. Also suppose that the pressure is zero. When it is equal to zero, the left end of the beam moves down. In this position, the baffle is away from the nozzle. Therefore, the nozzle backpressure is zero.

Now suppose that the input pressure increases to 5-psi. When it does, the baffle moves toward the nozzle. As the nozzle is covered, the backpressure increases. So, the pressure in the balancing bellows builds up. The increasing pressure in the balancing bellows causes an increased force on the right end of the beam. This balancing force tends to move the baffle away from the nozzle. The baffle will be positioned in respect to the nozzle so that backpressure, acting on the balancing bellows, results in a force equal to the input force. In this way, the output pressure is determined by the input pressure.

The previous example shows that the input force will be applied at a distance from the fulcrum. This is the distance at which the balancing force is applied. If the two bellows are the same diameter, the output pressure will almost equal the input pressure. If the input is known, then the output can be determined. Suppose that a pressure-to-quantity converter was installed on the output. This is known as a calibrated recorder or a gauge. Closer examination shows that the input pressure can be determined.

Figure 12-3 consists of the same components as Figure 12-2, except that the balancing bellows is twice the distance from the fulcrum as the input bellows. Operation of this mechanism is identical with the one that has equal lever lengths in Figure 12-2 except that the output of pressure is proportional, not equal, to the input pressure. In fact, the output pressure is equal to almost half the input pressure because the mechanism does not balance forces alone—it balances force times distance. The input force times the input distance (d_i) equals the output force times the balancing distances (d_b), or, the balancing

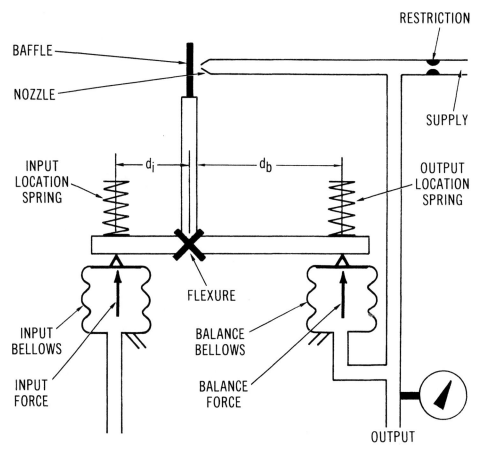

FIGURE 12-3   Moment-balance mechanism, unequal lever lengths.

distance is twice the input distance. The balancing pressure or the output pressure need only be half as great as the input pressure.

Force times a distance is referred to as a moment, which means that the output moment equals the input moment. Mechanisms that respond to forces acting on levers are called moment-balance mechanisms.

The mechanism just described is a pressure transmitter using the moment-balance principle. Most people will mistakenly describe this instrument as being force-balance. In this discussion, the moment-balance mechanism is a branch of the force-balance mechanism.

In this example, the two bellows were of equal size. Rarely is this case. It is possible to balance a mechanism having different pressures by changing the lengths of levers and by changing the bellows areas.

The mechanism of Figure 12-4 is quite different when compared with Figs. 12-2 and 12-3. This mechanism has a beam, a pivot, an input element, a balancing element, an opposing spring, and a detector. The first difference is that the detector is located on an extension of the beam. In the previous examples, it was fastened to the center of the beam. It is not important where the detector mechanism is located as long as it senses beam movement. The second difference is that the mechanism contains a spring that is opposing the balancing bellows.

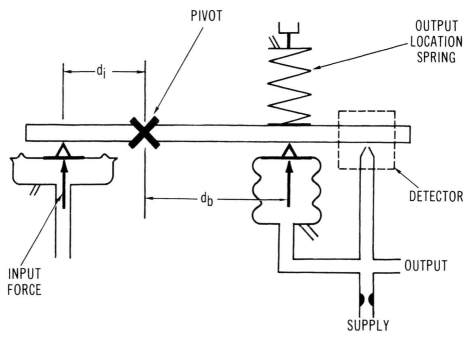

FIGURE 12-4   Moment-balance mechanism with output location spring.

Assume that the input pressure signal is zero. When this occurs, a 3-psi output signal is then required. The spring makes it possible to obtain the 3-psi output signal when the input is zero.

Here is how it works. Assume that the output pressure is zero, which means that the balancing force is also zero. The spring is compressed by turning the adjustment screw inward. This adjustment causes force to be applied to the beam. The beam then rotates towards the nozzle. The backpressure will level out. The pressure results in a balance force equal to the spring force. By adjusting the spring compression, the balancing bellows can be loaded with pressure. The output pressure, in this case, will be 3-psi. Notice that the input pressure is zero, yet there is a 3-psi output signal.

Of course, it is possible to get a variety of different outputs for zero input pressure. 3-psi was used in this example because it is a standard output signal level.

Alternate Arrangements

Figure 12-5 shows a moment-balance mechanism. The operation of this mechanism is identical with the one shown in Figure 12-4. The same relationships hold true. The difference here, is that the input and balancing elements are on the right-hand side of the

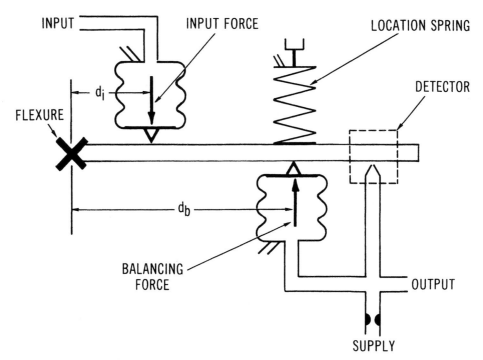

FIGURE 12-5   Moment-balance mechanism with input and output on opposite sides of beam.

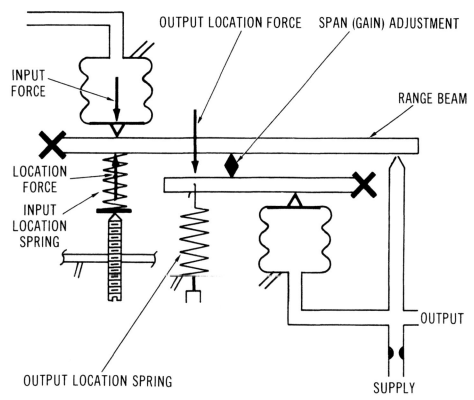

FIGURE 12-6   Moment-balance mechanism with double beam.

flexure. This is also unique in that the input element is above the beam. The input and balancing elements are found as many times on the same side of the beam as on the opposite side.

Figure 12-6 diagrams a moment-balance mechanism which is similar to Figure 12-5. In Figure 12-6, a range beam has been added to the mechanism. The range beam responds to large changes in the ratio of output pressure to the input force. The ratio is the balancing force. Addition of the range beam makes this a lever-type gain mechanism. Gain mechanisms will be discussed in greater detail in chapter 13.

Figure 12-6 shows a spring opposing the input element. The spring's function is to obtain a 3-psi output for an input that is not zero. There are instances when a 3-psi output conforms to a zero pen reading. The input might be a reading other than zero. For example, a 3-psi output might be needed when there is an input of 20-in of water.

MOMENT-BALANCE DIFFERENTIAL-PRESSURE TRANSMITTER ▬▬

The moment-balance principle is widely used in pneumatic measuring instruments. These instruments are usually transmitters. Transmitters sense variables and convert them to proportional pressure values. By examining the output pressure of a transmitter, it is possible to determine the value of a sensed variable. Differential-pressure transmitters that use the moment-balance principle will now be discussed.

Components

The components of a differential-pressure transmitter are:

1. A differential-pressure-to-force converter (the input).
2. A beam mechanism.
3. Balancing elements.
4. A detector mechanism.

Arrangements

Figure 12-7 shows a simplified diagram of a differential-pressure transmitter. This mechanism contains all essential components and elements. The diagram was shown and discussed earlier in this chapter. This diagram represents a moment-balance mechanism.

The input element converts differential pressure that is applied to a force beam. Immediately above the force beam is the range beam. Acting against the range beam is a balancing bellows. It converts the output air pressure to a force that is applied to the range beam. The force applied to the range beam acting through the beam contactor is applied to the force beam. The balancing force opposes the input force.

The input component consists of low and high pressure compartments and a diaphragm element. Two pressures acting on the diaphragm are converted to forces acting against each other. The difference between the two forces is applied to the left end of the force beam. Because both compartments are under pressure, beam moment must be taken out through a seal bellows. The instrument shown has the usual output and input span-

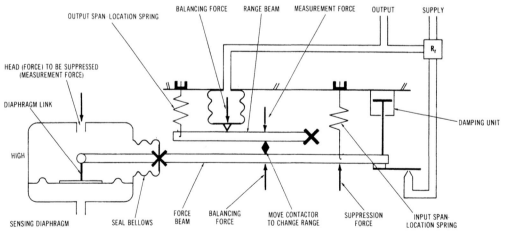

FIGURE 12-7   Simplified diagram of a differential-pressure transmitter. (*Courtesy of ABB Kent Taylor*)

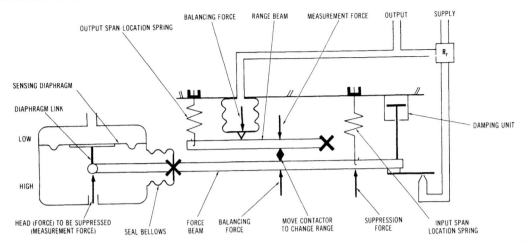

FIGURE 12-8   Same diagram as Figure 12-7, but with pressure housing reversed.

location springs. A damping unit is attached to the right side of the force beam. This unit is designed to reduce beam vibration. Mounted on the beam is a baffle/nozzle mechanism.

Compare Figure 12-7 with the more detailed diagram of Figure 12-9. Examination of these two diagrams will show that there is very little difference. Figure 12-9 is a schematic including details not available in Figure 12-7. Study these two diagrams. Notice that Figure 12-8 leaves out considerable detail when compared to the actual instrument. This diagram shows a differential-pressure mechanism for use on liquid level. Notice the position reversal of the diaphragm assembly.

Principle of Operation

Operation begins when an input is applied to the high and low pressure parts of Figure 12-9. Assume that high-pressure increases, while low pressure remains constant. If the high-pressure signal increases, the force beam will rotate about the fulcrum or pivot point. The baffle will then lift away from the nozzle. Nozzle backpressure will drop. The relay is reverse acting. Its action causes an increase in relay output pressure. As a result, balancing pressure increases which causes a change in balancing force. The resulting force acts on the range beam and through the beam contactor which causes it to move. This action causes the baffle to shift toward the nozzle. Hence, the instrument is brought to balance. This will be at the point where the balancing moment equals the increase in input moment.

Alternate Arrangements

Consider a second example of a differential-pressure transmitter. Figure 12-10 is a simplified diagram of this transmitter. It is shown in its mounted position. That is, with the force and range beams mounted vertically. The range beam is not fixed at either end of the vertical mounted mechanism. Both ends of the range beam rotates about a fulcrum point that is near the center of the beam. This fulcrum can be moved to change the span

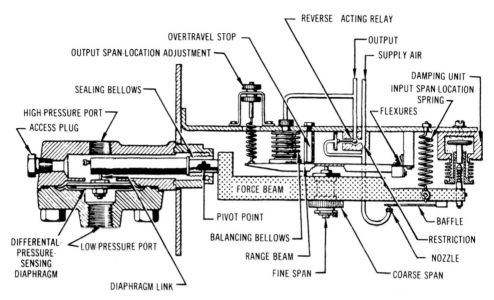

FIGURE 12-9   Schematic of a differential-pressure transmitter. (*Courtesy of ABB Kent Taylor*)

of the instrument. Span changing is accomplished here by moving the contactor between the range and force beams, which changes the ratio of range beam and force beam lengths.

Notice that the top end of the range beam is connected by a link attached to the force beam. The other end of the range beam has a spring assembly attached to it. Range beam movement is controlled by the balancing bellows attached below the pivot point. Operation of the entire assembly is very similar to that of the basic mechanism.

Principle of Operation

Assume now that there is an increase in the high-pressure input signal of Figure 12-10 which is applied to the lower right side of the instrument. It will cause the upper end of the force beam to move to the right and will permit the baffle to approach the nozzle. The nozzle backpressure will increase. The relay output will increase causing the balancing force to increase. An increase in the balancing force will cause the upper end of the range beam to move to the left. The balancing and measurement forces applied to the beam link must be equal. If they are, the positon of the baffle will be stable. The baffle must maintain this position, which will cause the balancing force to equal the measurement force.

Figure 12-11 is a detailed diagram of an actual differential-pressure transmitter. Compare Figures 12-10 and 12-11 with Figure 12-11. Notice in Figure 12-11 that the input diaphragm is a capsule filled with fluid. Any movement of the diaphragm surfaces results in an oil flow through a restriction within the capsule. It is in this manner that the transmitter is dampened.

The sealing element of the transmitter is a diaphragm. This element is labeled flexure seal in Figure 12-10 and sealing diaphragm in Figure 12-11. In the Taylor instrument of

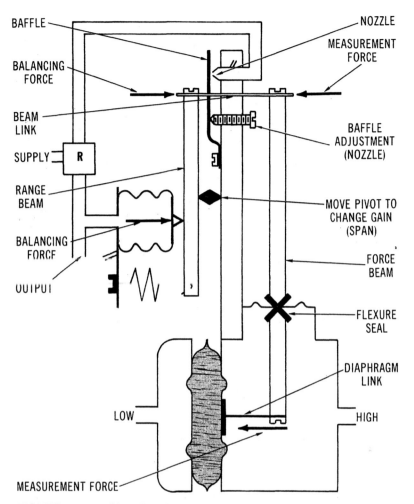

FIGURE 12-10   Simplified diagram of a differential-pressure transmitter.

Figure 12-9, it is a seal bellows. Functionally, these components are equivalent. It is worth noting that a sealing diaphragm causes some thrust on the beam. This thrust must be provided for or the sealing element would be damaged by internal pressure. In the Foxboro instrument, the sealing element also serves as a flexure. Thrust is taken care of by flexible straps mounted to the instrument chassis and the end of the beam. In the Taylor instrument, thrust is provided for by a flexure that also serves as the beam pivot.

Alternate Transmitter

The Bailey instrument of Figure 12-12 is an alternate differential-pressure transmitter that finds widespread usage. The cutaway view of this transmitter is used here because it shows some of the actual working parts of the instrument. This particular transmitter is used in liquid level measurement. Output signal pressures range from 3- to 15-psi or

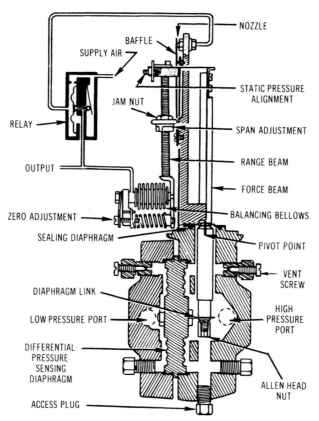

FIGURE 12-11   Schematic of a differential-pressure transmitter. (*Courtesy of the Foxboro Company*)

3- to 27-psi. These pressures can be transmitted up to 1000-ft to indicators and recorders, or 400-ft to controllers.

Principle of Operation

A simplified diagram of the Bailey transmitter is shown in Fig 12-13. Diaphragm motion causes the force beam and flexible vane to move according to the strength of the signal. Movement of the force beam alters the vane/nozzle relationship. This occurs in such a way that the nozzle backpressure in chamber B is changed. The action either opens the inlet or exhaust valve. When this occurs it produces a change in booster output pressure. Pressure is then applied to the restoring bellows where it repositions the restoring beam. The restoring beam is in contact with the force beam through the range adjustment. It then moves the diaphragm back to a position that balances the vane/nozzle. The booster output pressure is now proportional to the measured input level. The output is the transmitted signal. The signal is sent to an indicator, recorder, or controller at a distant location.

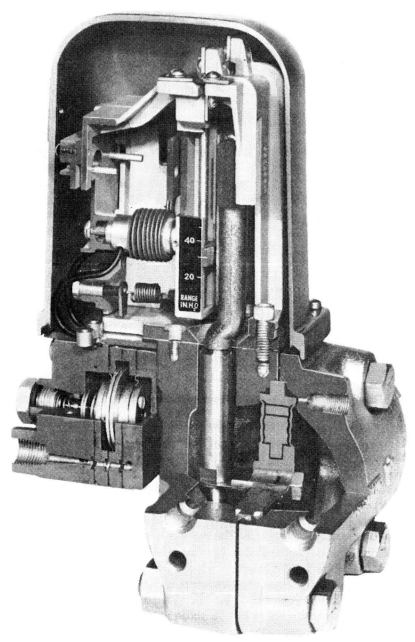

FIGURE 12-12   Cutaway view of a differential-pressure transmitter. (*Courtesy of Bailey Controls Co.*)

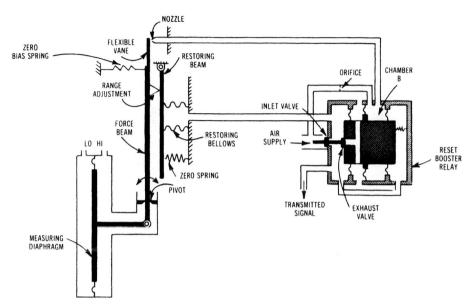

FIGURE 12-13   Simplified diagram of a differential-pressure transmitter. (*Courtesy of Bailey Controls Co.*)

ALIGNMENT

Alignment is an important consideration in the repair and calibration of an instrument. Before any instrument is calibrated, it is necessary that each of the components be in tune with each other. In Figure 12-7, some of the components are fixed while others are movable.

The input component housing, the input and output span-location springs, the balancing bellows, the damping unit, and the nozzle are all fixed near the chassis. The sensing diaphragm, the force beam, the baffle, and one end of the range beam can be changed relative to the instrument chassis. The problem of aligning consists of relating the movable components to the fixed ones. In moment-balance instruments, it is the relating of the two beams and the diaphragm.

In making an alignment, one component is chosen as a reference. In the case of the Taylor transmitter of Figure 12-7, the reference point is the force-beam fulcrum point. This point has to be aligned so that the thrust on the force beam passes through the fulcrum; also called the static-pressure alignment. If this thrust does not occur, there will be a torque on the beam. This movement will result in a calibration error. The instrument is designed so that the flexure can be adjusted, which permits the pivot point to be moved to the center of the force beam. The fulcrum point position would then be fixed.

The nozzle is then fixed relative to the chassis. Yet, the baffle must be within 0.0002-in of the nozzle. The baffle would then be aligned with the nozzle. To line up the force beam, an adjustment screw is used. The screw uses the baffle/nozzle as a positon indicator and establishes the position of the force beam.

Aligning the diaphragm uses force beam movement. Diaphragm movement must be small if it is to accurately convert differential pressure to force. It is, therefore, necessary that the diaphragm be centered around the force beam movement. This is accomplished by making the following adjustments:

After fixing the position of the force beam, the diaphragm is loaded with air pressure. The pressure pushes it against the housing. This establishes its position relative to the instrument chassis. The diaphragm is then fastened to the force beam by clamping it to the diaphragm link. The pivot point is now aligned relative to the chassis. The force beam is relative to the pivot point and nozzle. Also, the sensing diaphragm is relative to the force beam. The components still to be aligned are the balancing bellows, the damping unit, and the range beam.

The pivot end of the range beam is fixed near the chassis. It needs to be arranged so that the force beam is parallel for a 9-psi output pressure. This can be accomplished by supplying a 50 percent input signal. The zero spring is then adjusted so that a 9-psi output is obtained. The damping unit is aligned by the use of the force beam. The damping unit is aligned to the force beam by sliding the unit relative to the chassis. The reason for this is, so that there is no friction between the damping unit plunger and its case.

The transmitter is now aligned. The components of the instrument have been centered or positioned so that it can be calibrated. Only Taylor's differential-pressure transmitter has been considered thus far. The alignment of other instruments must be now be considered.

The basic reference point alignment may vary, depending on the instrument. In Foxboro instruments, the basic alignment reference is the diaphragm flexure seal. Unlike the Taylor instrument, the pivot point is fixed. The basic alignment here is accomplished in the manufacturing of the instrument. Still, the diaphragm, the baffle, and the range beam need to be synchronized. The diaphragm alignment is accomplished by clamping a link in a prescribed fashion. The baffle is aligned by the baffle adjusting screw. The beams are aligned by the beam link adjustment. Finally, the bellows are lined up by adjusting the output span-location spring.

Figure 12-14 is a diagram of the Fischer & Porter differential-pressure transmitter. This diagram appears to be basically the same as the others. However, this transmitter has an adjustable damping restriction and a temperature compensating bellows in the sensing diaphragm unit.

MOMENT-BALANCE TEMPERATURE TRANSMITTERS ━━━━━

The moment-balance principle is used by a variety of transmitting instrumets. Among these are temperature and pressure transmitters, moment-balance controllers, and valve positioners. Essentially, the difference in these instruments is in the input components.

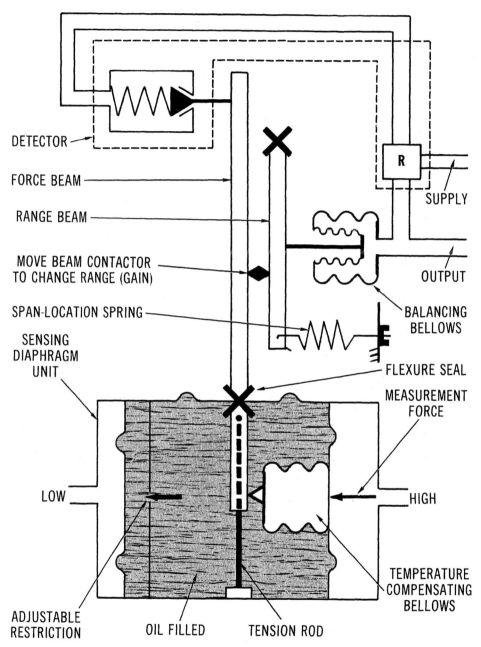

DETECTOR

FORCE BEAM

RANGE BEAM

MOVE BEAM CONTACTOR
TO CHANGE RANGE (GAIN)

SPAN-LOCATION SPRING

SENSING
DIAPHRAGM
UNIT

LOW

ADJUSTABLE
RESTRICTION

OIL FILLED   TENSION ROD

R

SUPPLY

OUTPUT

BALANCING
BELLOWS

FLEXURE SEAL

MEASUREMENT
FORCE

HIGH

TEMPERATURE
COMPENSATING
BELLOWS

FIGURE 12-14   Simplified diagram of a differential-pressure transmitter. (*Courtesy of Fischer & Porter Company*)

Components

The components of temperature transmitters are:

1. The temperature-to-force converter or input element.
2. A beam mechanism.
3. Balancing components.
4. A detector mechanism.

Arrangements

The arrangement of a temperature transmitter is similar to a differential-pressure transmitter. Figure 12-15 is a simplified diagram of a temperature transmitter. Notice the

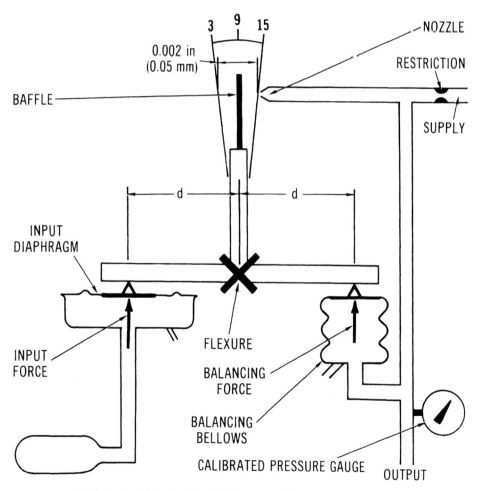

FIGURE 12-15   Simplified diagram of temperature transmitter.

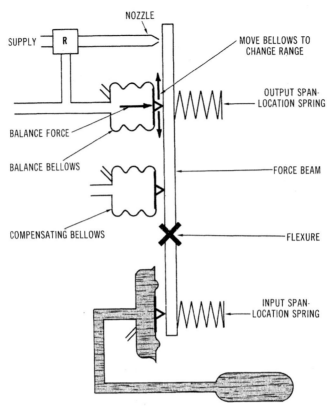

FIGURE 12-16   Simplified diagram of a temperature transmitter. (*Courtesy of The Foxboro Company*)

differences between the two types are mainly in the input components. For temperature transmitters, the input component is a closed system. This system consists of a bulb, capillary, and diaphragm element. The bulb converts temperature to pressure. The diaphragm unit converts pressure to force. This force, proportional to temperature, is applied to the force beam. The following arrangements are the same as those for a differential-pressure transmitter.

Figure 12-16 is a simplified diagram of Foxboro's filled-system moment-balance transmitter. In this diagram, there is a force beam. Operating against the force beam are the input element and balancing bellows. No range beam is used because it is a fixed-span instrument. An additional element and a pressure-and-temperature-compensating bellows operates on the force beam.

Principle of Operation

Earlier, it was stated that all parts of the filled system respond to temperature. Here, changes in temperature around the bulb need to be sensed. It is the temperature and pressure-compensating bellows that compensate for any error resulting from temperature.

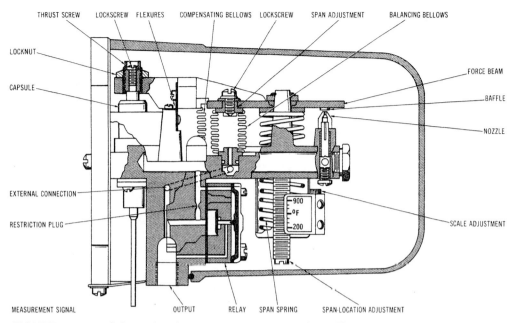

FIGURE 12-17   Schematic of a temperature transmitter. (*Courtesy of The Foxboro Company*)

The diaphragm produces a pressure-compensating problem. A diaphragm is a pressure-to-force converter. Atmospheric pressure acting on the diaphragm results in variations in force. Since atmospheric pressure varies, diaphragm forces may change even though the temperature remains the same. If the forces vary, there will be an error. As atmospheric pressure changes around the diaphragm, it also changes around the bellows. The compensating bellows is arranged to compensate for any shift in atmospheric pressure. The bellows is sized and positioned on the beam. This causes the two forces produced by the atmosphere to cancel each other.

For a given temperature rise, there is an increase in force from the diaphragm. This is shown in Figure 12-16. The force causes the beam to deflect to the left, which in turn, causes the baffle to move toward the nozzle. There is a proportional increase in nozzle backpressure, which causes relay output pressure to increase. The increase in output pressure acts on the balancing bellows and results in a force opposing the input or diaphragm force. Relay output and balancing force will increase until balancing occurs. If balancing pressure is measured with a gauge or a recorder, it will indicate the input pressure. This can be done by knowing the output pressure.

This instrument has a fixed span which is determined by the design of the filled system. In order to calibrate the instrument, it is necessary to make small span changes by having the balancing bellows mounted at an eccentric angle. As the bellows is rotated, its lever length changes which then changes the feedback moment.

Figure 12-17 is a schematic diagram of the temperature transmitter. Figure 12-18 is a pictorial diagram. Study these and relate them to Figures 12-15 and 12-16. As these comparisons are made, recognize the elements as they appear in the actual instruments.

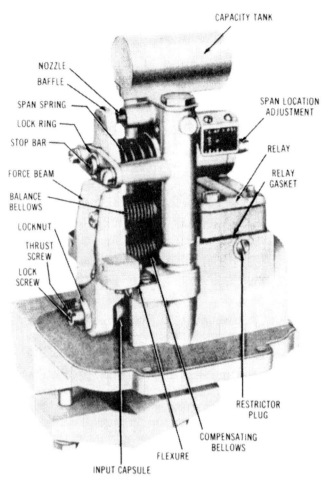

FIGURE 12-18   Pictorial diagram of a transmitter. (*Courtesy of The Foxboro Company*)

Alternate Arrangements

Figure 12-19 is a simplified diagram of a temperature transmitter. Notice that this instrument has the force and range beams suspended from the force plate. When the input element expands, it causes beam assembly movement through tension rather than compression. Instead of the input-element force pushing against the beam, a force plate is used. This plate pulls the beam's tension links.

Suppose now that there is an increase in the applied temperature. This causes a corresponding increase pressure on the force plate. Through connecting links, the force plate pulls against the force and range beams causing the left end of the range beam to rise. This movement passes through the link/lever mechanism to the baffle. The baffle then moves away from the nozzle. Nozzle backpressure drops to a small value. The relay, which is reverse acting, causes an increase in output pressure. Output pressure acting

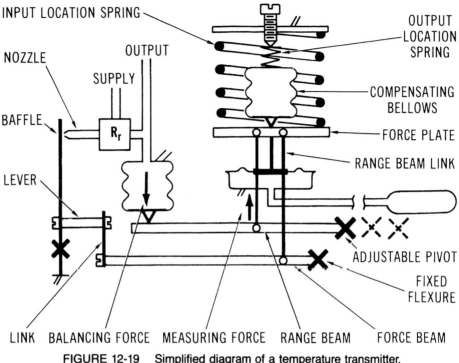

FIGURE 12-19 Simplified diagram of a temperature transmitter.

through the balancing bellows, results in a downward force on the range beam. This force tends to pull the force plate down. Recall that the original increase in temperature pushed the force plate upward. The output downward force then balances the input upward force. Balance then depends on baffle/nozzle clearance, the upward input force and downward output force. Figure 12-20 shows a schematic diagram of an ABB Kent Taylor temperature transmitter. A photograph of this transmitter is shown in Figure 12-21.

The operational span of the transmitter can be changed by altering the flexure point of the force beam relative to the force plate. Changing the flexure point alters the ratio of lever length. Additional span changes are possible by varying the stiffness of the span-location springs. Note that changing the span-location spring only changes the original loading of the input capsule. With a selection of different springs, it is possible to obtain a 3-psi output for a variety of temperatures.

Suppose now that an applied temperature of 30°C produces an output of 3-psi. Assume that a suppression spring is installed that is twice as stiff as the original spring. This will load-up the input capsule. 60° C would now be needed to produce an input moment equal to the balancing moment created by a 3-psi output.

As the span-location spring is changed, the force against the balancing bellows changes. The Taylor transmitter also has a compensating bellows. This bellows acts on the force plate rather than directly against the force beam.

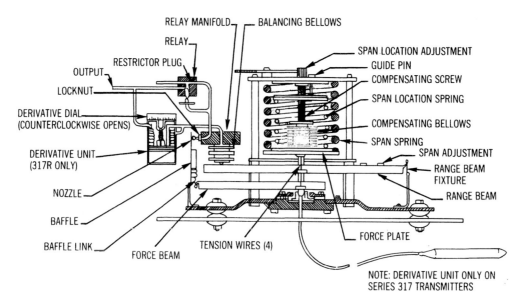

FIGURE 12-20   Schematic of a transmitter. (*Courtesy of ABB Kent Taylor*)

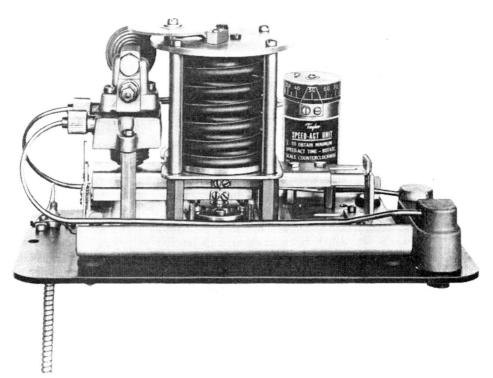

FIGURE 12-21   A transmitter. (*Courtesy of ABB Kent Taylor*)

MOMENT-BALANCE PRESSURE TRANSMITTERS ▰▰▰▰▰▰▰▰▰

The moment-balance principle of operation is widely used in pressure transmitters today. The operation of this instrument is similar to that of the temperature transmitter. The instrument presented here has a single beam with a center flexure point. The two resulting levers are equal in length.

Components

The components of a moment-balance pressure transmitter are:

1. A pressure-to-force diaphragm input element.
2. A beam mechanism.
3. A detector mechanism.
4. A balancing mechanism.

Arrangements

Refer to the moment-balance pressure transmitter of Figure 12-22. Input is applied to the pressure-to-force diaphragm element on the left side of the instrument. Figure 12-23 shows two input elements for a transmitter. The balancing bellows is mounted on the right side of the assembly. The input force and balancing force of the two bellows units is applied to the beam. The pivot or flexure point is in the center of the beam. This develops two levers of equal length. The baffle is connected to the center flexure point. Movement of the baffle changes the clearance of the nozzle.

Principle of Operation

Operation of the moment-balance pressure transmitter was discussed at the beginning of this chapter. Review the operational procedure of the mechanisms in Figure 12-2 and Figure 12-15.

SUMMARY

A pneumatic transmitter is used to transmit air signals from the input sensor to some distant or remote location. A transmitter employs a source of air, an air controlling element, and output tubing to carry air to a receiving location. An output pressure of 3- to 15-psi is common. This air signal can represent temperature, differential pressure, density, and liquid level.

The moment-balance principle is commonly used in pneumatic transmitters. The term moment refers to the resulting turning effect due to an unbalance force applied to a lever. The moment of force is equal to the force multiplied by the distance of the force from the fulcrum about which it turns. The components of a moment-balance transmitter

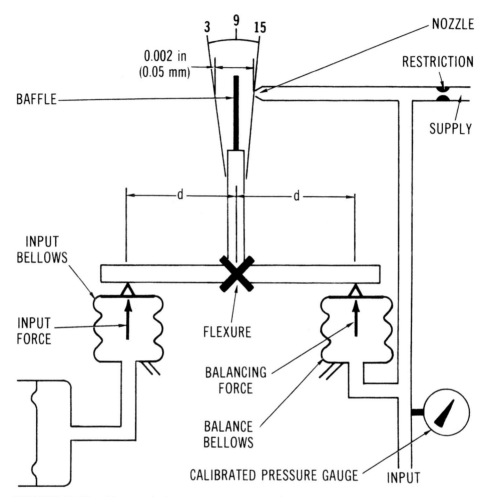

FIGURE 12-22   Moment-balance pressure transmitter.

are an input element to convert a process variable to force and a balancing element to convert air pressure to force.

Other than the difference in the input element, there are no differences between pressure and temperature transmitters that use the moment-balance principle. Notice that the differences between pressure and temperature transmitters are restricted to that portion of the input system exposed to the process. In a temperature transmitter, the component exposed to the process is a temperature-to-pressure converter. In the case of the pressure transmitter, the component exposed to the process is a pressure-to-pressure converter. In addition to its primary purpose of converting pressure, there is a sealing function. The purpose is to isolate process fluids from the instrument.

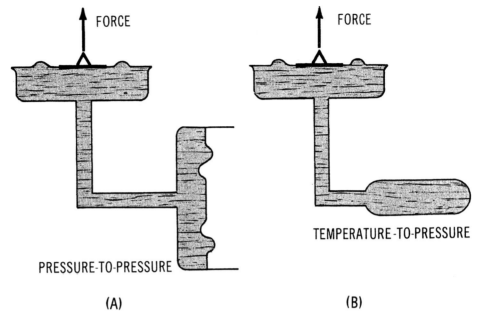

FIGURE 12-23   Input elements for transmitters. (A) Pressure element; (B) Temperature element.

ACTIVITIES

TRANSMITTER ANALYSIS

1. Refer to the operational manual of the transmitter used in this activity.
2. Identify the function of the transmitter used in this activity.
3. Turn off the transmitter or disconnect it from the system.
4. Identify the type of mechanism used in the transmitter.
5. With the operational manual as a reference, align the components of the mechanism.
6. Connect the transmitter to a source of air for operation.
7. Apply input air of a known value to the unit.
8. Check the operational span of the assembly by altering the value of input air.
9. Explain the operation of the transmitter used in this activity.
10. Disconnect the air source and reassemble the unit.

QUESTIONS

1. What are the components of a moment-balance transmitter?
2. What is meant by the term moment-balance?
3. How does the moment-balance mechanism of Figure 12-2 operate?

4. What are the components of a differential-pressure transmitter?
5. What is a representative output signal pressure from a transmitter?
6. What is alignment?
7. What are the components of a moment-balance temperature transmitter?
8. What are the components of a moment-balance pressure transmitter?

13
Moment-Balance Positioners

OBJECTIVES

Upon completion of this chapter, you will be able to:

- Define the term positioner.
- Explain the operation of a pressure-to-position converter.
- Identify the components of a positioner.
- List some examples of moment-balance positioners.
- List some applications of valve positioners.
- Explain how gain is achieved with a positioner.

KEY TERMS

Actuator.   That portion of a device which converts mechanical, pneumatic, thermal, or electrical energy into some mechanical motion.

Feed-forward.   A predictive form of control that feeds a corrective signal forward to the controller based on a measurement or distrubance in the operation of a system.

Gain.   The amount of increase in a signal as it passes through any part of a control system.

Integral.   A component that is essential for completeness or a built-in part that makes an item complete.

Louver.   An opening with one or more slanted and movable fins to allow the movement of air or light.

Position-to-pressure converter.   An instrument that accepts an input signal which changes the physical position of a component and translates this to a change in pressure.

Pressure-to-position converter.   An instrument that accepts an input signal which changes the pressure of a component and translates it into a change in physical position.

Valve actuator.   A device that changes mechanical, pneumatic, thermal, or electrical signal energy into the mechanical movement of a valve.

MOMENT-BALANCE POSITION DEVICES ▬▬▬▬▬▬▬▬▬▬

In previous chapters, moment-balance instruments were discussed. These instruments are designed to convert process variables to a proportional pressure. Essentially these devices are transmitters. Devices that convert position-to-pressure or convert pressure-to-position will now be discussed. Position-to-pressure converters are generally called setpoint transmitters. Pressure-to-position converters can be recorders or indicators.

There is an additional group of instruments known as pressure-to-position converters. These may be called valve positioners. They accept a low-volume air signal and convert it to a high-volume air output signal. This output is then applied to the valve actuator.

Components

1.   An input mechanism.
2.   A beam mechanism.
3.   A detector mechanism.
4.   A balancing mechanism.

Arrangements

The components are arranged in the typical moment-balance configuration. In this respect, the position-to-balance converter is identical with other moment-balance instruments. The difference arises out of the input components. In measuring instruments, the input components were devices that converted the process variable to a proportional force. Position-to-pressure converters employ an input mechanism that converts position to a proportional force. An essential element of this mechanism is a position-to-force converter. Recall that a spring is a position-to-force converter. The position of interest is applied to one end of the spring. The spring then converts that position to a force that is applied to the force beam.

Principle of Operation

In moment-balance instruments, the input force, acts through a lever. This is balanced by an output force acting through its lever. The detector mechanism determines when the two forces times distance are in balance. This refers to the moment of the mechanism. Figure 13-1 shows a moment-balance diagram for a position input. Figure 13-2 is a diagram of a Taylor transmitter. This setpoint transmitter is a position control device. The position of an index is converted to a proportional pressure. Setpoint transmitters are used in conjunction with controllers.

Now consider the operation of a setpoint transmitter. Suppose that the instrument is in balance with the indicator at some point on the scale. Let the setpoint knob rotate to

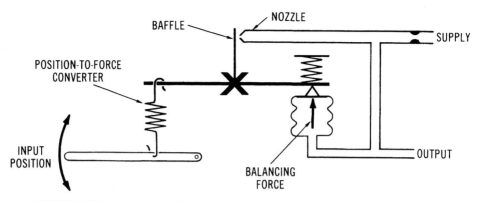

FIGURE 13-1   Fundamental moment-balance mechanism with position input.

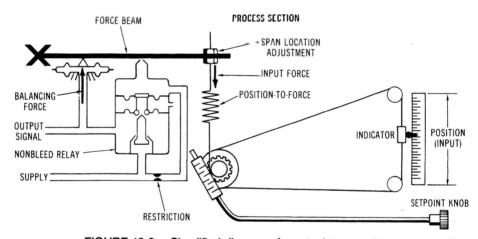

FIGURE 13-2   Simplified diagram of a setpoint transmitter.

move the indicator up scale. The purpose of the instrument is to change the output pressure to a new indicator position.

As the indicator is moved up scale, the lower end of the spring moves downward. This causes a downward force to be applied to the force beam. As the beam moves down, it covers the nozzle. This action builds up the nozzle backpressure, and the relay output pressure increases. An increase in relay output pressure, acting on the balancing diaphragm, causes an increase in balancing force. When the balancing moment is equal to the input moment, the balance action will stop. The baffle has assumed a position with respect to the nozzle that results in a backpressure or output pressure proportional to the setpoint pressure. It is also proportional to the setpoint index position.

Figure 13-3 is a more detailed diagram which shows the actual physical relationships. Figure 13-4 is a photograph of the instrument. Study these figures and relate them to the fundamental diagrams.

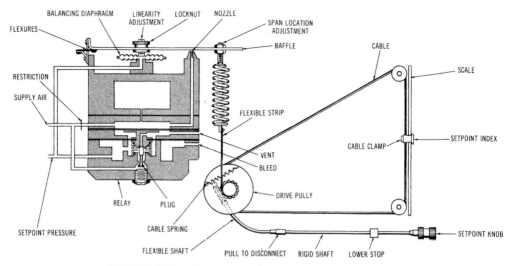

FIGURE 13-3   Schematic of a setpoint transmitter.

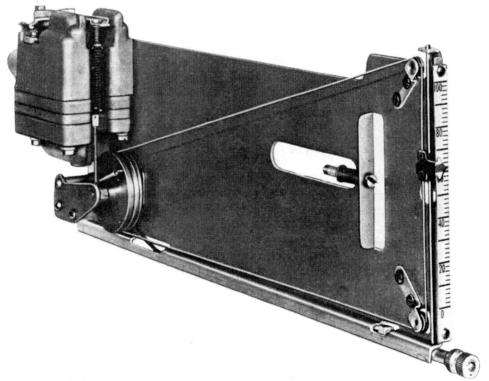

FIGURE 13-4   A setpoint transmitter. (*Courtesy of ABB Kent Taylor*)

PRESSURE-TO-POSITION CONVERTERS ■■■■■■■

In some respects, the pressure-to-position converter is opposite of the position-to-pressure converter. The instrument now being considered is the recorder mechanism of a transmitter. This mechanism, unlike the recorders of other manufacturers, is a self-balancing instrument. It is a feedback-type mechanism. The recorders discussed in the first part of the chapter are feedforward types.

A self-balancing recorder mechanism drives a pen that is equal to the input pressure. In this respect, the recorder is no different than the conventional feedforward mechanism. The difference between these mechanisms occurs when the pen does not assume its proper position. The air pressure driving the pen mechanism will change until the pen is at the proportional position.

Components

The essential components of the pressure-to-position converter are:

1. An input mechanism.
2. A beam mechanism.
3. A detector mechanism.
4. A balancing mechanism.

Arrangements

The input element, which is a pressure-to-force converter, operates against the force beam. The output of the pen drive mechanism acts through a position-to-force converter or spring. This output also acts against the beam mechanism. The difference between the input and the balancing forces is detected by a baffle/nozzle detector. Changes in baffle/nozzle position are fed to the indicator mechanism. Figure 13-5 is the fundamental diagram of a pressure-to-position converter. It differs from the transmitters because the

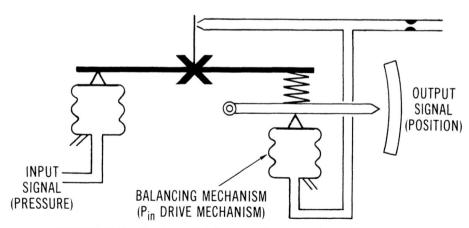

FIGURE 13-5   Fundamental diagram of pressure-to-position converter.

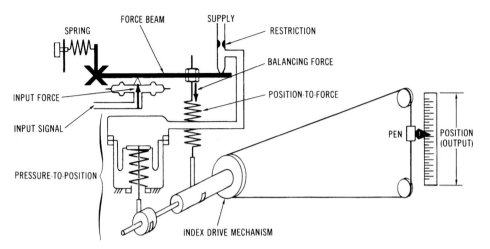

FIGURE 13-6   Simplified diagram of a recorder section of a converter mechanism. (*Courtesy of ABB Kent Taylor*)

output is not the balancing pressure. The output is the change of position of the balancing component.

Principle of Operation

In Figure 13-6, assume the mechanism is balanced for a midscale position of the indicator. The input pressure to the mechanism is then increased, which causes a force on the force beam that tends to cover the nozzle. Nozzle backpressure then results. The increase in backpressure loads up the pen driving mechanism. A downward movement of the rolling diaphragm element then occurs. The downward motion rotates the shaft, which drives the pen.

Notice that as the shaft rotates, the lower end of the spring moves downward. This action results in a force being fed back to the beam. The balancing force opposes the input force. The detector will change the pressure on the pen drive mechanism. The pressure keeps changing until the balancing moment is equal to the input moment. Figure 13-7 is a schematic diagram of the actual mechanism. Figure 13-8 is a photograph. Study these illustrations to understand mechanism operation.

FORCE-BALANCE VALVE POSITIONERS

A valve positioner ensures that the actuator stem takes a position equal to the input signal. The input signal almost always is the output of a controller. Physically, the valve positioner is mounted directly on the actuator. There is a linkage between the actuator positioner and the valve stem.

A positioner may be integral with the pneumatic mechanism mounted on the top. The term integral refers to a component that is essential for completeness of a whole item. Figure 13-9 shows an integral-mounted unit. A positioner may also be sidearm mounted. A sidearm mounted positioner is shown in Figure 13-10, page 218. Func-

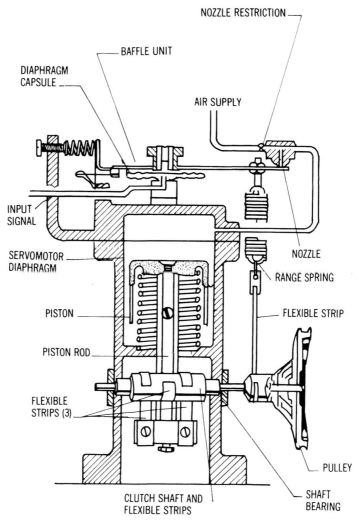

FIGURE 13-7   Schematic of the recorder section of a converter mechanism. (*Courtesy of ABB Kent Taylor*)

tionally, these positioners are equivalent regardless of the mounting unit. In addition to being used on valves, positioners have other functions. They can be used in piston actuators, louvers, and on variable-speed drive units.

Figure 13-11 (page 219) shows the sidearm positioner valve of Figure 13-10 with the pneumatic control removed. When force is applied to move the valve stem upward, it opens the valve. Removing this force causes the valve to return to its closed position.

Components

1.   An input element that converts air pressure to force.
2.   A force-beam system.

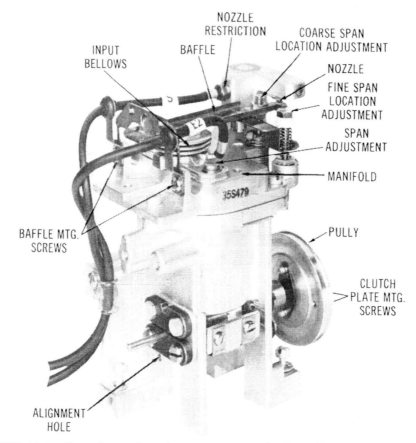

FIGURE 13-8   Recorder section of a converter mechanism. (*Courtesy of ABB Kent Taylor*)

3.   A balancing-lever system.
4.   A position-to-force converter.
5.   A detector mechanism.

It is important to recognize that the valve positioner considered by itself is not a functional entity. The valve positioner not mounted on the valve is missing an essential component. That component is the equivalent of the balancing bellows of the fundamental moment-balance mechanism. The balancing element required is furnished by the valve. Specifically, the valve diaphragm and spring convert pressure to position. Notice that the purpose of the valve diaphragm and spring is to drive the valve stem.

Arrangements

The valve positioner is mounted on the valve so as to have the valve-stem position applied to the force beam. Usually, though not always, the valve-stem position is fed back

FIGURE 13-9   An integral-mounted positioner.
(Courtesy of Fisher Control International, Inc.)

through a compound-lever system. This system is arranged so that gain can be changed. The valve-stem position, passing through the lever system, is applied to a position-to-force converter spring. This force is applied to the force beam opposing the input force. The detector is arranged to sense beam position, identifying any unbalance between the input and balancing moments.

Principle of Operation

Figure 13-12 is a fundamental diagram of an actuator positioner. Notice that it is identical to the diagram of other moment-balance mechanisms. An interesting feature of the valve positioner is that the balancing element is the diaphragm of the valve. Operation is identical to the basic moment-balance mechanism.

Figure 13-13 shows a simplified diagram of a Masoneilan valve positioner connected to a valve. This positioner is sidearm mounted as shown in Figure 13-14. See how this actuator/positioner operates. Compare Figure 13-12 with Figure 13-13.

Assume the positioner is in a balanced condition and there is an increase in the input signal. This increase will cause an increase in force that rotates the force beam counterclockwise. The detector responds to this movement, causing a decrease in output pressure. Decrease in output pressure, operating against the valve diaphragm, permits the valve-stem to lift. Valve-stem upward movement, acting through the lever system, causes

FIGURE 13-10 A sidearm-mounted positioner. (*Courtesy of Fisher Control International Inc.*)

the lower end of the spring to move downward. The increase in balancing force opposing the input signal force tends to move the beam clockwise. The detector will cause the output pressure to change the amount necessary to drive the valve stem to the position required to balance the input signal.

Notice that the output of the detector, which is an air-pressure signal, is not equal to the input-pressure signal. In fact, the output of the detector can be almost any pressure for a given input signal. It is this aspect of the positioner that makes it an important piece of equipment. The positioner develops the pressure needed to drive the actuator stem so that it is proportional to the input signal.

VALVE POSITIONER USES

Valve positioners are used for the following reasons:

1. To compensate for the forcing effects of fluids causing an unbalanced valve plug.
2. To minimize the friction effect.
3. To increase the speed of response of the control valve.
4. To split the travel of valves.

FIGURE 13-11   A sidearm positioner with the control removed. (*Courtesy of Fisher Control International Inc.*)

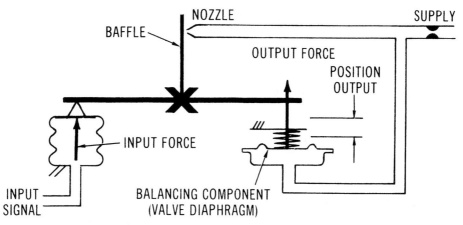

FIGURE 13-12   Fundamental diagram of actuator positioner.

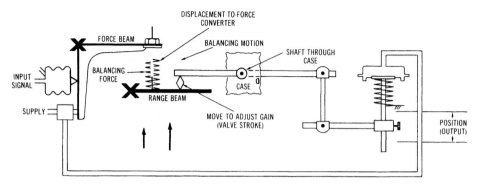

FIGURE 13-13   Simplified diagram of a valve positioner. (*Courtesy of Masoneilan International, Inc.*)

Consider the forcing effect of an unbalanced valve plug. Assume a fixed input signal pressure and a valve is operating in a fluid at 100-psi of pressure. When the valve is closed, there must be no pressure on the downstream side of the valve. Yet, acting underneath the valve is a pressure of 100-psi. If the valve plug area is 1-in, there is a force of 100-lb or 445-N. This force lifts the stem, permitting an unwanted increase in flow through the valve. Notice that the input signal has not changed. Flow through the valve should not change. The valve positioner will correct for this thrust on the valve stem.

In this example, the input signal is constant. Due to the change in pressure under the plug, the stem lifts. As the stem lifts, a force is applied to the positioner force beam. The detector responds to this change in force and operates to increase the pressure on the diaphragm, resulting in a downward force on the stem which drives it to its original position. Now there is a different pressure on the actuator.

Operation of the actuator positioner under high friction conditions is somewhat similar. Assume an increase in input signal, which would mean that the valve stem should move. Suppose it does not because friction is holding it back. If it does not move, there is no new balancing force opposing the input force. As a consequence, the output pressure decreases until the valve stem moves. It will drop to zero if the valve stem does not move. Usually, the detector will cause the air pressure on the actuator to change. The change is generally enough to make up for the friction forces holding the actuator stem movement back. If the input signal is in the opposite direction, the pressure on the valve will increase to the supply pressure.

A word of caution should be noted for this operation. Actuator positioners should not be used to compensate for unnecessary friction from improperly maintained control valves. There is a clear possibility that an actuator positioner or a high-friction valve will result in a valve that cycles. The result would be an unstable control operation.

Actuator positioners are primarily used to increase the speed of response. Recognize that the air delivery capacity of a controller is limited. Some actuators are large and require considerable amounts of air to cause a change in valve-stem position. In this situation, the valve moves slowly because it takes time to build up pressure within the actuator. If a valve positioner is used, the controller changes pressure in the input bellows

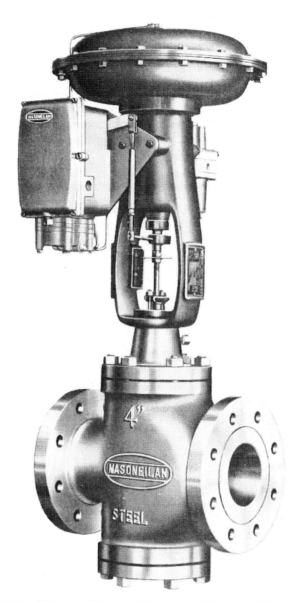

FIGURE 13-14   Valve positioners. (*Courtesy of Masoneilan International, Inc.*)

of the positioner. Therefore, the input force increases rapidly. The detector mechanism of the positioner is designed to furnish large amounts of air, which causes the valve to move rapidly in response to changes of input signal.

Valve positioners are used to split the range of valves. An input range of 3-to 15-psi normally causes the output to have full valve travel. For a 3-to 15-psi input, only part of full valve travel may be desired. A fairly common application has one valve moving from fully closed to fully open as the input signal changes from 3-to 9-psi. An input of 9-to 15-psi causes a second valve to move from fully open to closed. This means that the valve ranges have been split. Valve positioners can do range splitting very well. An adjustable gain mechanism is used to accomplish this operation.

GAIN MECHANISMS

Figure 13-15 shows an adjustable gain system consisting of parallel levers and a movable contactor. A parallel lever system is used on positioners. One lever of the system is the force beam. The second lever pivots through the instrument case.

Notice that the two levers are joined by a contactor, as in Figure 13-16. The contactor is arranged so that it slides along the levers. As the contactor moves from right to left, the range beam length is increased. The length of the input lever decreases. In other words, the input lever length is changed. This change in ratio of lever length alters the ratio of output-position to input-position. The parallel lever gain mechanism is a Z-type linkage using a very short link.

Now study the lever mechanisms of Figure 13-16. Notice in Figure 13-16A that the contactor is near the input-lever pivot point. In Figure 13-16B, the contactor is near the output-lever pivot point. See how moving the contactor relative to the pivot changes the gain. Assume now that the contactor is near the input-lever pivot point as in Figure 13-16A. This causes a relatively small movement of the contactor for a large input move-

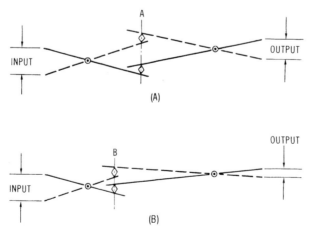

FIGURE 13-15   Adjustable gain system. (A) Contactor in position A; (B) Contactor in position B.

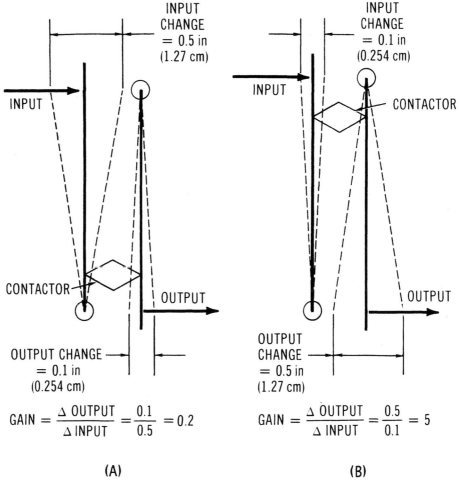

FIGURE 13-16 Changing gain in parallel lever system. (A) Lower gain position; (B) High gain position.

ment. When the contactor moves a small amount, it drives the output lever a small amount. The diagram shows the input moving 0.5-in. The contactor is located so that for this 0.5-in movement, the output lever will move 0.1-in. The gain is the change in output divided by the change in input. The Greek letter delta (Δ) denotes a change in value. The expression then becomes:

$$\text{gain} = \Delta \text{ output}/\Delta \text{ input}$$
$$= 0.1\text{-in}/0.5\text{-in}$$
$$= 0.2$$

Now examine Figure 13-16B. In this figure, the contactor is near the output-lever pivot point. Even though input lever and contactor move only a small amount, this movement applied near the pivot point on the output lever produces a relatively large output movement. If the input moves 0.1-in, the output may move 0.5-in depending on lever lengths. In this example, the gain will be:

$$\text{gain} = \Delta \text{ output}/\Delta \text{ input}$$
$$= 0.5\text{-in}/0.1\text{-in}$$
$$= 5$$

As the contactor is moved upward from one extremity to another, the gain is changed from 0.2 to 5. Of course, for intermediate contactor positions, gains between 0.2 and 5 can be obtained.

The parallel lever system is widely used. It is necessary that it be clearly understood. This mechanism has adjustable gain. The mechanism makes it possible to change the ratio of the output to the input.

Recognize that valve-stem travel is established by the range of the valve being used. The standard range is 3-to 15-psi for full travel. Other ranges are also obtainable.

The reason for splitting valve ranges is to cause a full change in valve-stem position for a 50 percent change in input signal. This can be done if a full change in valve stem position causes the balancing force to be half of its normal value. If it is half, it will balance a 50 percent change of input signal. A 3-to 9-psi input signal will cause full valve travel. Sometimes valve ranges are split in thirds.

MOMENT-BALANCE POSITIONERS ▬▬▬▬▬▬▬▬▬▬▬▬▬▬▬▬▬▬▬▬

Thus far, the moment-balance principle has been discussed as it applies to actuator positioners. This discussion has centered around the purpose of the actuator and its operation. The discussion will now be directed toward additional examples of actuator positioners. In particular, they are used on valves and piston operators.

The first example to be considered is a positioner used by ITT to operate its piston-type valve operator. This positioner is integrally mounted as compared to sidearm mounted. This component is very similar to positioners already discussed.

In addition to ITT positioners, there will be a discussion of devices used by Foxboro and Bailey. These instruments are used to control piston operators.

Components

The components of these positioners are the same as those discussed previously. They include:

1.   An input element to convert air pressure to force.
2.   A force-beam system.
3.   A balancing-lever system.

4. Position-to-force converter to change lever position to force.

5. A detector mechanism.

6. The valve diaphragm that converts pressure to position.

Arrangements

The ITT positioner, unlike others, is integrally mounted which means that the positioner mechanism is installed in the top of the valve housing. Figure 13-17 is a diagram and Figure 13-18 is a photograph of the ITT positioner.

Examine Figure 13-17 and compare it with the list of components and with the diagram. Locate the input element. It is a diaphragm that converts the air-pressure signal to force. This force, acting through a shaft, is applied to a force beam. A force proportional to valve position is applied to the other end of the force beam. The detector senses the force beam position through an extention of the shaft. The two forces are brought into balance by the detector.

Notice that there is no balancing lever system. Since the positioner is integrally mounted, feedback to the force beam can be made directly without additional levers.

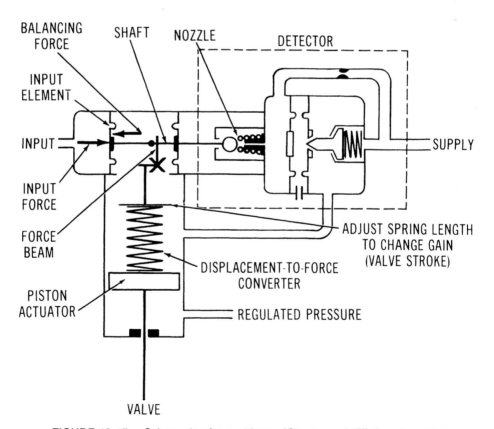

FIGURE 13-17   Schematic of a positioner. (*Courtesy of ITT, Conoflow Div.*)

FIGURE 13-18   An integral positioner. (*Courtesy of ITT, Conoflow Div.*)

Elimination of levers requires that different provisions be made for adjusting the gain of the positioner. These are different from those used when levers are present. Gain adjustments are made by adjusting the effective length of the spring.

The function of a spring is to convert a change in position to a change in force. For a given position, the resultant force will depend on the spring characteristic. Given a particular spring, it is possible to change its spring characteristic by changing its length, thus altering the output force for a given input by changing the effective length of the spring. Since the ratio of output to input is changed, the gain is changed.

The detector mechanism is a standard baffle/nozzle nonbleed relay combination. The mechanism consists of a port and ball combination rather than a nozzle and flat baffle. Functionally, they are equivalent.

Principle of Operation

Assume that the positioner is in a balanced condition and a new signal is applied. The input signal, acting on the diaphragm, is applied to the force beam. It moves the baffle relative to the nozzle, which changes the backpressure operating the relay. As a result, the relay output changes. This change in relay output is connected to the upper side of the piston operator. The new pressure causes it to move. This movement, acting through the spring element, changes the balancing force. When the balancing force approximates the input force, the piston actuator will come to a fixed position.

Notice that balancing is between the input force, valve position, and the relay output. As a result, the pressure on the piston will change. This is the pressure which will force it to a position which results in a balancing force equal to the input force.

The relay output pressure applied to the upper portion of the piston operator also loads the input diaphragm. If this alone were the case, the positioner would not be functional. Relay output pressure would directly oppose the input signal pressure. Notice that a second seal diaphragm is identical to the input diaphragm. The relay output pressure also loads the seal diaphragm. Since the diaphragms are identical, the resulting forces are also identical, causing the diaphragms to oppose each other. The resulting forces oppose each other and balance out. It is in this way that relay backpressure on the input diaphragm is eliminated.

To summarize, the ITT integral positioner is identical with positioners discused earlier. The differences are in the method of mounting, changing the gain, and baffle/nozzle construction. In this instrument, when the positioner is integrally mounted, gain is altered by changing the effective length of the spring. The baffle/nozzle detector is a ball/port combination.

A second example of a position device is the Foxboro piston operator. Foxboro calls it a Poweractor. This can be used to operate valves, butterflies, louvers, variable-speed drives, or any other mechanical equipment that requires positioning.

Components

The components of the Poweractor of Figure 13-19 are typical of the valve positioner components already discussed. Notice the force beam and the range lever furnishing the adjustable gain. The detector is a double-acting pilot valve. The position-to-force spring is driven by the piston.

Arrangements

The Poweractor is sidearm mounted. Connection to the piston is made through and extension on the piston shaft. Connected to the end of the extension on the piston shaft is the balancing spring. This spring converts piston position to force.

Principle of Operation

In positioners, it was assumed that a change in the input signal was needed to start the operation. In this example, assume that the input signal remains constant. The load on the piston, however, is subjected to some changes which could be caused by lumps of material under the roll that the piston positions. These lumps tend to move the piston upward. As the diagram of Figure 13-20 shows, the piston moves upward while the balancing force is reduced. This permits the force beam to rotate upward. The detector

FIGURE 13-19   A poweractor. (*Courtesy of The Foxboro Company*)

senses the upward movement of the force beam. It acts to increase the air pressure on the upper side of the piston. It also reduces the air pressure on the lower side of the piston positioner, causing increased force to move the piston downward. Recall that material lumps caused the piston to move upward, thereby driving the piston to a position that balances the input signal.

Piston positioners respond the same as other positioners when the input signal is changed. For that matter, all positioners respond the same when loading changes.

The last example of moment-balance positioning devices to be discussed is a Bailey piston positioner. This positioner is similar to the Foxboro Poweractor. There is a difference between these two positioners. The Bailey unit is arranged so that it has a nonlinear relationship between piston position and input signal. Therefore, it is possible to get a nonproportional piston position for an input pressure signal. Remember that in other devices the relationship between piston and input signal was proportional. In other words they were linear.

The nonlineraity feature can be extremely valuable. It is widely used when the relationship between flow and valve position is nonlinear. If the positioner is nonlinear in the same way that the flow-to-position is nonlinear, input signal and flow can be made linear.

Components

The components of a Bailey positioner are the usual ones, with one exception. Between the piston position feedback and the position-to-force converter is a nonlinear link. This link is a cam. Nonlinear links were discussed in an earlier chapter. The nonlinear link, or

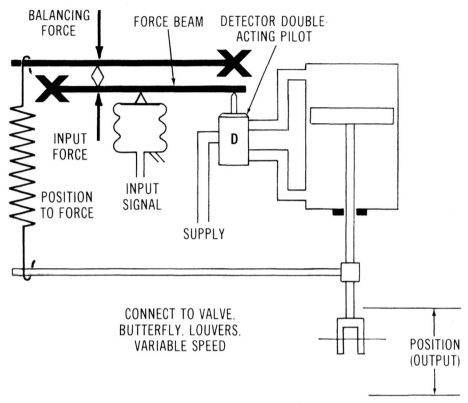

FIGURE 13-20   Simplified diagram of a Poweractor. (*Courtesy of The Foxboro Company*)

cam, is driven by a two-gear train. Mounted on one of the gears is a cam which is followed by the range lever.

Arrangements

A Bailey positioner is sidearm mounted. It usually has a piston and lever system forming a package. The device to be positioned is attached to the takeoff beam, as shown in Figure 13-21. The takeoff beam is also the lever providing piston-position feedback to the positioner. It is comparable to the lever mechanism that rotates in other valve positioners.

Principle of Operation

Assume now that the positioner is balanced at a signal of about 3-psi, and that this signal changes to 4-psi. The new signal will result in a force on the force beam and tends to rotate the force beam upward. The detector senses this upward movement. It increases pressure on the lower side of the piston and decreases it on the upper side. These changes in pressure cause the piston to move upward. The takeoff beam will rotate clockwise. The motion, acting through the lever and gears, rotates the cam counterclockwise. The cam

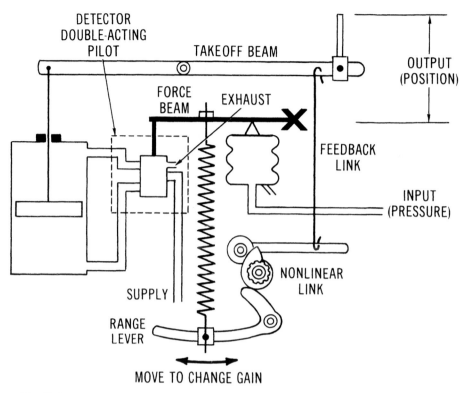

FIGURE 13-21   Simplified diagram of a positioner. (*Courtesy of Bailey Control Co.*)

pushes the range lever downward causing an increase on the force beam. Cam movement continues until it balances the increased input signal.

Suppose the postioner were balanced at 12-psi and the signal increased to 13-psi. This pressure changes an amount similar to that discussed previously. Operation of the positioner and piston is identical to that just described. However, the range lever now operates against a different proportion of the cam. See Figure 13-22.

Suppose that for the 1-psi change between 3- and 4-psi, the cam rotates 22.5°, or half of the full angularity rotation of the cam. The rotation will result in a change in the cam follower of 1/8-in. A 1/8-in change will result in the lower end of the spring changing 1/4-in. Now examine the changes obtained when the cam operates at 12-psi. When the cam rotates from 12-psi to the 13-psi position, the cam follower moves 1/4-in, or twice as much as before. For a 1/4-in movement, the lower end of the spring will move 1/2-in. Yet, the input signal changed the same 1-psi.

The signal change of a positioner is generally the same as the balancing force. A change in balancing force at the 12-psi level is twice the change in balancing force at the 3-psi level. This difference in balancing forces for the same input-signal change will result in different piston-position changes. Hence, the relationship between piston position and signal becomes nonlinear. The specific relationship is determined by the profile of

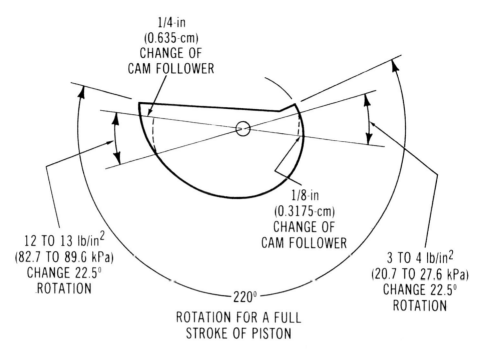

1/4-in
(0.635-cm)
CHANGE OF
CAM FOLLOWER

1/8-in
(0.3175-cm)
CHANGE OF
CAM FOLLOWER

12 TO 13 lb/in^2
(82.7 TO 89.6 kPa)
CHANGE 22.5°
ROTATION

3 TO 4 lb/in^2
(20.7 TO 27.6 kPa)
CHANGE 22.5°
ROTATION

220°
ROTATION FOR A FULL
STROKE OF PISTON

FIGURE 13-22   Cam relationships of a positioner. (*Courtesy of Bailey Controls Co.*)

the cam. A variety of cams is available. Cams can be designed to obtain any of the required relationships between piston position and input signal. The Bailey positioner was studied because it is used on a piston actuator. The same positioner can be used on a variety of actuator-driving devices. Figures 13-21 and 13-22 show the relationship of elements and components in the Bailey positioner.

SUMMARY

There are two common classes of devices in pneumatic instrumentation that are used to influence position. Position-to-pressure converters are called setpoint transmitters. An additional group of instruments are pressure-to-position converters. The valve positioner is an application of this instrument.

In a position-to-pressure instrument, physical changes in the input are converted to a force used to regulate pressure at the output. This type of instrument has an input, a beam, a detector, and a balancing mechanism.

Pressure-to-position converters are the opposite of position-to-pressure converters. There is an input, beam, detector, and balancing mechanism. The input is pressure-actuated and the output produces a physical change in position.

Force-balance positioners are essentially valve-positioning devices. These instruments are responsible for actuating valve position proportional to the input signal.

Integral positioners have their works mounted in the top. Sidearm positioners have their works mounted on the side.

There are a number of unique positioners on the market today. These instruments differ from the standard device by mounting methods, gain-changing mechanisms, and baffle/nozzle construction.

ACTIVITIES

POSITIONER ANALYSIS ▬▬▬▬▬▬▬▬▬▬▬▬▬▬▬▬▬▬

1. Refer to the operational manual of the positioner used in this activity.
2. Identify the type of positioner used in this activity.
3. If the positioner is connected to a system, do not disassemble it. If the positioner is not attached to a system, remove its outside housing.
4. Locate the position of the converter mechanism.
5. Describe the operation of the mechanism.
6. If the positioner is attached to a functioning system, describe the function of the instrument.
7. Explain the operation of the positioner.
8. Reassemble the instrument.

QUESTIONS

1. What is meant by the term pressure-to-position converter?
2. What is a setpoint transmitter?
3. What are the components of a pressure-to-position converter?
4. What is the principle of operation of a pressure-to-position converter?
5. What is a valve positioner?
6. What are some uses of valve positioners?
7. How does a gain mechanism function?
8. What are some of the unique differences in positioners today?

14
True Force-Balance Instruments

OBJECTIVES

Upon completion of this chapter, you will be able to:

- Distinguish between moment-balance and true force-balance instruments.
- Identify the mechanism of a true force instrument.
- Explain the operation of a true force-balance mechanism.
- Explain how a true force-balance instrument is calibrated.

KEY TERMS

Convolutions.   A structure that has rolled or curled elements in its construction. Part of the construction of a diaphragm.

Counterforce.   A force that moves in a direction opposite to or in the reverse direction of another force.

Effective diameter.   A size consideration that is somewhat less than the actual diameter of a diaphragm.

Element.   A basic or rudimentary part of a larger assembly.

Span.   A measure of the extent, stretch, reach, or spread between two limits.

Stack.   To arrange in a pile with one element on top of the preceeding element.

Stack-type instrument.   An expression that describes the structure of diaphragm elements used in a true force-balance mechanism.

Thrust shaft.   A central rod around which a number of diaphragms are arranged in a stack.

True force-balance.   An operating principle that employs a stack of diaphragms arranged around a central shaft so that each element lines up with the center of the mechanism.

TRUE FORCE-BALANCE

Moment-balance instruments are characterized by forces acting at distances from pivot points. A force times a distance is a moment. The moment-balance principle was applied to instruments used in recording, transmitting, and positioning.

Moment-balance mechanisms are closely related to force-balance mechanisms. There is another class of mechanisms that respond to the true force-balance principle used in true-force recorders, transmitters, positioners, and controllers. The true-force-balance principle is also used in many relays.

True force-balance instruments are characterized by pressure-to-force elements only. No levers are used in these mechanisms except for the input lever positioners.

True force-balance instruments frequently are called stack-type devices. This expression shows that the elements are stacked on top of one another, similar to a stack of pancakes. It is done so that force and counterforce act along the same axis.

Components

1.   An input diaphragm.
2.   A balancing diaphragm.
3.   A thrust shaft.
4.   A detector mechanism.

Notice the absence of any force beams or lever systems.

Arrangements

The pressure-to-force converting elements are arranged on the same axis so that the center of the element lines up with the center line of the instrument. The range-location springs are also arranged on the same axis. This construction results in the stacking of elements. Each element is clamped to the thrust shaft. The detector mechanism is arranged to sense the position of the thrust shaft. The outer edges of the diaphragm are clamped between ring-shaped pieces. These space the diaphragms and form the compartments.

Principle of Operation

Figure 14-1 shows that the input signal is applied to the upper compartment located against the diaphragm. The diaphragm converts input pressure to a force which moves the thrust shaft downward towards the nozzle. The nozzle backpressure builds up. The back-

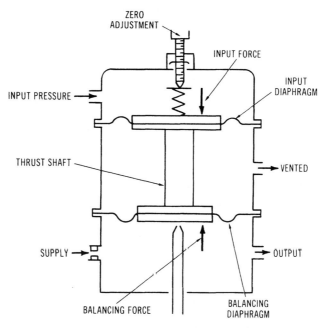

FIGURE 14-1 True force-balance mechanism.

pressure is applied against the balancing diaphragm. Acting on the balancing diaphragm, it is converted to a balancing force. The balancing force directly opposes the input signal force.

Nozzle backpressure will change to the pressure required to balance the input-signal pressure. When the forces are balanced, the thrust shaft also balances in front of the nozzle. It causes the required nozzle backpressure to produce a balancing force.

OUTPUT SPAN LOCATION

When true force-balance devices are used as transmitters, the output signal should be the standard 3- to 15-psi. The input signal is proportional to a measured variable, which means that for a zero input signal, the output signal should be 3-psi.

Assume now that the input signal is zero. The nozzle backpressure signal must also be zero. Yet, a 3-psi output for a zero input is needed.

Suppose that the thrust shaft is loaded with a spring. When the spring is compressed a force is applied to the thrust shaft. This force moves the shaft towards the nozzle. As a result of this movement, the nozzle backpressure will increase. The backpressure acting on the diaphragm forces the thrust shaft upward against the force of the spring. When the balancing force equals the spring force, the thrust shaft will come to equilibrium. Simple adjustment of the output span-location spring can be used to obtain a 3-psi output with a zero input signal.

SPAN CHANGING ■■■■■■■■■■■■■■■■■■■■■■■■■

The mechanism just described is a one-to-one relay. That is, the output signal divided by the input signal equals one, which means that the gain is one. Suppose a gain value different than one is required. In moment-balance instruments, lever lengths can be changed. In true force-balance instruments there are no levers.

The output force of a diaphragm is equal to the input times the area of the diaphragm. By changing diaphragm areas, it is possible to get different force values for an applied pressure.

Figure 14-2 is a diagram of a true force-balance mechanism. In this unit, the diaphragm has twice the area of the balancing diaphragm. The operation of this device is the same as that of Figure 14-1. The primary difference is that now 1-psi of input pressure is balanced by 2-psi of balancing pressure.

Different ratios of diaphragm diameters are used to obtain different output and input values. By changing this ratio, it is possible to change the ratio of output to input signals. Changing diameter ratios is the method used to make large changes in gain.

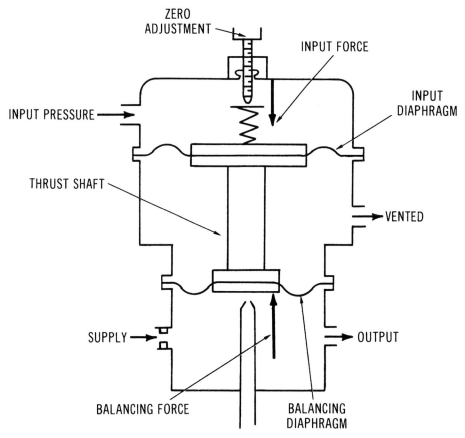

FIGURE 14-2   Force-balance mechanism with a diaphragm area ratio of 2 to 1.

For small gain changes, there is a special group of diaphragms. These diaphragms are designed to make minor changes in the input/output ratio by changing diaphragm diameter. When speaking of diaphragm diameter, there is no concern with the extreme outside diameter, which is the point where the diaphragm clamps to the ring components. It is called the effective diameter. The effective diameter is smaller than the actual diameter. It varies as the center portion of the diaphragm moves relative to the housing.

Figure 14-3 shows the center of the diaphragm moving downward. The convolutions of the diaphragm have a tendency to turn or roll. The section of convolutions closest to the housing become longer than those near the center, reducing effective diaphragm diameter which, in turn, reduces diaphragm area. This property is used for correcting minor differences between output and input signals. Value differences of 2 percent or less can be controlled by this procedure. In other words, moving the center of the diaphragm relative to the housing is a span adjustment.

The amount that the ratio of diameters can be changed is quite limited. This value is perhaps less than 2 percent of the input span. Unlike moment-balance transmitters, force-balance transmitters have ranges that are fixed. In moment-balance transmitters, the input span can be changed. Values of 25- to 250-in of water can be changed by moving the contactor of the lever gain mechanism.

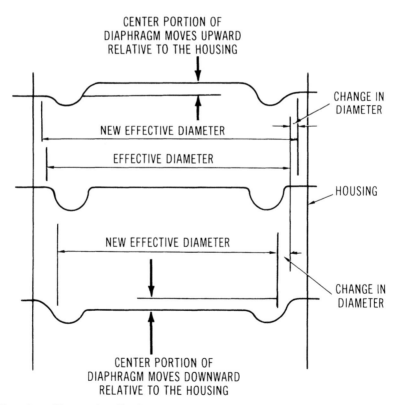

FIGURE 14-3   Change in effective area of diaphragm at different operational pressures.

Large span changes can be made in a true force-balance transmitter. It is necessary to make changes in the ratio of the sensing diaphragm to the balancing diaphragm. This is accomplished by using balancing diaphragms of different diameters. The sensing diaphragm remains the same.

FORCE-BALANCE TRANSMITTERS

An operational essential of the true force-balance mechanism is that the forces involved directly oppose each other. A feature of this mechanism is the lack of levers. The elements are arranged so that they directly oppose each other. In the moment-balance principle, the force involved acts at a distance from a fulcrum point. The true force-balance principle can be applied to differential-pressure transmitters. The Taylor transmitter of Figures 14-4, 14-5, and 14-6 is an example of this instrument.

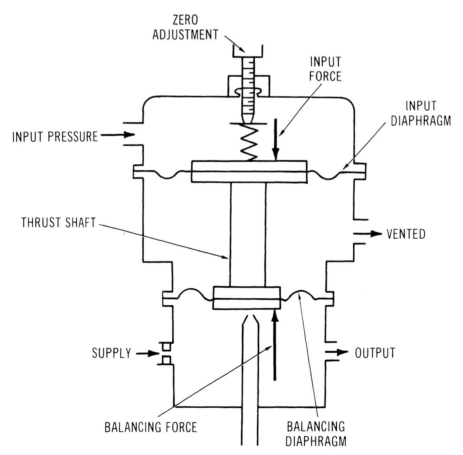

FIGURE 14-4   Simplified diagram of a transmitter. (*Courtesy of ABB Kent Taylor*)

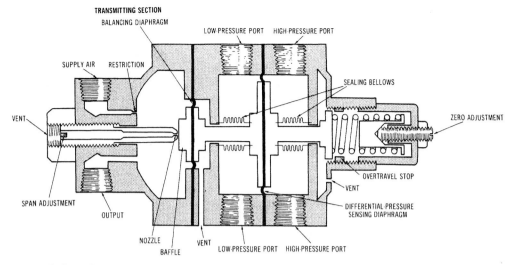

FIGURE 14-5 Schematic of a transmitter. (*Courtesy of ABB Kent Taylor*)

FIGURE 14-6 Cutaway photograph of a transmitter. (*Courtesy of ABB Kent Taylor*)

Components

1. The differential-pressure sensing diaphragm.
2. High-and low-pressure compartments.
3. A thrust shaft.
4. A balancing diaphragm.
5. A nozzle backpressure compartment.
6. A detector mechanism.

Arangements

The sensing diaphragm and balancing diagphragm are mounted on the thrust shaft. The thrust shaft is arranged in the housing so that it can move a small amount. Movement is sensed by a baffle/nozzle located in the lower end of the shaft. The shaft end serves as the baffle. The nozzle is threaded into the housing. Turning the nozzle moves the tip of the nozzle relative to the housing. Nozzle backpressure loads the balancing diaphragm.

Mounted at the upper end of the thrust shaft is an output span-location or zeroing spring. One end of the zero spring rests against the thrust shaft. The other end rests against a threaded screw. The load on the spring can be changed by adjusting the zero screw.

Two seal bellows are provided to form the high-pressure and low-pressure compartments. The seal bellows are identical and are arranged to perform a sealing function only.

The sensing diaphragm is considerably larger than the balancing diaphragm. This means that a full range of balancing pressure is required to balance a small range of differential pressure.

Principle of Operation

Refer to Figure 14-4. Assume that the instrument is balanced for some input differential pressure. Assume further that there is an increase in differential pressure. The increase in pressure acts on the input diaphragm, causing an increased force that moves the thrust shaft downward toward the nozzle. Covering the nozzle increases backpressure and the output.

An increase in nozzle backpressure, acting on the balancing diaphragm, causes an increasing upward force. When the increasing upward force balances the change in downward force, the thrust shaft reaches equilibrium. When the forces are balanced, the output pressure is proportional to the differential pressure.

CALIBRATION ADJUSTMENTS ■■■■■■■■■■■■■■■■■■■■■

Suppose the output pressure of an instrument does not exactly coincide with the applied input pressure. If not, the instrument is not in calibration. How could it be calibrated?

Assume that as the input differential pressure changes from zero to 100-in of water, the output changes from 2- to 14-psi. Examination of these numbers shows the output span of 12-psi to be correct. The span location, however, is off by 1-psi. To correct for this error in calibration, the span-location screw is adjusted downward which increases the spring force on the thrust shaft. The increase in spring force must be balanced by the nozzle backpressure. The span-location screw is adjusted inward until the nozzle backpressure is 3-psi. A change of 1-psi at 2-psi will change the 14-psi to 15-psi.

Consider an example of calibration being an error in the span. That is, for a 100 percent change of input, the output does not change 100 percent. To be specific, assume that an input signal changes from zero to 100-in of water. This causes the output signal to change from 3- to 14-psi. The output should have changed from 3- to 15-psi, meaning that the span is short by 1-psi.

To increase the span, it is necessary to change the ratio of the sensing and balancing-diaphragm areas. Recall that the effective area of a diaphragm changes as the center portion moves relative to the housing. The problem, then, becomes one of moving the diaphragm relative to the housing. If the thrust shaft could be moved relative to the housing, the central portion of the diaphragm will also move.

Looking at the mechanism may lead us to believe that adjusting the span location screw would move the thrust shaft downward. This is only partly true. Increasing the loading of the spring will move the thrust shaft a few thousandths of an inch. Therefore, movement of the shaft by adjusting the screw is less than an inch. Also notice that by changing the thrust shaft a small amount changes the output. The output pressure changes from 3- to 15-psi. Therefore, the span adjustment does not move the thrust shaft a significant amount.

Another way of moving the thrust shaft must be found. Examine the nozzle. Notice that it is threaded into the housing. Suppose the nozzle is turned away from the baffle 1/32-in. As the nozzle is moved, the nozzle backpressure will decrease. As the nozzle backpressure decreases so does the balancing force. If this force decreases, the thrust shaft will move to the same distance from the nozzle. This is the same distance as it was before the changing of the nozzle. By turning the nozzle away from the baffle, the thrust shaft is moved. This move is relative to the housing by 1/32-in. The central portion of the diaphragm has moved 1/32-in relative to the housing, which then changes the effective area of each of the diaphragms.

Observe the diaphragm convolutions of Figure 14-4. The convolutions are curled edges of the two diaphragms. Notice that the sensing or input diaphragm convolution points down. The balancing convolution points up, causing the thrust shaft to move. Downward movement causes the effective area of the input diaphragm to become smaller. It becomes smaller as the effective area of the balancing diaphragm becomes larger. Thus, the ratio of the diaphragm area decreases. Because of the reduced ratio, a large nozzle backpressure is needed to balance a differential pressure if the span is to increase by the 1-psi that it lacked.

As the nozzle is moved, the loading of the zero spring is changed because the moving of the nozzle also moves the thrust shaft. If the shaft is moved, the spring force changes slightly. As a result, there is an obligation to alternately adjust span and zero. This is not the usual situation.

Figure 14-5 is a detailed diagram of the ABB Kent-Taylor transmitter. This diagram is a considerably simplified drawing of the actual mechanism. The actual mechanism thrust shaft consists of a series of threaded members and O-ring seals. These two items clamp the diaphragm and sealing bellows. Care must be taken to assemble the thrust shaft and install it within the housing. Figure 14-6 is a cutaway photograph of the mechanism. Study Figures 14-4, 14-5, and 14-6 and understand them thoroughly.

Force-balance instruments employ a type of construction that is susceptible to internal and external leaks. Therefore, these instruments should be carefully checked for leaks.

FORCE-BALANCE POSITIONERS ▬▬▬▬▬▬▬▬▬▬

Moore Products Co. and Fisher Controls Co. positioners will be discussed here. Both of these positioners are the usual sidearm-mounted type. Sidearm-mounted positioners require an external linkage system that feeds back the valve position to the positoner.

In force-balance devices, forces directly oppose each other. In moment-balance instruments, the forces act at distances from a fulcrum point. Do not confuse linkage that is used to route a movement with linkage that is part of the positioner.

Components

The components of a force-balance positioner are:

1. An input device that converts the input signal to force.
2. A balancing mechanism consisting of the actuator and a feedback element that converts the actuator position to force.
3. A thrust shaft.
4. A detector mechanism.

Arrangements

All elements, other than the external lever system, are arranged on a center line. The input force is applied to one end of the thrust assembly. The balancing force is applied to the other end. The detector is arranged to sense any change in position of the thrust assembly.

For a specific arrangement, consider the Moore positioner. Figure 14-7 is a photograph of the positioner and Figure 14-8 is a schematic diagram. The Moore positioner consists of a position-to-force converter or a spring. This converter is arranged to directly load the input-signal element or a bellows. The actuator position is fed through a parallel-lever gain mechanism. The thrust assembly is a disk of metal mounted between the spring and the bellows.

The thrust assembly used by the Moore positioner is elementary compared to the ABB Kent-Taylor transmitter. It is also elementary when compared to the thrust assembly of force-balance controllers and relays. The reason that the thrust assembly is simple in the Taylor device is that only two forces or signals are interacting. If more than two forces were involved, as is the case of the others, a more complicated thrust assembly would be required.

OPERATION OF THE MOORE POSTIONER ▄▄▄▄▄▄▄▄▄▄▄▄▄▄▄▄

Assume that the positioner is in balance. The input force is balanced by a force that is a function of the valve position. Now let the input signal increase. The increased signal will result in an increased force being applied to the thrust assembly. The force tends to lift the assembly. The pilot valve responds to the lift. This causes the supply port to open wider and the exhaust to be reduced. This action increases the output pressure of the pilot valve. The valve moves the actuator down.

Downward actuator movement is fed back through a parallel-lever gain mechanism. The gain mechanism output is applied to the top of the balancing spring and to the assembly. At this point, the two forces equal each other. The thrust assembly will come to an equilibrium. Equilibrium is reached when air pressure drives the actuator to a positioner, resulting in a force equaling the input-signal.

FIGURE 14-7   A positioner (*Courtesy of Moore Products Co.*)

The only role the bellows plays is to provide a flexible air passage. The passage channels the exhaust air from the pilot valve to the atmosphere. Air escapes through small holes in the thrust assembly.

OPERATION OF THE FISHER POSITIONER

Figure 14-9 is a diagram of the Fisher positioner. A photograph of the positioner is shown in Figure 14-10. The positioner is sidearm-mounted. The actuator position is fed back to the positioner through a parallel-lever system. In this respect, it is the same as the Moore positioner. Note that one of the levers forms a 90° angle. This is the same as the 180° angle lever used by Moore and others. This lever is mounted at 90° because the thrust assembly is mounted at a 90° angle relative to the actuator travel. The thrust assembly

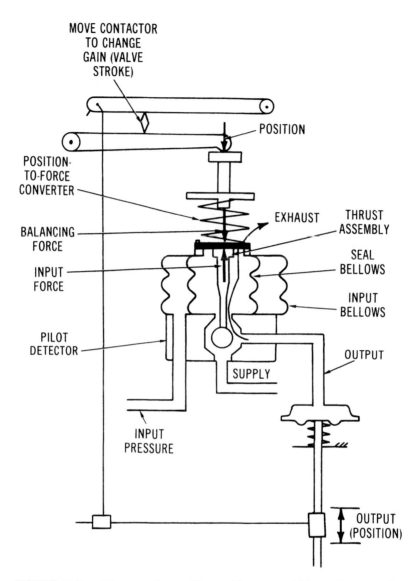

FIGURE 14-8   Diagram of a positioner. (*Courtesy of Moore Products Co.*)

appears to be different from that used by Moore and Fisher. The shaft is arranged so that the input signal force and the balance force are applied at each end.

The input signal is applied to a bellows-in-a-can assembly. The force passes through the thrust shaft and is applied to a balancing spring. The detector is arranged to detect any movement of the thrust shaft. The detector is a nonbleed pilot valve. A nonbleed valve is constructed so that the supply and exhaust seats face in the same direction. It also

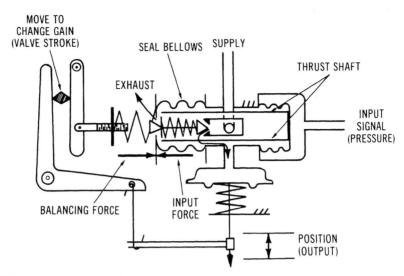

FIGURE 14-9   Diagram of a positioner (*Courtesy of Fisher Control International, Inc.*)

has a moveable exhaust port. A sealing bellows is required so that the exhaust port can move. The output of the pilot valve is taken from between the two seats.

Assume now that the postioner is in a balanced position. If it is, both the supply and exhust ports will be closed. Recall that this is true for all nonbleed valves. Now increase the input signal pressure.

An increase in the input signal causes the assembly to move to the left. This lifts the supply plug from the supply port and permits the supply air to pass through the port, which will result in an increase in pilot-valve output pressure. Output pressure, acting on the actuator, moves it downward. The movement feeding through the lever system compresses the spring. It causes an increase in balancing force. The increase in balancing force directly opposes the increase in input-signal force. When these forces equalize, both the supply and exhaust ports will be closed. The pressure applied to the actuator will remain constant. The positioner and actuator are therefore in equilibrium. This develops a new position proportional to the input-signal pressure.

Notice the output of the pilot valve. It is applied to the actuator. It also fills the inside of the input-signal bellows and the seal bellows. These two bellows are identically sized. The forces across the bellows cancel each other. The change in pilot-valve output pressure does not introduce any changing force on the thrust assembly.

SUMMARY

In sidearm-mounted positioners a lever system is used. Nevertheless, the balancing force directly opposes the input signal force. The two forces are applied to opposite ends of the thrust assembly. The detector is arranged to detect any small displacement of the thrust assembly.

FIGURE 14-10 A positioner. (*Courtesy of Fisher Control International, Inc.*)

The thrust assembly of force-balance instruments takes on a variety of physical forms. Notice that all the forces involved converge on the thrust assembly. The thrust assembly must be arranged so that it can respond to small differences in these forces. The detector discovers these differences and changes forces to bring them to balance.

The usual construction is to mount the thrust assembly on a diaphragm. A bellows can also be used in this assembly.

ACTIVITIES

TRUE FORCE-BALANCE INSTRUMENTS

1. Refer to the operational manual of the instrument used in this activity.
2. Identify the function of the instrument.
3. If the instrument is not attached to an operating system, it can be calibrated.
4. Supply a source of air to the instrument and monitor the value of the input and output.
5. With the aid of the operational manual, what is the desired operational span of the instrument?
6. Adjust the input air supply to zero. What is the value of the output?
7. Adjust the input air to 3-psi. What is the value of the output?
8. Alter the zero adjust screw to produce an appropriate output.
9. Adjust the input air to 15-psi. What is the value of the output?
10. Does the instrument have a gain factor?
11. How does a change in input compare with the change in output pressure?
12. If the span of the instrument is correct, no calibration is needed. If the span is low or high, it can be adjusted by altering the span adjustment screw.
13. Disconnect the air source from the instrument.

QUESTIONS

1. What is meant by the term true force-balance?
2. How does true force-balance differ from moment-balance?
3. What is a stack-type of instrument?
4. What are the components of a true force-balance mechanism?
5. How is the force-balance principle applied to a transmitter?
6. How does a force-balance positioner operate?
7. What are the components of a force-balance positioner?

15
Motion-Balance
Principle and
Applications

OBJECTIVES

Upon completion of this chapter, you will be able to:

- Define the term motion-balance mechanism.
- Identify three basic types of motion-balance mechanisms.
- Explain the operation of a motion-balance mechanism.
- Identify the fundamental elements of a type-3 mechanism.

KEY TERMS

Angle motion-balance.   A type or variation of the motion-balance principle that responds to two signals applied to opposite ends of a floating lever.

Linear motion-balance.   A variation of the motion-balance principle which has no levers and responds to two signals that are directly opposed to each other.

Motion-balance mechanism.   A mechanical structure that responds to two signals that are applied to opposite ends of a floating lever.

Self balancing.   The property of an instrument in which the absence or change of signal causes equilibrium to be reached or restored.

Type-1 mechanism.   An angular motion-balance assembly that employs a floating lever which has measurement and balancing signals applied to each end of the lever.

Type-2 mechanism.   An angular motion-balance assembly that consists of a floating lever and a balancing lever.

Type-3 mechanism.   An angular motion-balance assembly that consists of a floating lever and two fixed pivoting levers in its structure.

INTRODUCTION

In previous discussions it was noted that all instruments can be divided into two major branches. They are feedforward and feedback, or self-balancing. Self-balancing instruments, in turn, can be divided into two more categories. They are motion-type and force-type. In Chapters 10 to 14, force-type instruments were shown that are available in two forms. These are true force-balance and moment-balance.

In this and following chapters, motion-balance instruments will be considered. As with force-balance instruments, motion-balance devices also appear in two forms. The first is similar to the moment-balance and is called angle motion-balance. The second form, similar to true-balance, is called linear motion-balance.

Angle motion-balance instruments are by far the most common motion-balance instruments. Transmitters made by Moore and ABB Kent-Taylor are examples of two linear motion-balance instruments.

MOTION-BALANCE MECHANISMS

Our discussion of motion-balance instruments shall proceed in the same way as force-balance instruments. Motion-balance instruments have a fundamental mechanism. This mechanism is incorporated in the design of a variety of instruments.

Elements

Elements of the fundamental angle motion-balance mechanism are:

1.   An input lever that pivots on a fixed pivot.
2.   A balancing lever that pivots on a fixed pivot.
3.   A floating lever that rotates on two floating pivots.
4.   An error detector.

Arrangements

The measurement signal is in the form of a motion to the input lever. The balancing motion drives the other fixed lever. The floating lever is arranged between the two fixed levers. The ends of the floating levers move as measurement and balancing motion occurs. The detector is positioned to detect any displacement of the floating lever.

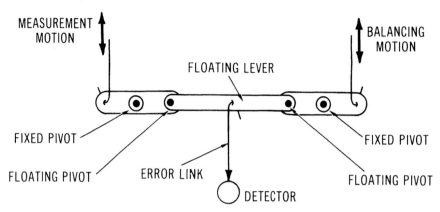

FIGURE 15-1   Fundamental angle motion-balance mechanism.

Figure 15-1 is a diagram of a fundamental angle motion-balance mechanism. This mechanism is used to convert a measurement to a proportional pressure. In addition to transmitters, this mechanism is used in controllers and valve positioners. Differences in applications arise out of the way the input motion or the balancing motion originates and is used. Motion-balance linkage is perhaps the most important instrument mechanism.

The motion-balance lever mechanism appears in a wide variety of instruments. It is a common mechanism that is used in many unusual ways. For example, it is used as a motion divider in Fischer & Porter's electronic rotameter transmitter. It is also used as an adjustable ratio mechanism. Whenever there is a need to relate two signals to get one output, any variation of a motion-balance mechanism may be used. A study of motion-balance mechanisms used in controllers is discussed in later chapters.

Principle of Operation — Type-3 Mechanism

Operation of the fundamental angle motion-balance mechanism will now be discussed. Assume that the mechanism is in balance and there is an increase in measurement signal. The increased signal will rotate with the fixed measurement lever counterclockwise. This lifts one end of the floating lever. The floating lever rotates about the balancing floating pivot. As it rotates, the center portion of the floating lever is raised. The detector responds to this displacement of the floating lever by decreasing the output pressure. The output pressure is applied to the balancing bellows. This decrease in output pressure causes a change in balancing motion. The motion rotates the balancing lever counterclockwise. It lowers the balancing floating pivot and with it the balancing end of the floating lever. Recall that the measurement signal raised the measuring floating pivot. This, in turn, raised the center portion of the floating lever. Now the balancing motion lowers the other end of the floating lever, which tends to lower the center of the floating lever.

The detector will change its output pressure. This change will continue until the balancing equals the measurement motion.

The diagram shown is symmetrical. The detector takeoff is at the center of the mechanism. This is highly idealized, but it does demonstrate how the input and balancing motions are made equal; that is, equal within the range of the detector. It also shows how the detector causes these motions to be made equal. Figure 15-2 shows the balancing action. Notice, the difference in lever position for balance at 3-, 9-, and 15-psi.

The output of this instrument is an air pressure. This was also true of the force-type instruments. The air pressure is proportional to the input measurement motion, meaning that the output is proportional to the input. Remember that a transmitter converts a measurement to a proportional pressure.

The mechanism just described is only one type of motion detector. It is called type 3 and is the most complicated because three levers are involved.

Type-2 Mechanism

Figure 15-3 shows a motion-balance mechanism that uses two levers. In the discussion of Figure 15 1, the measurement motion was applied to a measurement fixed lever. This, in turn, shifts the floating lever. If the measurement motion were connected to only one end of the floating lever, the fixed lever would not be needed. Two levers characterize the type-2 motion balance mechanism.

Assume the measurement motion lifts the floating lever. The detector responds to the new position of the floating lever. The detector also changes its output. It then rotates the balancing lever counterclockwise. The right end of the floating lever lowers. Remember that the original motion was to lift the left end of the floating lever. The balancing motion lowers the right end. The lowering will continue until the center of the floating lever is brought back to its original position.

Type-1 Mechanism

Consider a third motion mechanism. One of the fixed levers is eliminated by directly connecting the measurement motion to the floating lever. It is also possible to eliminate a second fixed lever by connecting the balanced motion to the other end of the floating lever. Figure 15-4 shows a type-1 motion-balance mechanism. The detector of this mechanism, like the other two, responds to a change in the center of the floating lever.

Suppose that measurement motion lowers the left end of the floating lever. The detector responds to this change by altering the output pressure. It acts through a pressure-to-motion converter. This action lifts the balancing end of the floating lever. The mechanism will come to equilibrium when the balancing motion equals the measurement motion.

ANGULAR MOTION-BALANCE TRANSMITTERS ■■■■■■■■■

Three types of angular motion-balance mechanisms have been discussed. It was noted that two signals are involved in the operation of this mechanism: measurement and balance. They are applied in opposite directions to opposite ends of the floating lever. The detector is arranged so that it responds to a small displacement of the floating lever.

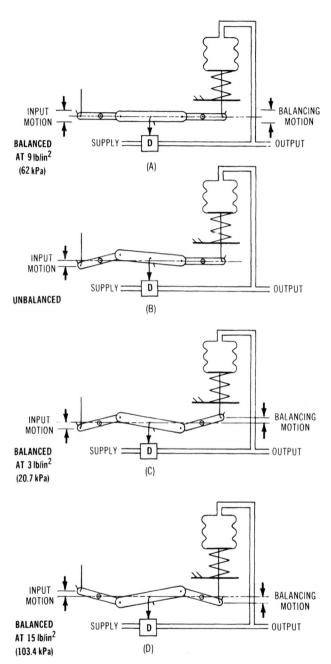

FIGURE 15-2   Balancing action of type-3 angle motion-balance mechanism. (A) Balanced a 9-psi; (B) Unbalanced; (C) Balanced at 3-psi; (D) Balanced at 15-psi.

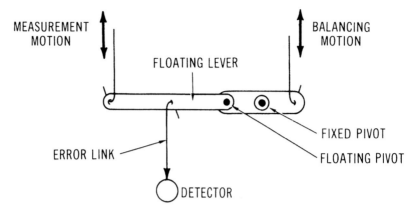

FIGURE 15-3   Type-2 angle motion-balance mechanism.

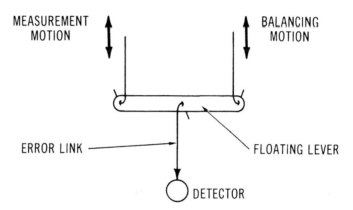

FIGURE 15-4   Type-1 angle motion-balance mechanism.

Operation of the fundamental mechanism was described as being symmetrical. The actual mechanism is, however, not very symmetrical in its operation. Frequently, the levers are two-sided and folded 90°. It is necessary that one be able to identify the levers regardless of their physical appearance. This is not always easy to do. Some motion-balance lever systems can be quite involved when compared to force-balance lever systems. Lever identification can be made by performing the following steps in order:

1. Locate the floating lever.
2. Locate the levers driving the floating lever.
3. Locate the fixed and floating pivot points.
4. Locate the the input motion and balancing motion.

If this procedure is followed, there should be no difficulties in identifying mechanism levers.

Elements of A Type-3 Mechanism

1. A floating lever and two floating pivots.
2. An input lever and its fixed pivot.
3. A balancing lever and its fixed pivot.
4. A detector mechanism.

Arrangements

The floating lever is positioned between the two fixed levers. The points of contact between the floating lever and the ends of the two fixed levers form the floating pivots. A detector mechanism detects any displacement of the floating lever. The output of the detector is fed back to a pressure-to-motion converter. This is generally a bellows/spring combination. The pressure-to-motion converter drives the balancing lever.

Principle of Operation

Figure 15-5 is a diagram of a motion-balance mechanism used in the Foxboro transmitter. First, identify the levers. Then locate the fixed and floating pivots.

Notice that the measurement lever is made of two one-sided levers mounted on the same shaft. The balancing lever is shaped so that there are scooped-out extensions. Do not let this odd construction hide the fact that the balancing lever is a 180° two-sided lever. The scooped-out ends provide clearance and mechanical stops for other elements.

Identify the fixed pivot point of the balancing lever. Observe where the balancing motion is fed into the balancing lever. Locate the point of contact between the floating lever and the balancing lever. In this diagram, it is a pinned connection between the two

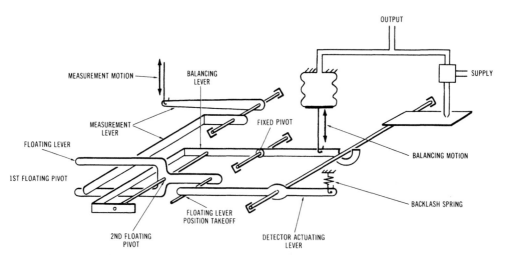

FIGURE 15-5   Motion-balance mechanism in a transmitter. (*Courtesy of The Foxboro Company*)

levers. The point of contact between the measurement lever and the floating lever forms the floating point.

The detector mechanism is more elaborate that the mechanism indicated in the diagram. In spite of its complexity, its function is simply that of detecting displacements of the floating lever. A detector actuating lever rotates if there is displacement of the lever. As it rotates, it changes the baffle/nozzle detector clearance. This lever system in no way changes the fact that the baffle/nozzle alters dispacement of the floating lever.

The specific arrangement of mechanisms results from the need for making mechanisms compact. There is also the need for changing the direction of motion.

MOTION-BALANCE LEVER MECHANISM

Assume that the mechanism is in a balanced condition. The measurement signal then changes in a downward direction. It can go either up or down, even though the diagram shows it centered. This change in motion rotates the measurement lever counterclockwise. It then lowers the mesurement end of the floating lever. The floating lever pivots on the balancing lever. This rotates the actuating lever clockwise and uncovers the nozzle. The nozzle backpressure drops off as does the relay output backpressure. The pressure-to-motion converter responds to the dropoff of relay output pressure which lifts and rotates the balancing lever counterclockwise. The balancing end of the floating lever drops.

The dropping of the balancing floating pivot permits the detector actuating lever to rotate counterclockwise. Remember that the measurement motion caused it to rotate clockwise. The detector actuating lever will continue to rotate clockwise. The rotation lasts until the clockwise motion balances the original measurement motion. Notice that the measurement and balance motions are balanced when the motions of the two floating pivots are equal.

Figure 15-6 is a more detailed diagram showing the mechanism. Figure 15-7 is a photograph of this mechanism. Study these diagrams to be sure that you can identify the essential elements of the mechanism as they appear in the photograph.

Elements of A Type-1 Mechanism

For the second example of an angular motion-balance transmitter, one manufactured by Masoneilan International will be used. This is of the liquid-level displacer-type and is shown in Figure 15-8.

Earlier, the operation of the displacer torque-tube instrument was considered in detail. How angular motion, due to change of level, is converted to a proportional output pressure will be considered.

The essential elements of the mechanism are:

1. An input motion link.
2. A floating lever on two floating pivots.
3. A detector mechanism.
4. A balancing motion link.

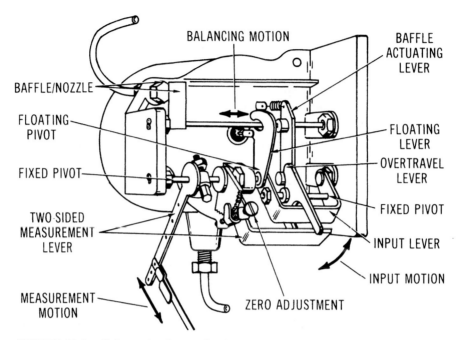

FIGURE 15-6   Schematic of a mechanism. (*Courtesy of The Foxboro Company*)

FIGURE 15-7   A mechanism. (*Courtesy of The Foxboro Company*)

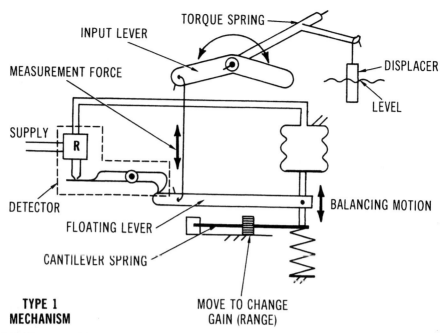

FIGURE 15-8 Motion-balance mechanism in a transmitter. *(Courtesy of Masoneilan International, Inc.)*

Observe the measurement signal coming into one end of the floating lever. The balancing motion is fed into the other end of the floating lever. The elements identified so far are a floating lever and two floating pivots. The balancing motion is directly connected to the floating lever. Therefore, this is a type-1 mechanism.

Even though there is a lever involved in the detector mechanism, its function is transmitting dispacement to the floating lever to the detector. Further recognize that the detector senses any dispacement in the floating lever. The specific part of the floating lever connected to the detector is unimportant.

Also notice that the balancing mechanism consists of a bellows, spring, and a variable spring element. The effective length of the spring is changed. The amount of motion obtained for a given change in balancing pressure can be varied as the length of the spring is varied. This changes the gain of the transmitter.

Principle of Operation

Assume that the mechanism is balanced and the level decreases about the displacer. As a result of this change in level, the input lever rotates clockwise. As a result, the measurement motion moves up. It lifts the measurement end of the floating lever. This, in turn, causes the baffle to move away from the nozzle. The nozzle backpressure falls off. This falloff in pressure causes an upward balancing motion. This is the motion that rotates the floating lever counterclockwise about the measurement end of the floating lever.

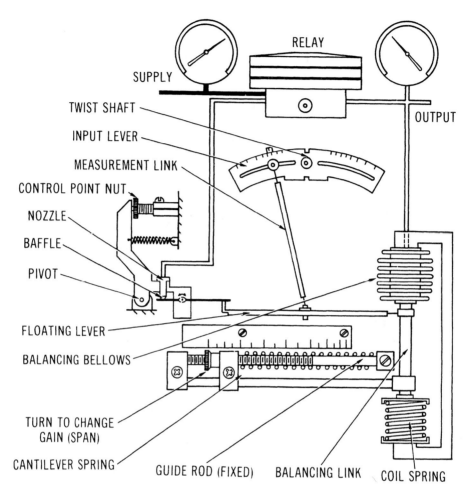

FIGURE 15-9   Schematic of a mechanism. *(Courtesy of Masoneilan International, Inc.)*

The balancing counterclockwise motion causes the baffle to move towards the nozzle. Remember that the original measurement motion caused the baffle to leave the nozzle. The balancing motion will continue until it almost balances out the measurement motion. At that point, the baffle will have taken the position with respect to the nozzle. This results in a detector backpressure that is equal to the measurement motion. The measurement motion has then been converted to a proportional output pressure, meaning that level has likewise been converted to a proportional output pressure.

Figure 15-9 is a schematic drawing and Figure 15-10 is a photograph of the transmitter. Study these diagrams and locate the essential elements of the mechanism.

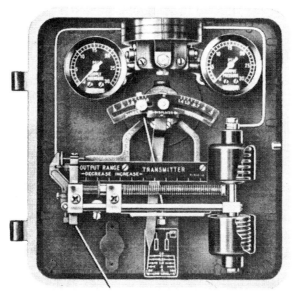

FIGURE 15-10   A mechanism. *(Courtesy of Masoneilan International. Inc.)*

SUMMARY

There are three types of angle-balance mechanisms. In all three mechanisms, a change in measurement motion results in a change in balancing motion. The balancing motion is due to the change in detector output pressure. The output pressure acts on a pressure-to-motion converter. The detector output pressure changes until the balancing motion is equal to the measurement motion.

The detector output pressure is the output of the instrument. The instrument output pressure is proportional to the measurement input motion. The motions involved are larger, possibly as much as 1/2-in. The motions of force-balance instruments are approximately the motion required to fully range the detector.

Two types of motion-balance mechanisms are used in transmitters. The key to understanding these instruments is not very involved. One must first locate the floating levers. Next, the fixed levers and the floating pivot points are located. There will be either one, two, or no fixed levers depending on whether the mechanism is type-1, -2, or -3. Fixed levers drive the mechanism.

The elements of these mechanisms are used to transmit the position of the floating lever to the detector. Knowing what elements to look for make it easy to recognize these elements as they appear in the instrument. Identify the elements before attempting to analyze the operation of the mechanism. If this is done, it reduces confusion.

Notice that the measurement motion and balancing motion as such are completely independent of each other. In no way is the measurement motion actually "felt" by the balancing motion. It is the detector responding to these movements that causes the balancing action. Compare this action with moment-balance mechanisms. The measurement force opposes the balancing forces because they were applied against a force beam.

ACTIVITIES

INSTRUMENT ANALYSIS

1. Refer to the operational manual of the instrument used in this activity.
2. Identify the type of instrument used in the activity.
3. If the instrument is not connected to an operating system, remove its cover or outside housing.
4. Identify the type of mechanism employed by the instrument.
5. Explain how the mechanism responds to a change in signal applied to its input.
6. Connect an air source to the input of the instrument. Monitor the input and output air of the instrument.
7. Adjust the input air to 0-psi. Monitor the output.
8. Apply several different values of input air while monitoring the output.
9. Notice the reponse of the mechanism to each value of input air. Can you see the self-balancing effect?
10. Remove the air source from the instrument and reassemble its housing.

QUESTIONS

1. What is the meaning of the term motion-balance?
2. What are the components of a type 3 angle motion-balance mechanism?
3. What are the construction differences between the three angle motion-balance mechanisms?
4. What is meant by the term self-balancing?
5. What is the key in identifying different angle motion-balance mechanisms?

16
Angle
Motion-Balance
Positioners

OBJECTIVES

Upon completion of this chapter, you will be able to:

- Define the term angle motion-balance.
- Compare the operation of motion-balance and moment-balance mechanisms.
- Explain the operation of an angle-gain mechanism.
- Identify the components of an angle motion-balance mechanism.

KEY TERMS

Angle-gain mechanism. An assembly of components that changes the gain or level of air pressure between the input and output of an instrument by rotation of a baffle/nozzle assembly.

Baffle. A plate or strip of material used to deflect, check, or regulate the flow of air through a device.

Baffle striker. A hammer-like device that is forced against the baffle of an air control mechanism.

Cam. An oblong-shaped mechanical component that produces a reciprocating motion when a roller or pin is moved against its edge.

Cantilever. A projecting beam or member supported at one end only.

Cantilever spring.   A spring device that is supported at one end and responds to the action of a bellows.

Force beam.   A bar or strip of material that has some form of effort or force applied to it in the operation of a balancing mechanism.

Nonlinear.   A condition in which a variable value conforms to an irregular shaped pattern or characteristic.

Plane.   A surface of such nature that a straight line joining two of its points lies wholly on the same surface.

Quadrant.   A 90° degree section or sector of a circle.

THE ANGLE MOTION-BALANCE MECHANISM ▬▬▬▬▬▬▬

In this chapter, positioners that use an angle motion-balance mechanism will be discussed. The function of any valve positioner is primarily the same as any other valve positioner. The reason is to ensure that the actuator position is proportional to an input signal. It is the actuator position that balances the input signal.

Components

1.   An input element to convert an air pressure to motion.
2.   A motion-balance lever mechanism.
3.   Balancing components (the actuator).
4.   A detector mechanism.

Notice that an essential component of the actuator positioner is the actuator. It is the actuator that furnishes the balancing motion.

Arrangements

Angle motion-balance positioners are sidearm mounted only. The positioner is mounted on the actuator so that actuator motion can be fed back to the positioner. The element used is generally a two-sided lever. This lever rotates as the actuator stem moves. The lever is one element of the motion-balance lever system.

The input signal is fed to a pressure-to-postion converter such as a bellows/spring. This assembly drives one end of the floating lever. The other end of the floating lever is driven by the balancing lever. It, in turn, drives the actuator. A detector is arranged to describe any displacement of the floating lever. The output of the detector drives the actuator.

Figure 16-1 is an angle motion-balance positioner manufactured by Foxboro. It uses a two-lever motion-balance mechanism. The input signal is directly applied to one end of the floating lever.

Principle of Operation

Assume that the positioner and actuator are in balance and that the input signal increases. The increase in input signal will lift the floating lever. The floating lever pivots on the

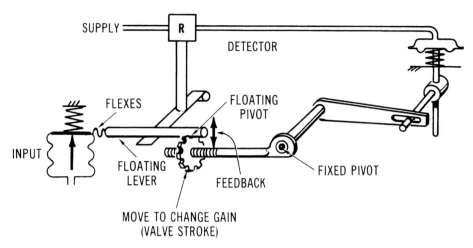

FIGURE 16-1   Simplified diagram of an angle motion-balance positioner.

balancing lever. As it lifts, the baffle covers the nozzle. The reverse-acting relay output decreases and permits the actuator stem to move upward. This upward movement rotates the balancing lever counterclockwise. It lowers the balancing end of the floating lever. The floating lever then pivots on the input. The downward movement continues until it balances the input motion.

COMPARING MOTION AND MOMENT BALANCE

As a rule, positioners are always used with actuators. Actuators are motion devices. The purpose of the actuator is to convert an air-pressure signal to a corresponding position. For each range of input signals, there is a corresponding actuator motion.

If a force-type mechanism is used, it is necessary to convert the actuator motion to a corresponding force. This was done in force-balance instruments by placing a spring between the actuator and the positioner. The input signal was applied to a pressure-to-force bellows. In this way, the input signal and the balancing signal become forces applied to a force beam. The beam reacts to any difference in forces. The detector identifies beam movement and initiates any corrective action.

Consider now the angle motion-balance mechanism. Since it is a motion-balance mechanism, the actuator motion is directly usable. There is no need for a conversion element. Hence, there is no need for a spring to be placed between the actuator and the positioner. Actuator position is directly applied to the floating lever. The input signal must be converted to motion. This is achieved by feeding the input signal to a pressure-to-position converter such as a bellows/spring assembly. The two signals involved are then motions. They do not react on each other. The balancing motion follows the input motion due to the action of the detector. These two motions are applied to opposite ends of the floating lever. The detector identifies any movement of the floating lever.

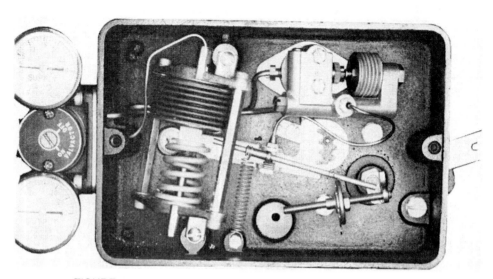

FIGURE 16-2   A positioner. (*Courtesy of The Foxboro Company*)

Angle motion-balance and force-balance positioners are arranged to detect small displacements of the floating lever. This causes balancing so that the actuator will assume a position proportional to the input signal. The adjustable gain mechanism in both instruments is the parallel-lever type. As the contactor between the parallel levers is moved, the ratio of levers changes; hence, the gain of the positioner is changed. See the positioner of Figure 16-1.

The floating lever of the Foxboro mechanism has a dual function. It operates as a floating lever and as the second lever of the parallel-lever gain mechanism.

Figure 16-2 is a photograph of the positioner. Study this picture and be able to recognize components as they appear in the actual instrument.

TYPE "C" POSITIONERS

Angle-balance positioners that do not use a parallel-lever gain mechanism will now be considered. The type "C" positioner is an angle motion balance positioner. It is functionally identical to the positioner just described. The type "C" positioner consists of the same components as the angle motion positioner. However, the specific arrangements of these components are quite different. The gain mechanism is not the parallel-lever mechanism. It is described as an angle-gain mechanism. For the moment, do not be concerned with the angle-gain mechanism.

The input device is a pressure-to-position converter consisting of a bellows and a folded cantilever spring. The cantilever spring does double duty: It is part of the input device and part of the balancing lever.

See the positioner of Figure 16-3. The balancing lever is a two-sided lever pivoted through the case. The actuator side of the balancing lever is a straight line between the

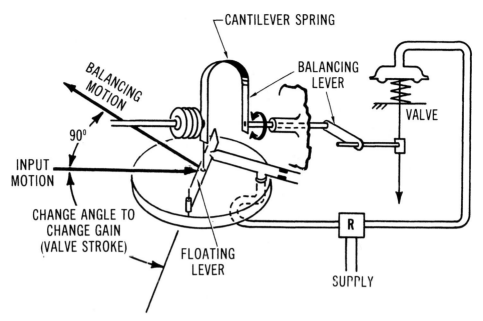

FIGURE 16-3   Balancing mechanism in a type "C" positioner. (*Courtesy of The Foxboro Company*)

bellows pivot point and the baffle actuating ball. Do not let the form of the cantilever spring be confused with the balancing lever.

The balancing lever is arranged so that one end is part of the floating lever. Notice that as the input signal changes, the cantilever spring compresses. This permits one end of the balancing lever to slide along the axis of rotation. The input position is directly applied to the floating lever. Actuator position passes through the two-sided balancing lever. One side of this is also the floating lever.

The components of a type "C" positioner have an unusual arrangement. They are also different in another respect. Notice that the two signals involved are at right angles to each other rather than parallel. Since the signals are at rights angles, it is necessary that the mechanisms be arranged to take this into account. The result of this relationship is that the motions, rather than operating in one plane, operate in two planes. The diagram shows the two planes of operation. Mechanisms arranged to operate in two planes are ideally suited to use angle-gain mechanisms.

Assume now that the positioner and actuator are in balance and that the input signal increases. The input signal moves the baffle striker in a direction that permits it to cover the nozzle. The resulting increases in nozzle backpressure cause an increase in relay output pressure. The relay output pressure moves the actuator stem downward which rotates the balancing lever in a direction that causes the baffle to uncover the nozzle. The original motion caused the baffle to cover the nozzle.

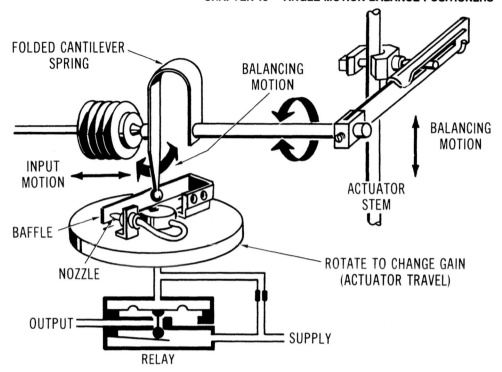

FIGURE 16-4   Balancing action in a type "C" positioner. (*Courtesy of The Foxboro Company*)

Balancing motion uncovers the baffle in a type "C" positioner. The movement continues until the backpressure results in a balancing motion that approximates the input motion.

Figure 16-4 is a pictorial diagram of the type "C" positioner. Figure 16-5 is a photograph of the mechanism. Study these figures to recognize the location of essential components.

THE FISHER POSITIONER

The Fisher Controls Company positioner is quite similar to the Foxboro positioner just described. The input motion and balancing motion are at right angles to each other. The floating lever is quite unusual in construction. A two-dimensional lever is used. All levers so far have been single dimensional. Do not be confused by this construction. The Fisher positioner has a two-sided lever with the two sides operating in different planes. See Figure 16-6.

Figure 16-7 is a diagram of the Fisher mechanism. The balancing lever is operated by a cam at the actuator end. Remember from the discussions of cams that lever length changes with angular rotation of the cam. Note that cam placement is between the actuator and the floating lever. This makes it possible to obtain a nonlinear relationship

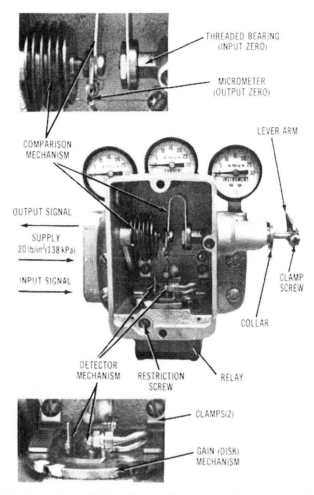

FIGURE 16-5   Type "C" positioner. (*Courtesy of The Foxboro Company*)

between actuator travel and the input signal. In this respect, the Fisher positioner is similar to the Bailey piston positioner.

In the operation of the Fisher positioner, do not be confused by the presence of the cam. The floating lever, although pivoted by a tension wire, is free to float. This balancing motion is applied to one end of the floating lever. The input motion is applied to the other end of the floating lever. Each end acts as a pivot for the other.

The detector is arranged to respond to any movement of the floating lever. The detector takeoff point can be rotated from the input end of the floating lever to the balancing end of the floating lever. It is in this way that the gain of the positioner is changed. This angle-gain mechanism is similar to the gain mechanism of the Foxboro type "C" positioner.

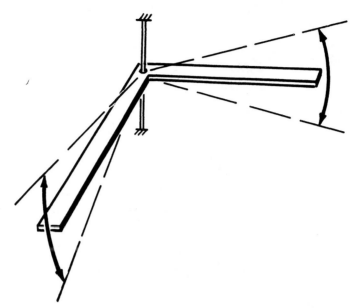

FIGURE 16-6   Floating lever action in a positioner. (*Courtesy of Fisher Control International, Inc.*)

We will now take a look at the operation of the Fisher positioner. First assume that the positioner and actuator are balanced. The input signal then increases. This increase lifts the floating lever. The tension wire pivot and the balancing cam act as the pivot point. The lifting of the floating lever causes the baffle to cover the nozzle. The relay output pressure increases, moving the actuator downward. This movement rotates the balancing lever counterclockwise, lowering the end of the floating lever. The tension wire and bellows act as the pivot point. Remember that the input signal raised the floating lever which, in turn, causes the baffle to cover the nozzle.

The baffle then uncovers the nozzle. The baffle remains uncovered until the detector output pressure changes. The amount changed must be sufficient to change the actuator position and an amount equal to the input position.

Figure 16-8 is a photograph of the actual positioner. Study this figure to be certain that you are able to recognize the essential components as they appear in the mechanism.

ANGLE-GAIN MECHANISM

Angle-gain mechanisms are used by Foxboro and Taylor to change the gain of their positioners. In addition to their use on positioners, they are used in controllers and transmitters. Recall that the function of an adjustable gain mechanism is to make it possible to change the ratio of output to input.

The parallel-lever gain mechanism has been studied. The input and output signals are parallel when used on this gain mechanism. Remember that in an angle-gain mechanism

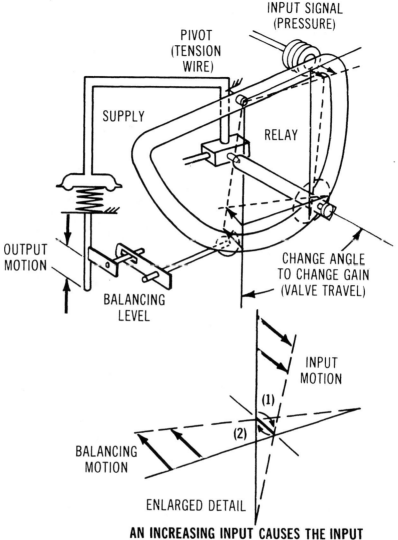

FIGURE 16-7　Balancing mechanism in a positioner. (*Courtesy of Fisher Control International, Inc.*)

FIGURE 16-8   A positioner. (*Courtesy of Fisher Control International, Inc.*)

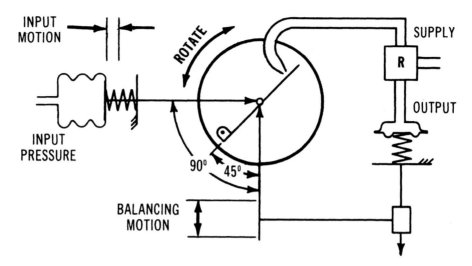

FIGURE 16-9   Angle-gain mechanism; 45° setting.

the two signals are at right angles to each other. In the discussion of positioners using the angle-gain mechanism, exact operation of the mechanism was not discussed. In this chapter, the angle-gain mechanism will be carefully studied. It is important that this operation be understood. It is a key mechanism in the operation of a number of instruments.

Components

1.   A rotating base piece.
2.   Two signal motions.

Arrangements

The baffle/nozzle mechanism is mounted on the base piece. The base piece is mounted so that it can be rotated around its center. The two signal motions are at right angles to each other and arranged to converge on the baffle.

Principle of Operation

Figure 16-9 is a diagram of the angle-gain mechanism as used in valve positioners. As diagrammed, the baffle forms an angle of 45° with the two signals.

Assume now that an increase in input pressure occurs. This increase results in an input motion that causes the baffle to move away from the nozzle. The nozzle backpressure drops off. The relay output pressure decreases and the actuator moves upward. This upward motion of the actuator is fed back to the baffle. It causes the baffle to move toward the nozzle. Recall that the original motion moved the baffle away from the nozzle. The baffle moves toward the nozzle. It continues until there is an output pressure that drives the actuator to a postion proportional to the input pressure.

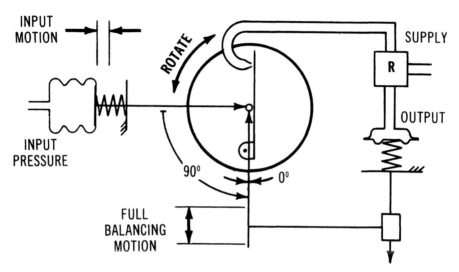

FIGURE 16-10   Angle-gain mechanism; 90° setting to input signal.

Assume that for a 3- to 15-psi change in input pressure signal, the input motion is 1/2-in. Because the angle is 45° the output motion would also have to be 1/2-in. The ratio of output to input is a gain of one.

Input and Baffle at 90°

So far the operation of the positioner has been described. Now consider how the gain mechanism changes the gain of the actuator for a given input-pressure change. This makes it possible to change the actuator output with respect to the input pressure change. Figure 16-10 is the same as the mechanism shown in Figure 16-9. However, the baffle and nozzle have been rotated on the base piece. It now forms a 90 angle with the input signal and a 0° angle with the balancing signal.

When the baffle is in the position shown in Figure 16-10, it causes the positioner to operate. Assume that the input pressure increases. The increased input pressure will move the baffle away from the nozzle. Since the baffle is 90° relative to the input signal, a very small change in input signal will fully uncover the nozzle. The nozzle backpressure and the relay output will drop to zero, causing the actuator to drive to its upper stop. This results in a full range balancing motion. However, this full range balancing motion does not change the baffle position relative to the nozzle. The balancing motion is parallel to the baffle. The detector mechanism cannot respond to the change in balancing motion. Yet, a very small change in input signal will cause the actuator to stroke through its full range. Recall that the baffle forms a 45° angle with the input and balancing signal. This balancing motion has just as much effect on the detector as the input motion.

What is the gain when the baffle forms an 90° angle to the input signal? A very small change in input signal will cause a full change of output. Assume that the input signal changes enough to fully uncover the nozzle and stroke the actuator through its full range. Further assume that the actuator stroke is 1/2-in divided by 0.002-in, or 250.

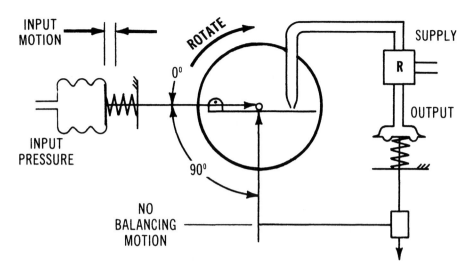

FIGURE 16-11   Angle-gain mechanism; 90° setting to balancing signal.

Figure 16-11 shows the baffle positioned so that it forms a 90° angle with the balancing signal. When the baffle is in this position, the input signal can change from 3-to 15-psi with no effect on the baffle. This is because the input signal motion is now parallel with the baffle. For a full change of input there is a zero change in balancing action. The ratio of output to input is zero divided by 1/2. Zero divided by 1/2 is zero. The gain of the instrument is zero when the baffle forms a 90° angle to the balancing motion.

We have shown the baffle relative to the input signal in three positions: 45°, 90°, and 0°. As the baffle is rotated from 0° to 90°, the gain of the positioner changes from zero to 250. This change in gain is quite large and adequate. Theoretically, it is possible to obtain gain changes from zero to infinitely large values by rotating the baffle relative to the signal a full 90°. But some small input motion is required to fully cover and uncover the nozzle. The gain is limited by the sensitivity of the detector. As a practical matter, gains that exceed 300 may make the instrument unstable, causing a continuing cycle.

Action

In addition to the wide changes in gain that can be obtained, the mechanism possesses an additional feature. Suppose that it continues to rotate the baffle relative to the input signal. This causes the baffle to form a 45° angle relative to the signals.

Figure 16-12 shows the baffle at a 45° angle. Compare Figure 16-12 to Figure 16-9. Notice in both figures that the baffle forms a 45° angle. It is, however, in a different quadrant.

Follow the operation of the mechanism when the baffle is as shown in Figure 16-12. Assume that the input pressure increases. This increase in pressure will cover the nozzle. The nozzle backpressure will build. The relay output will increase. The actuator, responding to this increased air pressure, will lift. Notice that the actuator is arranged so that for an increasing air signal, the actuator stem lifts.

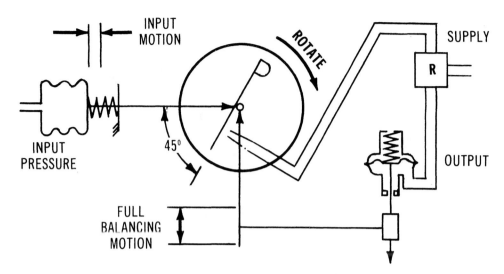

FIGURE 16-12   Angle-gain mechanism; action reversed compared to Fig. 16-9.

The actuator of Figure 16-9 is arranged so that for an increasing air signal, the actuator stem moves downward. The actuator actions are reversed. As the relay output increases, the actuator moves down. Hence, the balancing motion moves in a direction that causes the baffle to move away from the nozzle. Recall that the original motion moved the baffle toward the nozzle. These two motions will come to balance when the balancing motion approximates the input motion.

As was the case in Figure 16-9, the gain of the positioner is one. By rotating the baffle to a different quadrant, however, the action of the balancing motion relative to the input motion has been reversed. The practicality of being able to do this makes it possible to easily match a positioner to actuators having different actions. In the case of controllers, the ability to rotate the baffle to different quadrants makes it possible to change the action of the controller.

Figures 16-13 and 16-14 show the baffle for 0° and 90° positions relative to the input motion. The gain will be 250 and zero, respectively. However, a balancing motion will be in a reverse direction to the balancing motion of Figures 16-10 and 16-11. The action of the positioner is therefore reversed.

Review the operation of the Fisher positioner discussed earlier in the chapter. Study the angle-gain mechanism, as shown in Figure 16-8. Notice that the baffle "sees" all the balancing signal when it is positioned so that the takeoff point lines up with the input signal. The gain can be changed from zero to 250. If the baffle takeoff is positioned beyond the cam and into the other quadrant, the positioner action is reversed.

By repositioning the baffle 180°, it is possible to change the gain from 250 to zero direct-acting, to zero to 250 reverse-acting.

An advantage of the angle gain mechanism is that it can make large changes in gain and can reverse its action. This is done by going beyond the point where feedback enters the mechanism.

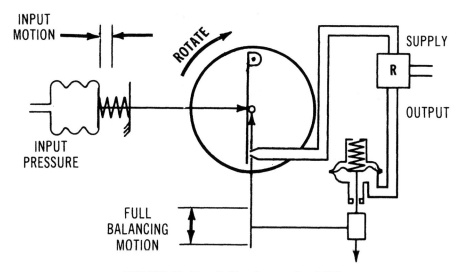

FIGURE 16-13 Setting for a gain of 250.

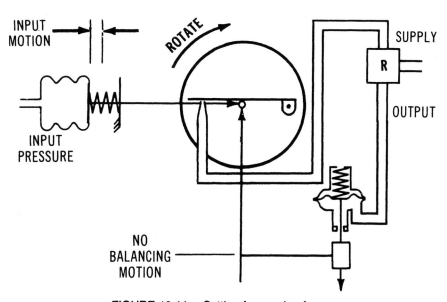

FIGURE 16-14 Setting for a gain of zero.

SUMMARY

In this chapter, three positioners have been discussed. The first positioner is a conventional motion-balance linkage of the two-lever type. The Foxboro type "C" positioner and the Fisher positioner are angular moment-balance positioners.

Angular moment-balance positioners use a two-sided lever in their construction. This lever rotates as the actuator stem moves. Input signals are applied to one end of the lever mechanism.

Actuators are motion devices. The purpose of the actuator is to convert an air pressure signal to a corresponding position. A force-type mechanism is used to convert actuator motion to a corresponding force. This is done in force-balancing instruments.

Angle-gain mechanisms are used to change the gain of positioners. This mechanism changes the ratio of output to input, makes large changes in gain, and can reverse its action.

ACTIVITIES

INSTRUMENT ANALYSIS

1. Refer to the operational manual of the instrument to be used in this activity.
2. Remove the cover or housing of the instrument.
3. Locate the input element of the instrument.
4. Describe the operation of the input element.
5. Locate the motion-balance lever mechanism.
6. Describe the operation of the lever mechanism.
7. Does the instrument have a gain mechanism?
8. Connect an air source to the input of the instrument. Monitor the input and output air of the instrument.
9. Adjust the air source to produce 3-psi.
10. Can you see the response of the mechanism? Does the change in input cause a change in output?
11. Adjust the air to a 5-, 10-, and 15-psi while monitoring the output and mechanism response.
12. Explain how the mechanism responds to a change in air pressure.
13. Reassemble the instrument and disconnect the air source.

QUESTIONS

1. What is the primary function of a valve positioner?
2. What are the fundamental components of an angle motion-balance mechanism?
3. Explain the meaning of the term angle motion-balance.
4. How does the angle-gain mechanism operate?
5. What is the function of the rotating base piece of an angle gain mechanism?

17
Linear
Motion-Balance
Instruments

OBJECTIVES

Upon completion of this chapter, you will be able to:

- Define the term linear motion-balance.
- Identify the elements of a linear motion-balance mechanism.
- Explain the operation of a linear motion-balance transmitter.

KEY TERMS

Balancing bellows. An assembly that employs a bellows to achieve the balancing operation of a mechanism.

Chase. The act of following, trailing, rapidly pursuing, or trying to overtake something that is moving.

Chasing action. The moving response of the nozzle and baffle of a linear motion-balance mechanism.

Conduit. A pipe, channel, or tile for protecting electrical wires, air flow, or a stream of hydraulic fluid.

Differential movement. The difference in displacement that occurs as a result of force applied to two different elements.

Flexible conduit.   An elastic pipe or tube that is capable of bending or being reshaped.

Relay-valve detector.   A sensing element that responds to the physical change of valve position in a relay mechanism.

LINEAR MOTION-BALANCE PRINCIPLE ▬▬▬▬

In preceding chapters, angle motion-balance instruments were discussed. Recall that motion-balance instruments are available in two forms. These are angle motion-balance and linear motion-balance. The linear motion-balance mechanism is comparable to the true force-balance mechanism. It has no levers because the signals involved are directly opposed to each other. Additional levers may be required to bring the motion to the balancing mechanism. The detector mechanism responds to any difference in the motions involved.

The number of linear motion-balance instruments is much more limited than the number of angle motion-balance instruments. There are only a few common types of linear motion-balance instruments. Linear motion-balance instruments will, however, become more widely used in the future.

In this chapter, the linear motion-balance transmitters of Moore Products Company and ABB Kent Taylor Instruments will be studied.

Elements

The essential elements of the linear motion-balance mechanism are:

1.   An input element to convert the input pressure signal to a proportional motion.
2.   A balancing element to convert the output pressure signal to a proportional motion.
3.   A detector mechanism.

Arrangements

The input element is a pressure-to-motion converter. It is arranged to drive one-half of the detector. The balancing element must be a pressure-to-motion converter. It is arranged to drive the second "half" of the detector. The input element, the detector, and the balancing element are all arranged on the same center line. There are no pivots or levers. Hence, the motions involved are linear. Figure 17-1 is a simplified diagram of the linear motion-balance mechanism. The baffle is driven by the input bellows. The nozzle is driven by the balancing bellows.

Principle of Operation

Assume that the mechanism is in balance and the input signal is increased. The increase in pressure acting on the bellows will cause the baffle to move toward the nozzle. The nozzle backpressure will increase. This increase in backpressure, acting on the balancing bellows, will expand the bellows and will move the nozzle upward. The original motion

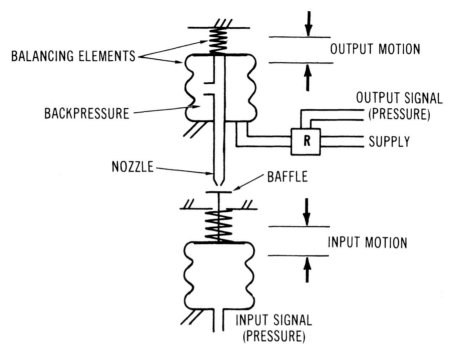

FIGURE 17-1   Simplified diagram of a linear motion-balance mechanism.

causes the baffle to move upward. The nozzle will move until its motion approximates the input baffle motion.

In the previous discussions of detectors, the nozzle or the baffle remained fixed. As the detector is used in a linear motion-balance instrument, the nozzle and the baffle move. The differential movement between the two detector elements causes the change in backpressure.

In no way does detector movement change the operation of the detector. The important point here is to notice that the two elements of the detector move relative to each other. The baffle/nozzle was shown for convenience. Other detectors can also be used.

THE MOORE LINEAR MOTION-TRANSMITTER ■■■■■■■

Figure 17-2 is a simplified diagram of the Moore linear motion-balance transmitter. Figure 17-3 is a cross-sectional view of the same instrument. This mechanism uses a baffle/nozzle, relay-valve detector mechanism. The relay valve is a standard nonbleed device. Input motion can originate from a wide variety of sources. The diagram shows input as angular rotation of a bellows-type differential-pressure meter body.

The baffle/nozzle is similar to a pilot valve. In fact, Moore calls it a pilot. Observe that as the valve plug is moved, only one restriction is varied. Recall that as the pilot valve moved, two restrictions were simultaneously varied. Since only one restriction var-

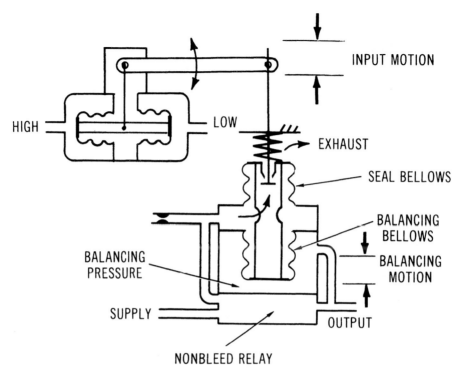

FIGURE 17-2   Simplified diagram of a linear motion-balance transmitter. (*Courtesy of Moore Products Co.*)

ied with the movement, this part of the detector can properly be called a baffle/nozzle. It appears quite different from the typical baffle/nozzle construction.

Let us see how it operates: Assume that the instrument is in balance and that there is a change in measurement that causes a downward input motion. This downward movement uncovers the nozzle. It causes the nozzle backpressure to drop, hence, decreasing the relay output pressure. The relay output pressure is fed back into a compartment containing the balancing bellows. As the pressure in this compartment is reduced, the bellows expands. As it expands, it carries the nozzle with it. This downward movement of the nozzle toward the baffle results in a covering of the nozzle. Recall that the original motion was a downward baffle movement that uncovered the nozzle. These two motions will be brought to balance when they approximate each other. Th baffle will position itself relative to the nozzle. The resulting backpressure drives the balancing bellows an amount that approximates the measurement motion. Therefore, the output pressure is proportional to the measurement.

Both elements of the detector need to move. It is therefore necessary that the nozzle portion of the detector be mounted on a flexible element. The element used by Moore to accomplish this is a bellows. Recognize, however, that the bellows functions as a flexible conduit.

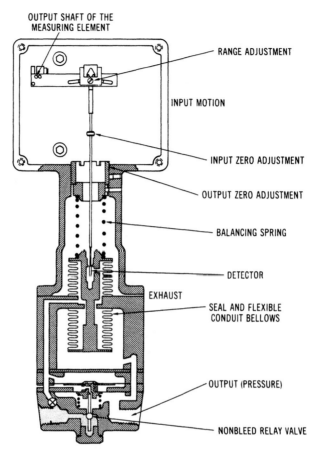

FIGURE 17-3   Cross-sectional view of a linear motion-balance transmitter. (*Courtesy of Moore Products Co.*)

The nozzle backpressure acts on the inside of both the balancing and sealing bellows. The sealing bellows and the balancing bellows have equal effective areas. The force developed by nozzle backpressure acts on the sealing bellows. This is cancelled out by the force developed by the same pressure acting on the inside of the balancing bellows. The resulting motion caused by the force acts on the balancing bellows.

TAYLOR LINEAR MOTION-BALANCE TRANSMITTERS ▬▬▬▬

A second example of a linear motion-balance instrument is the Taylor transmitter. Figure 17-4 is a simplified diagram of the mechanism. A baffle/nozzle relay-valve detector is used. The relay valve is nonbleed. In construction, the relay is different from most relay valves. It has the exhaust port mounted on a bellows rather than a diaphragm. This bellows makes it possible to obtain large movements of the exhaust port. If a diaphragm

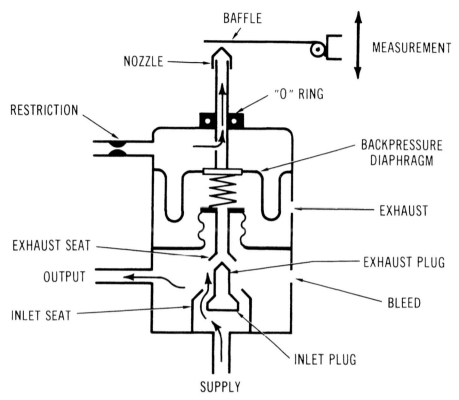

FIGURE 17-4   Simplified diagram of a linear motion-balance transmitter. (*Courtesy of ABB Kent Taylor*)

were used, the exhaust port movements would be small. The bellows, which supports exhaust port movement, also acts as the balancing element. The input element is a bellophragm. Between input motion and the balancing motion is a spring.

The bellophragm supports the nozzle and, as the bellophragm moves, so does the nozzle. The bellophragm moves as the nozzle backpressure changes. The nozzle backpressure, acting on the bellophragm, loads the relay valve through a spring. It is this spring that converts the large nozzle movements to a force that drives the relay. It is this aspect of the mechanism that distinguishes it from other relay valves.

Input motion is applied through the baffle. The baffle can be driven by a variety of input measuring assemblies. In many cases, the input is an angle motion. However, it is the linear portion of this angle motion that is detected by the nozzle. Actuator positioners use the true force-balance mechanism. Do not let the lever system that is used to route motion to the mechanisms be confusing.

Consider now the operation of the Taylor transmitter. Assume the mechanism is in equilibrium and that there is a change in measurement that causes the baffle to lift. The nozzle backpressure will decrease. The bellophragm will move upward. This reduces the force being applied to the bellows supporting the exhaust port. The exhaust port will open and that pressure will decrease. The relay output will also decrease. This decrease, acting on the balancing bellows, will cause the relay to balance at a new pressure. This is determined by the nozzle backpressure.

Notice that both elements of the detector are driven. The baffle is moved by the measurement signal. The nozzle is driven by the balancing output pressure. The differential movement is what establishes the output pressure. As is the case with all motion mechanisms, the baffle "chases" the nozzle or vice versa. This chasing action is much easier to see in the linear motion-balance instruments.

SUMMARY

A linear motion-balance instrument falls in the category of motion-balance instruments.

The essential components of this mechanism are an input element, a balancing element, and a detector. The input element converts input pressure to a proportional motion. The balancing element converts the output pressure signal to a proportional motion. The detector responds to any difference in motion.

There are no pivots or levers involved in the makeup of linear motion-balance instruments. In practice, this type of instrument responds to changes in pressure applied to a bellows system. As a result, instruments of this type have a linear output.

ACTIVITIES

LINEAR MOTION-BALANCE TRANSMITTERS ■■■■■■■■■■■■

1. Refer to the operational manual of the transmitter used in this activity.
2. Remove the cover or housing froom the instrument.
3. Identify the input and output of the instrument.
4. Note the location of the bellows assembly.
5. Connect a source of air to the input and monitor the input and output air pressure.
6. Adjust the input to produce 3-psi.
7. Note the response of the bellows assembly to this change in air.
8. How does the output air of the instrument respond?
9. Adjust the input to values of 5-, 10-, and 15-psi while observing the response of the mechanism.
10. Explain the response of the instrument to a change in air pressure.
11. Reassemble the instrument.

QUESTIONS

1. What is the difference between angular motion-balance and linear motion-balance?
2. What are the essential elements of a linear motion-balance mechanism?

3. Describe the function of the two bellows mechanism in control of the flapper/
    nozzle of a linear motion balance transmitter.
4. Explain the meaning of the term linear motion-balance.
5. What is a bellophragm?

18
Control Valves

OBJECTIVES

Upon completion of this chapter, you will be able to:

- Define the term final control element as used in a pneumatic system.
- Identify the components of the actuator of a control valve.
- Identify the components of a valve.
- Identify different types of control valve plugs.
- Describe the maintenance that is performed on the diaphragm of an air actuator.
- Describe the maintenance that must be performed on the valve body and subassembly of a control valve.
- Explain how a control valve is adjusted.

KEY TERMS

Angle valve. A type of globe valve having openings at right angles to each other. Generally, one opening is on the horizontal plane and one is on the vertical plane.

ANSI. The abbreviation for American National Standards Institute.

Bonnet. A metal cap or covering for a valve.

Bridgewall. The exterior housing of a valve.

Gland. The moveable part of a stuffing box in which packing is compressed or a device for preventing leakage of fluid past a joint in machinery.

Globe. A type of valve which has a ball, spherical, or rounded surface in its structure.

Final control element. That part of a system which is responsible for altering the output of a process.

ISA.   The abbreviation for Instrument Society of America.

Lapping.   A smoothing or polishing process involving an embedded abrasive material which is rubbed against a surface.

Meter.   A particular segment of a valve assembly which is responsible for measurement or movement.

Packing.   The material which is placed inside of a housing or container to prevent leakage.

Parabolic plug.   A bowl-shaped element located inside a valve to achieve control.

Piston plug.   A cylinderical vessel which moves back and forth in a housing to achieve control.

INTRODUCTION

The end point of measuring and controlling instruments is the final control element. The final element is usually a diaphragm actuator driving a control valve. In the work on positioners, it has been shown that the actuator is part of the positioner. If control is to be obtained, the control valve must be in good operating condition. The control valve is subject to damage, wear, corrosion, and erosion. Therefore, control valves require a fair amount of maintenance.Most process control problems could be eliminated by better care of the final element.

This chapter will concentrate on air-operated control valves driven by actuators of the diaphragm type. Piston-operated actuators are still available but not as widely used. These instruments shall not be discussed. Air-operated control valves shall be studied because they are widely used in actuators.

Components

The control valve consists of two major components:

1.   The actuator.
2.   The valve.

The actuator is made up of a:

1.   Flexible diaphragm.
2.   Spring and spring tension adjustment.
3.   Plate, stem, and locknut.
4.   Housing.

The valve is made up of the:

1.   Body.
2.   Plug.
3.   Stem.
4.   Pressure-tight connection.

Arrangements

The actuator is arranged in the following manner: The diaphragm is bolted to a dished metal head, forming a pressure-tight compartment. This controller output pressure is connected to this compartment. An actuator arrangement is shown in Figure 18-1.

The motion of the diaphragm is opposed by a spring.

The valve stem is attached to the diaphragm. Any diaphragm movement results in the same valve-plug movement.

Both the meter and plug can be direct or indirect. The action of the actuator may be such that either an increase or a decrease in air may lift the stem. The design of some actuators permits them to be reversed. The plug may be attached to the stem so that a lifting stem closes or opens the valve. Some plugs are simply reversible on the stem.

The valve action desired is based on what position the valve should take on an air failure. On one process, it may be desirable to have the valve wide open when the air fails. An example of this would be in a cooling process. On other processes, it may be

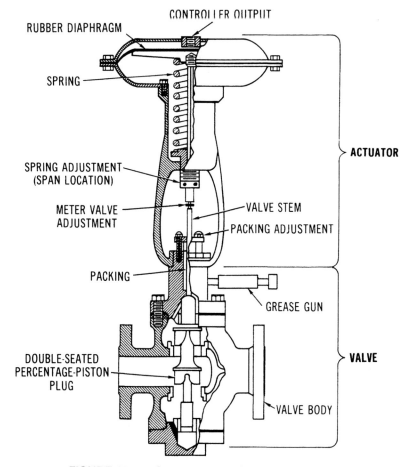

FIGURE 18-1   Control valve with mounted actuator.

better to have the valve closed, as in a heating process. When the process is decided and the type of plug is established, the actuator may be specified.

Operation

The controller output serves as the input to the control valve if no positioner is used. The actuator converts the controller output to a valve opening.

Suppose the controller output increases. The increase in pressure will compress the spring, allowing the diaphragm to move. The diaphragm movement is relayed to the valve plug through the stem.

Some control valves are designed so that a change in flow is a percentage of the amount at that particular valve position. For example, suppose the valve is three-quarters open and 35-gpm is flowing. The valve stem then opens an additional 10 percent of its travel. The flow should increase 10 percent of 35, or to 38.5-gpm.

Now suppose the valve is one-quarter open and the flow is 5-gpm. Suppose that the stem is again moved open the same 10 percent of its travel. In this case, the change in flow would be 10 percent of ±5, or 5.5-gpm.

This characteristic is called an equal-percentage. This feature is very desirable of a valve. The output results in a change in flow. This change is proportional to the amount flowing at the time the variation occurred. Control valve plugs are designed to give the percentage characteristic. Some formed plugs are shown Figures 18-2, 18-3, 18-4.

Other valve plugs are used. The most common is the linear plug. It is designed so that the opening changes in a line with the stem position.

The valve plug can be either parabolic or piston. The parabolic plug is designed to get the percentage characteristic. The piston plug has ''V'' notches cut into the piston walls to obtain the percentage response.

Alternate Arrangements

Some control valves use a piston instead of the rubber diaphragm. Instead of a spring, the valve can be arranged to use air pressure that acts as a spring.

The pressure-tight connection may be a bellows that expands and contracts with valve-stem travel.

CONTROL-VALVE MAINTENANCE ▬▬▬▬▬▬▬▬▬▬▬

Part 1. Diaphragm Air Actuator

Air-to-Push-Down Maintenance. The air-to-push-down diaphragm actuator is shown in Figure 18-5. This particular device works on simple menchanical principles. It is a completely separate unit independent of the valve body. When air pressure is applied to the top of the diaphragm, the actuator stem is pushed down. The motion or force is opposed by the compression of the spring.

Valve Spring Selection. The spring is selected so that the actuator will move when the air pressure reaches a predetermined point. It will complete its rated travel with an air

PERCENTAGE
PISTON

PERCENTAGE
PARABOLIC

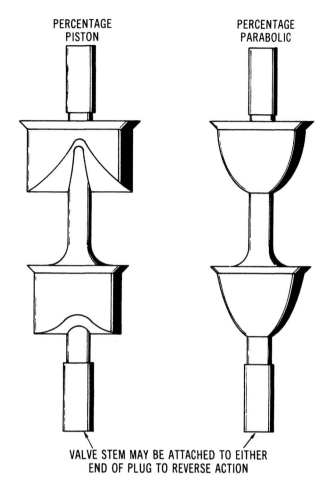

VALVE STEM MAY BE ATTACHED TO EITHER
END OF PLUG TO REVERSE ACTION

FIGURE 18-2   Double-seated valve plugs.

pressure on the diaphragm. The range of an actuator is the air pressure range in pounds-per-square-inch for the rated stroke. One ISA standard is 3- to 15-psi. Another range now widely accepted is 6- to 30-psi. The rate of the spring is selected for a 3- to 15-psi nominal range. This permits the stem to start its travel when the air pressure reaches 3-psi. It will complete the rated stroke when the pressure reaches 15-psi, +5 percent. From the standpoint of maintenance, the exact operating air-pressure range is not important as long as the rated valve stroke can be obtained without exceeding the maximum available loading pressure.

How well an actuator performs depends on its response to very small changes in air pressure. Alignment of moving parts is required for proper response. For good alignment, it is necessary to guide the actuator stem.

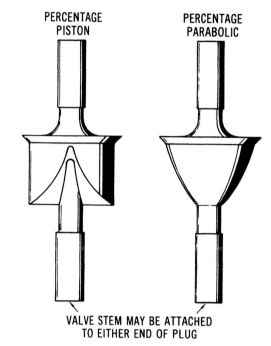

PERCENTAGE PISTON

PERCENTAGE PARABOLIC

VALVE STEM MAY BE ATTACHED
TO EITHER END OF PLUG

FIGURE 18-3   Reversible single-seated valve plugs.

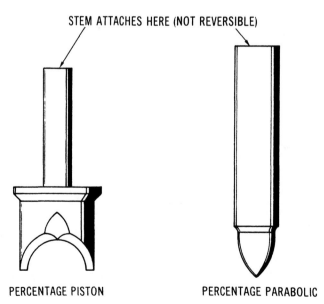

STEM ATTACHES HERE (NOT REVERSIBLE)

PERCENTAGE PISTON   PERCENTAGE PARABOLIC

FIGURE 18-4   Nonreversible single-seated valve plugs.

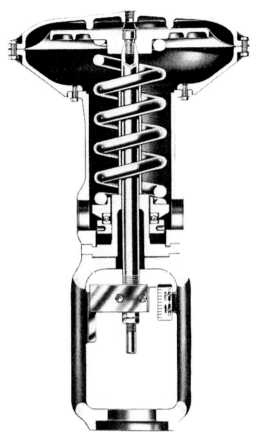

FIGURE 18-5   An air-to-push-down actuator. (*Courtesy of Fisher Control International, Inc.*)

Refer to the actuator of Figure 18-5. When the diaphragm conforms to the diaphragm plate, it serves as a flexible upper guide to the stem. A self-lubricated lower bearing, located in the adjusting screw, serves as a lower guide.

Diaphragm Replacement.   The only part normally requiring replacement is the diaphragm. To replace the diaphragm, relieve all spring compression by turning the adjusting screw counterclockwise. This will prevent the upper case from popping up when the cap screws are removed. This step is very important on actuators with a high spring compression. Remove the upper diaphragm case, actuator stem nut, washer, and diaphragm. Install the new diaphragm and reassemble the unit.

In performing this operation, there are several steps. First, connect an air supply with a gauge and regulator to the air connection on the actuator. Then turn the adjusting screw clockwise to slightly compress the spring. Apply air pressure. By feeling the stem, note the pressure when the stem begins to move. Readjust the spring compression until the stem begins to move at the minimum value of the air-pressure range stamped on the serial plate.

At times it is necessary to make an emergency diaphragm replacement. A flat-sheet stock diaphragm material can be used for actuators up to approximately 18-in outside case diameter.

It is important that the stroke moves without restriction. This is achieved by stretching the diaphragm bolt-circle diameter. It should be about 10 percent greater than the case-bolt circle diameter.

Air-to-Push-Up Diaphragm Actuator.   Refer to the air-to-push-up actuator of Figure 18-6. There is some difference between this actuator and the one just discussed. Note that the spring, diaphragm, and diaphragm plate are inverted. This is done so that when air pressure is applied to the diaphragm, the actuator stem moves upward. This actuator is used where design does not permit inverting the valve body and plug. Examples include angle valves, top-guided valves, and noninvertible designs.

Diaphragm Replacement.   To replace the diaphragm, relieve all spring compression by turning the adjusting screw counterclockwise. Remove the upper diaphragm case assembly which includes the spring barrel, spring, spring bottom, actuator stem nut, and diaphragm plate. Install the new diaphragm and reassemble. Readjust the spring setting using the same procedure as for the air-to-push-down unit. The diaphragm serves as a flexible upper guide. The packing box assembly serves as the lower guide.

The gasket of the lower diaphragm case/yoke and the packing box around the actuator stem prevent operating air leakage. Since this packing box is subjected only to low air pressure, the maintenance problem is not severe. The packing-box gland should be set up lightly. Assume that repacking is required. The actuator must be removed from the body subassembly in order to insert the preformed ring packing. The replacement ring should be covered with a thin coat of light cup grease before assembly to make it possible to form a seal with minimum bearing pressure.

Part II. Valve Bodies and Subassemblies

Four valve-body types in common use are shown in Figure 18-7A through 18-7D. For clarity, they are shown as simplified sectional drawings.

Double-Seated, Guided Valves.   Figure 18-7A is a typical top-and-bottom-guided, double V-port valve. The cast-globe body is widely used for general service applications. It is available in sizes up through 16-in. The ANSI ratings are through 600-psi and in certain sizes through 1500-psi. This type of valve can be furnished with other plug types, including the solid-turned parabolic, linear, quick-opening, and V-port. It is also available in single-seated designs.

Single-Seated Valves.   The single-seated valve includes a variety of small single-seated globe and angle bodies. They are designed for general service in sizes 1-in and smaller. This valve is a natural extension of Figure 18-7A. It is available in ANSI ratings through 600-psi. The trim is available in an interchangeable set of five nominal sizes of 1/8-, 1/4-, 3/8-, 1/2-, and 3/4-in.

Screwed-Bonnet Valves.   Figure 18-7C shows a heavy-duty screwed-bonnet valve. This valve is made in globe and angle types through 2-in and with a 1-in maximum or-

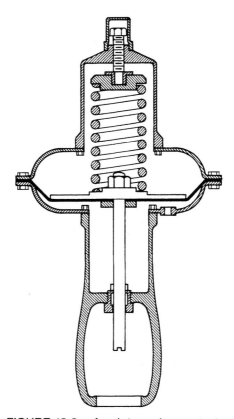

FIGURE 18-6   An air-to-push-up actuator.

ifice. The globe body is rated at 3000-psi at 450°F. The angle body is rated at 6000-psi at 450°F. The valve plug has a heavy single top guide and simple turned construction. The seat ring is a threaded venturi type with a heavy hex head, permitting it to be tightened against copper or other metallic gasket materials.

Split-Body Angle-Valve.   Figure 18-7D is a split body with an integral bonnet. The seat ring is clamped between the two body pieces. A ground metal-to-metal joint is used in place of gaskets. The seat ring is reversible. If the original seating surface is damaged, the ring can be inverted, which will produce a new seating surface. This valve is produced in sizes up through 8-in in ANSI ratings through 600-psi. The split-body design is also available in a straight through flow pattern.

Angle Valve.   An angle valve is shown in Figure 18-7D. The assembly is made in sizes through 6-in for heavy-duty service. These bodies were designed for use in severe operating conditions. Some service conditions may necessitate continuous or intermittent flushing of the body, a common procedure in some oil refinery applications. A connection is provided directly under the bonnet flange. The flushing medium passes through ports around the plug bushing, which prevents the formation of coke or the accumulation

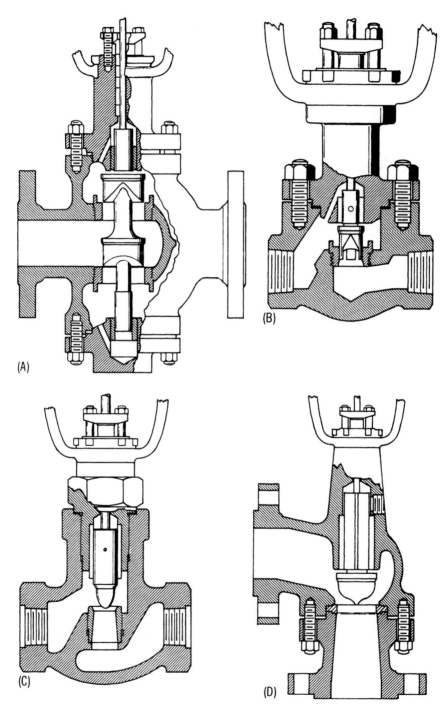

FIGURE 18-7   Valve bodies and subassemblies. (A) Top-and-bottom-guided V-port valve;
(B) Top-and-skirt-guided-percentage-piston valve; (C) Top-guided, heavy-
duty, screwed-bonnet valve; (D) Top-guided, split body, angle valve.

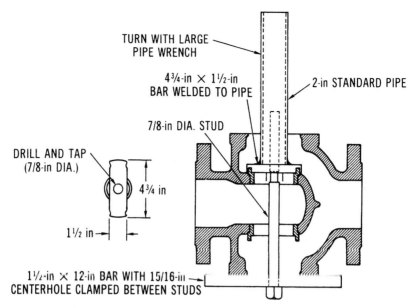

FIGURE 18-8   Seat-ring wrench you can make yourself.

of abrasive particles between the bushing and the plug. The seat ring is a venturi type. It is held in place by a flared, screwed retainer.

Bonnet and Blindhead.   The top-and-bottom guided invertible body style is shown in Figure 18-7A. It has a removable bonnet and blindhead. The valve is piloted in body recesses to very close clearances to assure positive valve-plug alignment. Be careful when removing these parts. Be certain that pilot sections, plug guides, and gasket surfaces are not damaged.

When reassembling the bonnet and blindhead, install new gaskets lightly coated with a good sealing compound. In an emergency, the original gaskets can be reused. A new gasket can be cut from high-grade sheet stock.

When making up the bonnet and blindhead connections, the nuts should be tightened in diametrically opposite pairs. Check the valve plug assembly periodically by moving it through its entire stroke to make certain that it is running freely. Successful, trouble-free operation of top-and-bottom guided valves depends on good alignment throughout the whole assembly.

Air-fin bonnets are provided to reduce packing-box temperature where fluids hotter than 400°F are being controlled. The effective cooling rate of the fins allows the use of standard packing material with service temperatures up to 800 F.

Plain extension bonnets are used for low-temperature service, usually below 32°F. The extension permits the use of standard packing material. The packing box is moved upward from the valve body to a higher ambient temperature condition, which also allows for the use of additional insulation when required.

Seat Rings.   Most standard control valves are furnished with renewable seat rings made of 18-8 stainless steel. The seat rings are precision-machined and piloted at maximum ring diameter into a recess in the valve body to ensure true alignment and accurate centering. A relatively heavy ring shoulder is provided to minimize distortion when the ring is set up tight in the body. The seating surfaces on the ring and in the body bridge-wall are kept narrow and given a special smooth finish to ensure a tight joint. Set up threaded seat rings using special fixtures and a lubricant used sparingly.

Making a Seat-Ring Wrench.   Threaded seat rings are set up very tightly when initially installed. After years of service, or when they are to be replaced, it often becomes difficult to remove them. One problem in removing the rings is to find a wrench that will not jump the lugs when force is applied. A suggested design for a wrench is shown in Figure 18-8. In this design, the wrench is prevented from rising by a tie rod running through the body. The application of heat may assist in the removal of some stubborn rings. Once the seat ring is removed, carefully inspect and clean the seating surface. Then insert the new ring. Rings should be coated with a light application of sealing compound. Sealing compound should also be used as a lubricant when making up the threads.

Grinding Seats.   With control valves using metal-to-metal seats, there is seldom a requirement for dead-tight shutoff. In double-seated valves, tight shutoff is virtually impossible in normal operation. Of all the seat designs, the beveled seat for both plug and seat ring is the type most commonly used.

The seating surfaces of the plug and rings are machined. They are seldom smooth enough to make a tight seal. In double-seated valve designs, the distance between the seats on the plug and seat rings may vary slightly. In order to correct these two conditions, a hand operation called "lapping" is necessary. The contact seating surfaces are kept narrow within the limits of good machining practice to give a combination that is easy to lap.

Lapping is started by daubing the compound on the seat. The compound is placed in several spots equally spaced around the periphery of the heat ring. Put the compound on both seats at once when fitting a double-seated plug. Be careful not to get compound on the skirt of a V-port or on the side of a turned-type plug. Compound at this point will wear the lateral surfaces. Insert the plug in the seat ring carefully until it is seated, then rotate it by using short oscillating strokes. After 8 or 10 strokes, lift the plug slightly from the seat and repeat. This intermittent lifting is important to keep the compound evenly distributed.

On large valves where the weight of the plug is substantial, it is advisable to mix a small quantity of lubricant, such as graphite, with the lapping compound. The lubricant mixture will slow down the cutting rate and prevent tearing of the seats. On very large valves, it is helpful to support the plug on a hoist to prevent the entire weight from resting on the seats.

Valve Packing.   Packing-box maintenance is one of the primary chores of routine control-valve service. The packing box must be tight, yet have low friction. Improper care, or improper lubrication or packing material, can result in leakage or excessive fric-

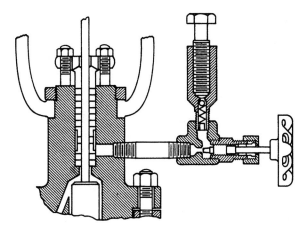

FIGURE 18-9   Packing box and lubricator.

tion. In control valves, the packing box is relatively deep with a lubricator and a lubricant ring normally provided. Figure 18-9 shows a typical packing-box section.

A typical packing material is Teflon-coated yarn. This material is a split ring packing for ratings up to 6000-psi and 400°F. Other packing materials, either solid or split-ring types, are available, including solid Teflon.

For valves operating at temperatures above 400°F, service is improved when an air-fin bonnet is provided. The air-fin bonnet is used to reduce the temperature in the packing-box. For extremely low temperatures, extension bonnets are used to protect the packing box.

Valve Lubricants.   The majority of packing boxes are designed for use with a lubricator. On cast-iron assemblies, the lubricator alone is furnished. It is designed with a ball check as a safety measure to prevent back flow of process fluid while the lubricator is being filled. On steel valves, a steel isolating valve is provided between the lubricator and the bonnet. The isolating valve gives positive protection at higher operating pressures and temperatures.

Keep the lubricator filled with the specified lubricant. It should be turned in firmly but not tightly. The lubricant is intended as an aid to the packing. It does not serve as packing itself. Under normal service conditions, it should not be necessary to lubricate the valve more than once every two weeks or so. At the time of service, one or two turns of the lubricator nut should be sufficient.

There are several standard lubricants available to meet the wide range of operating conditions under which the valve may be used. These lubricants are divided into:

1.   Those suitable for petroleum and allied chemicals.
2.   Those suitable for water solutions.

Tightness of the packing is maintained by a combination of lubrication and proper compression. It is necessary to take up on the packing gland until all of the available compression is used. It will then be necessary to repack.

In ordinary service, packing life is exceptionally long and there is no need for lubrication. In some cases, the valve can be furnished without the lubricator and lubricating ring. When it is finally necessary to repack, it may be justifiable to merely back off the packing gland. One or two rings of split packing can be inserted on top of the old packing.

When it is necessary to repack, the valve must be disassembled. The stem locknuts and indicator disk must be removed. Turn the plug stem out through the packing box. Remove the packing-box flange and gland. Remove the old packing and lubricant ring. More packing rings are installed to fill the packing box.

Part III. The Control Valve Adjustments

Valve Reassembly.   To assemble the valve body and actuator after service, follow this procedure.

1. Place the actuator on the body subassembly and secure with the clamping nut.
2. Place the packing-box flange over the plug and stem and follow with the first locknut, the indicator plate, and the second locknut.
3. Proceed with seating of the valve plug to the valve-seat rings.

For valves that are normally open, turn the plug stem into the actuator as far as it will go. Use the stem locknuts as a means of turning the stem. Lock the plug stem to the actuator stem with the locknuts. Reset the indicator plate until the indicator disk shows the maximum rated opening. Apply enough air pressure to the actuator to move the plug stem down. Continue to apply air pressure until the indicator shows the valve to be SHUT. Turn the plug stem out of the actuator stem until the plug is seated. Relieve the diaphragm pressure. Rotate the plug stem about one-quarter of a turn out of the actuator stem. Lock the plug stem to the actuator stem with the locknuts.

For valves that are normally closed, the plug must be seated. To do this, the plug stem is turned into the actuator stem. Apply air pressure to the diaphragm, which will move the plug off the seat. Rotate the plug stem one-quarter turn into the actuator stem. Relieve the diaphragm air pressure. Lock the plug stem to the actuator stem. Reset the indicator so that it shows the valve to be closed.

Normally, closed valves are assembled by turning the plug stem out of the actuator stem until the plug is seated. Reset the indicator plate so that the indicator disk shows SHUT. Apply enough air pressure to the actuator to lift the plug off the seat. Turn the plug stem one-quarter turn out of the actuator stem and relieve the diaphragm pressure. Lock the plug stem to the actuator stem with the locknuts.

SUMMARY

The end point or final element of a pneumatic system is usually a control valve. This element is normally diaphrapm actuated. Due to damage, wear, corrosion, and erosion this valve requires maintenance.

There are two kinds of control that can be applied to a valve. Some control valves are designed to rate percentages. Any flow change would be a percentage of the amount flowing at a valve position. This is called the equal-percentage characteristic. Another characteristic involves a linear plug. A linear plug is designed so that the opening changes linearly with the stem position.

Control-valve maintenance is divided into three major parts. The diaphragm air actuator is one major part of valve maintenance. In this regard, performance of the valve spring must be tested or evaluated. In addition, maintenance also requires periodic diaphragm replacement.

A second major part of control-valve maintenance deals with valve bodies and subassemblies. A typical maintenance problem is seat grinding. A special lapping compound is used to improve the seating surface. After this procedure, valve packing and lubrication must be added.

The third area of control maintenance deals with valve adjustments. The valve must first be completely reassembled before proceeding. Secondly, enough air pressure is applied to move the plug off its seat. Then the stem is adjusted one-quarter turn in or out of the actuator, depending on the valve design.

ACTIVITIES

CONTROL VALVE EXAMINATION ━━━━━━━━━━━━━━━━━

1. Examine the structure of a final control element.
2. Locate the actuator.
3. Where is the diaphragm of the actuator located?
4. How is the actuator attached to the valve assembly?
5. Explain the operation of the actuator.
6. Identify the valve mechanism.
7. Does this unit have a pressure-tight assembly?
8. Describe the structure of the valve assembly.
9. Is there any exterior adjustment of the control element?
10. Where is the diaphragm air supplied to the assembly?

FINAL CONTROL VALVE OPERATION ━━━━━━━━━━━━━━━

1. Refer to the operation manual of the final control element used in this activity.
2. Attach an air source to the input of the final control element.
3. Look through the body of the assembly to locate the valve plug. In some units, the valve plug may not be visible. An air or water source may be used to determine the status of the valve plug.
4. Is the valve open or closed?
5. Apply 5-psi of air to the control valve input.
6. Does the application of air cause any change in the plug of the valve?

7.   According to the operational manual, apply the maximum value of air to the input. Do not exceed 15-psi for most units.

8.   Does this change in air alter the position of the valve plug?

9.   Does this amount of air cause full or partial control of the valve plug?

10.   Disconnect the air source from the assembly.

QUESTIONS

1.   What is the final element of a pneumatic system?

2.   What are two kinds of control that can be applied to a valve?

3.   What is the definition of the equal-percentage characteristic?

4.   What is the characteristic which involves a linear plug?

5.   What are the three parts or areas of control-valve maintenance?

6.   In dealing with valve adjustments, why must air-pressure be applied to the plug stem?

7.   What type of compound is used to improve the seating surface?

19
Fundamental Controllers

OBJECTIVES

Upon completion of this chapter, you will be able to:

- Define the terms feedforward and feedback.
- Distinguish between indicating and nonindicating controllers.
- Define the term controller.
- Explain the operation of a narrow-band proportioning controller.
- Describe how external feedback is accomplished in a controller.
- Explain how a controller accomplishes gain.

KEY TERMS

Blind. A mechanical condition that has difficulty in sensing an operation. A controller that does not read or identify process variable information.

Derivative. An operation that shows a relationship between the time that a controlled variable occurs and the final control element changes.

External feedback. A signal that is returned to a control element outside of the device or instrument.

On/off control. A two-position control procedure that is either on or off for a period of time.

Measured variable. A controller term that refers to a measured process condition.

Mode.   A method, procedure, or manner of doing something such as the method of control.

Process.   Collective functions performed in and by the equipment in which a variable is controlled.

Process variable.   Any process such as temperature, flow, liquid level, or pressure that changes its value during the operation of a system.

Proportioning.   To adjust a part or a thing having the same or a constant ratio with respect to its proportion.

Setpoint.   A value to which a control is adjusted to maintain a desired process variable setting.

Upstream pressure.   Pressure that occurs in front of or before being applied to a device or instrument.

Variable.   A process condition of temperature, pressure, flow, or level that is subject to change that can be measured, altered, or controlled.

Vessel.   A hollow or concave container that houses a device.

INTRODUCTION

Controllers are more complicated than most of the instruments we have discussed. The complexity arises out of two factors: First, the output of the controller is dependent upon the relationship of two inputs. Second, the output of some controllers is time-dependent.

Instruments considered up this point have had one input and one output. These instruments were generally not time-dependent. That is, they were designed to repeat the input as fast as possible. Ideally, the output signal coincided in magnitude and in time with the input signal.

A controller is an instrument that identifies input signal values and produces an output that is used to alter the operation of a system. To understand a controller, let us consider the operation of a household heating system. The thermostat of this system serves as the controller. The temperature of the thermostat is manually set to some desired value, generally called the setpoint adjustment or value. It represents one input of the system. The second input is a measured temperature value, which can be called the measured variable, process variable, or controlled variable. This information is developed by the thermostat. The thermostat accepts the two input signals and compares the values. If the setpoint is 70°F and the measured variable is 70°F, no changes will occur. If the measured variable is less that the setpoint value the thermostat develops an output signal, this signal tells the gas solenoid valve to turn on in order to produce more heat. When the measured variable rises above the setpoint value the gas valve is signaled to turn off. In this heating system, the temperature is automatically maintained near the setpoint value. The thermostat serves as the controller of the automated heating system.

CONTROLLER TYPES

A typical indicating controller is shown in Figure 19-1. The controller is designed to respond to temperature, pressure, level, or flow, as well as differential pressure, vacuum, and absolute pressure. It will accept remote signals when used with a transmitter. It may also be attached locally to a system to monitor direct signals. Setpoint and process-variable values are displayed on a circular scale with pointers mounted on concentric shafts. Supply and output pressures are indicated by the gauges near the bottom of the instrument. An interior view of the same instrument with a different operating range is shown in Figure 19-2.

Another type of indicating controller, partially removed from its case, is shown in Figure 19-3. All operator control functions are available on the front panel of this instrument, including the setpoint, valve settings, and switching functions. Process information applied to the controller is indicated by pointers, also visible from the front panel. Controllers of this type are used to monitor a single process variable. Two or more controllers can be connected together to monitor different process variables at the same time.

FIGURE 19-1   A typical indicating controller. (*Courtesy of AMETEK, Controls Div.*)

FIGURE 19-2   Interior view of an indicating controller. (*Courtesy of AMETEK, Controls Div.*)

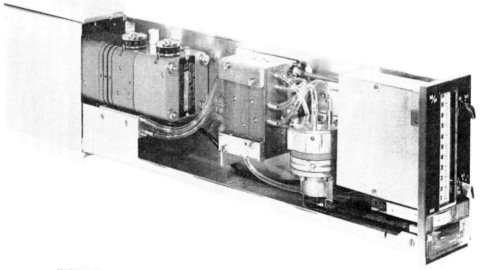

FIGURE 19-3   An indicating controller. (*Courtesy of Moore Products Co.*)

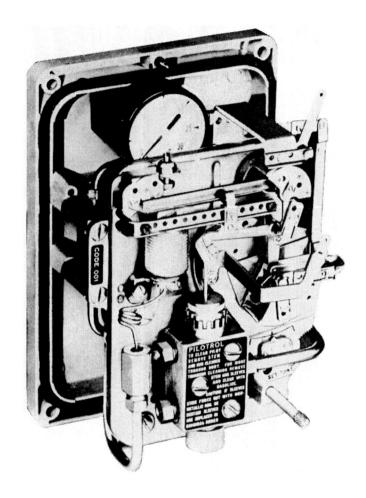

FIGURE 19-4   A nonindicating controller. (*Courtesy of Bailey Controls Co.*)

A nonindicating controller is shown in Figure 19-4. This instrument, called a "Pilotrol Controller" by its manufacturer, receives process signals in the form of changing pressure values. It then manipulates this information according to setpoint values and selected switching functions. A controller of this type is often called a receiver. Its output

FIGURE 19-5   Interior view of an indicating controller. (*Courtesy of Moore Products Co.*)

can change a final control element such as a valve positioner. Receiver outputs drive a recording indicator when records of process information are needed.

An internal view of an indicating temperature controller is shown in Figure 19-5. The indicating hand extends from the lower center and angles to the left. Temperature changes are sensed by the liquid-filled capillary tube. It is then applied to the bellows located in the lower right corner. This particular controller is used to open and close a gas pilot valve. The valve delivers pressure to the furnace according to the setpoint position.

The indicating scale, which has been removed from the instrument shown in Figure 19-5, displays both pressure and temperature. Setpoint adjustments are made through the half-round shaft that extends from the lower center of the unit.

NARROW-BAND PROPORTIONING CONTROLLERS

The first controller to be considered is a feedback narrow proportioning-band controller. For the time being, do not be concerned with the term narrow-band.

Components

1. A measuring input.
2. A reference input called a setpoint.
3. Feedback components.
4. A detector mechanism.
5. A lever system.

Arrangements

Figure 19-6 is a block diagram showing the relationships between the two components. Notice that there are two inputs to the controller. They are the measurement input and the setpoint input. There is only one output, which means that the output depends upon the relationship of two input signals. The measurement input is a signal that is proportional to the measurement. It can be, and frequently is, the output of a transmitter. The setpoint input is a signal begun by the operator. The operator is saying to the controller, "This is the value that I need." If the controller is working properly, the measurement signal will equal the setpoint signal.

Take a closer look at the block diagram of Figure 19-6. Locate the block labeled comparison mechanism. Notice that there are two inputs to the box and one output to the detector. The function of this box is to compare the measurement signal with the setpoint signal. If there is an output from the comparison box, it is called an error signal. The detector is used to identify error signals. The detector output pressure is fed back into the feedback components. These are represented in the box labeled feedback mechanism. The output of the feedback box is fed back to the detector. The purpose of the detector is to determine any unbalance between input signals. In this case, it would be the error signal and the feedback signal. If there is a difference, the detector operates to balance the two signals. The feedback signal is also the output signal. That is, the feedback signal is the signal that is applied to the final control element. This is normally a control valve.

Figure 19-7 is a controller mechanism. It shows how the elements and components are arranged in a functional controller. Notice that the box labeled *comparison mechanism* has two bellows opposing each other. A measurement signal is applied to one of these bellows. The setpoint signal is applied to the second. The two bellows drive a lever. Halfway down the lever, and pointing towards it, is a nozzle. The nozzle and this part of the lever forms a baffle/nozzle detector mechanism. At the lower end of the lever is the

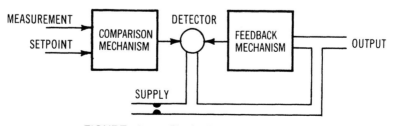

FIGURE 19-6   Block diagram of controller.

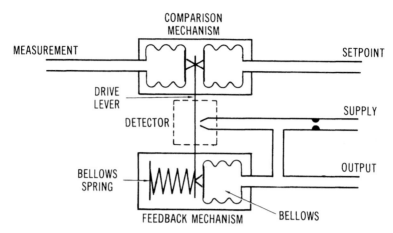

FIGURE 19-7   Diagram of controller mechanism.

feedback box. Inside this box is a bellows/spring. The nozzle backpressure is connected to this bellows. The output of the controller is this backpressure; that is, if a relay is not in use. If a relay is used, it is placed in the nozzle backpressure line. The controller output is the output of the relay. The introduction of a relay does not change the function of the controller.

Principle of Operation

Suppose that the instrument is in equilibrium. In this case, the two inputs are not changing, so there is no change in the output. Now suppose that the measurement signal increases. The increase in the measurement signal moves the lever towards the nozzle. The result of this movement is that the nozzle backpressure increases in value, and the increase is transmitted through the feedback bellows which then moves the lever further away from the nozzle. The nozzle backpressure stabilizes when the feedback movement equals the original input movement. The controller will have come to a new balanced position.

We will now examine the operation of a controller leaving the measurement input a constant value and changing the setpoint signal will now be considered. A decrease in the setpoint will permit the baffle to move toward the nozzle. The nozzle backpressure will increase. The increase in backpressure, acting through the feedback bellows, will move the lever away from the nozzle. When the feedback movement equals the changing setpoint signal, the controller comes to equilibrium at a new nozzle backpressure, hence, output. Notice that the output can be changed by altering either the measurement signal or the setpoint signal.

To express the idea in a different way, suppose the measurement and setpoint pressures are both increased. There will then be no change in baffle/nozzle position because no movement occurs. It should be recognized that the comparison mechanism responds to two signals. It also generates an output that is proportional to the difference between the two signals. If there is no difference, the output error signal is zero. If the measurement is greater than the setpoint, there is a plus error signal. If the measurement is less than the setpoint signal, there is a negative error signal.

EXTERNAL FEEDBACK ■━━━━━━━━━━━━━━━━━━━━━━━━━━━━━

The controller, much like the valve positioner, is not a functionally complete instrument. The controller is installed relative to a transmitter and a control valve. Figure 19-8 shows a controller installed in some process piping. The measurement begins at an orifice. Connected across the orifice is a differential-pressure cell. The orifice/differential-pressure cell combination converts flow to an air pressure. This air-pressure is the measurement signal. The setpoint signal is also an air pressure. The output of the controller is applied to a control-valve actuator. The valve is installed in the same process line as the orifice.

The operation of an installed controller will now be considered. Suppose now that the controller is in equilibrium, which causes the control valve to be in a fixed position. If the pressure upstream from the control valve increases, more flow will pass through the control valve. The increase in flow increases the differential pressure across the orifice. The increase in differential pressure is converted to an increase in transmitter output. This increase in output acts on the measurement bellows of the controller comparison mechanism, which causes the baffle to move toward the nozzle. The nozzle backpressure will increase. Backpressure acting on the feedback bellows will reposition the baffle and cause a new output. Output pressure will cause the control valve to move in a direction to reduce the flow. The reduction in flow will cause a reduction in transmitter output

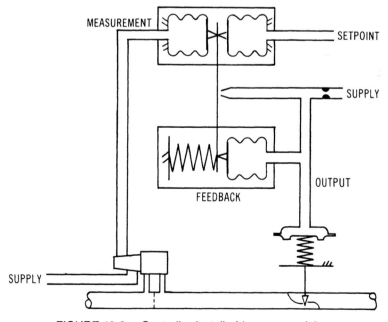

FIGURE 19-8   Controller installed in process piping.

pressure, thus producing a new measurement signal. The signal then changes the baffle/nozzle to a new position which, in turn, produces a new output pressure. These events continue until the flow comes to equilibrium.

It should be recognized that the operation of an installed controller occurs in steps. The balancing action is continuous. That is, as the measurement changes, the feedback changes. As the feedback changes, acting through the flow, the measurement also changes. It can be observed that there are two balancing, or feedback actions. One of the feedback actions is within the controller. It is brought about by the feedback bellows. The second balancing action is external to the controller. It is brought about by a change in measurement acting on the measuring bellows. This causes a new controller equilibrium point. Recognize that there is an external feedback action; however, do not let this be confusing. The controller output is required to balance the difference between the measurement signal and the reference or setpoint signal.

In the following discussion of controller mechanisms, the external feedback loop will be omitted in order to simplify the discussion. External loop changes do not alter controller action. In the discussion of controller theory, external feedback will be discussed. This discussion will show the reaction between controller settings and the processes being controlled.

CONTROLLER GAIN

The amount of output change for a given error signal depends upon the construction of the controller. More specifically, it depends upon lever length and detector position relative to the lever. In a mechanism diagram, the diagram shows the detector centered on the lever. This means that a change in feedback bellows position is equally effective as a change in either measurement or setpoint bellows position. The relationship between input and output is called gain. If nozzle position were changed on the lever, it would change the ratio of output to input. Therefore, the gain could be changed. The notion of changing the gain of a controller is extremely important. The mechanisms for changing controller gain are the same as those encountered in transmitters and valve positioners. There are two additional methods of changing the gain of a controller, as well.

In the operation of a controller, it was mentioned that there is an external feedback loop as well as an internal feedback loop. The controller diagrammed shows a comparison mechanism and a feedback mechanism. The fundamental operation of the controller was described as being a balance between two signals—the error signal and the feedback signal. The feedback mechanism is a simple proportional mechanism. The comparison mechanism is composed of bellows elements. The inputs to the comparison mechanism are energized by pneumatic signals. The detector is a simple baffle/nozzle detector.

Components

It is possible to have a controller made of only a few components. The controller of Figure 19-9 consists of:

1. A measuring input.
2. A detector.
3. A control lever.

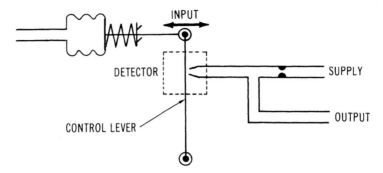

FIGURE 19-9   Simple controller consisting of measurement input, detector, and control lever.

Arrangements

Figure 19-9 shows a controller connected to a simple process application. The controller contains the three basic parts listed under the components heading. Note that the comparison mechanism is quite primitive compared to the fundamental controller diagrammed in the first part of the chapter. At first it appears that there is only one input. Second, notice that there is no internal feedback loop. There is, however, an external feedback loop.

The mechanism shown in Figure 19-10 can be described as a blind pressure controller. It is a pressure controller becauses it measures pressure and is installed to control pressure in a vessel. It is called blind because it does not indicate or record the measured variable.

Operation

Suppose the measurement range is zero to 10-psi, or zero to 69-kPa. Suppose further that the control valve has to have 9-psi, or 40-N, of force on the diaphragm to obtain 5-psi of pressure in the vessel. Therefore, the nozzle must be 0.001-in away from the baffle. This is true because of the sizing of the nozzle and restriction.

Suppose that the upstream pressure is increased. There would be an increased flow through the control valve. The pressure in the vessel would increase as would the pressure on the controller bellows. This increased pressure would move the baffle closer to the nozzle. The baffle/nozzle would increase the pressure on the control valve. The valve would start closing, which would reduce the flow and pressure to the vessel. A reduced pressure acting on the bellows would cause the baffle to move away from the nozzle, reducing nozzle backpressure and permitting the valve to open. Eventually, the nozzle backpressure, valve and vessel pressure, and measurement would be equal. The new balance point would cause the valve to be slightly more closed than it was prior to the increase in pressure.

Suppose now that the pressure in the vessel needs to be 8-psi. Notice that the detector mechanism is mounted so that the nozzle can be moved relative to the baffle. Suppose further that the nozzle is moved away from the baffle. Let us now follow the action of the controller.

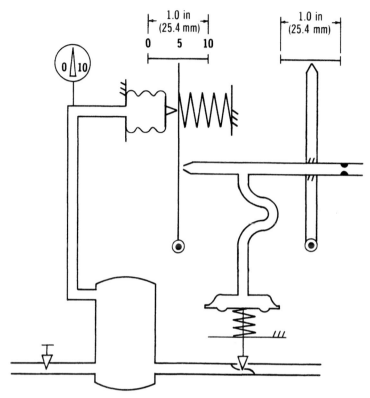

FIGURE 19-10   Simple controller connected to a process.

Moving the nozzle away from the baffle will cause a decrease in nozzle backpressure which, in turn, will cause the control valve to open. Control valve opening permits more pressure to build up in the vessel. This increase in pressure, acting on the controller measurement, will move the baffle toward the nozzle. This tends to increase the nozzle backpressure, which closes the valve. The controller and process will come to equilibrium at some new pressure greater that 5-psi. There is no way of knowing what pressure this will be. The controller is blind. If a pressure-indicating gauge were offered, examination of the gauge could be made. This would permit the nozzle to be readjusted to obtain the desired pressure. The position of the nozzle constitues the second controller input. It is a way of obtaining a comparison between the actual and the desired pressure.

OTHER ASPECTS OF CONTROLLER GAIN

Suppose that a controller has a change in measurement pressure that causes a bellows deflection of 1-in. Also suppose that a change of 0.002-in in baffle/nozzle clearance will change the nozzle backpressure from 3- to 15-psi. Suppose further that the nozzle is centered on the baffle. It will now be decided how much measurement pressure is needed to cause a full change in nozzle backpressure and valve position.

Assume that measurement motion is applied to the baffle at twice the distance the nozzle is from the pivot. This measurement must move 0.004-in to change the baffle/nozzle clearance 0.002-in. The pressure change necessary to obtain this 0.004-in is 0.4-psi. In other words, a 0.4-psi change will cause the valve to move from fully open to fully closed. The 0.4-psi of water pressure is so small that the controller would continue to cycle from one limit to another. It would alternately open and close the valve. This action can be described as on-and-off control. Yet, this expression of control can be somewhat misleading. It would be preferable to describe the controller as having a very high gain. It could also be considered as a very narrow proportioning-band controller. It is very important to determine the exact gain of the controller.

Gain is the ratio of output to input. In the preceding example, the output changes 100 percent when the input changes 0.4 percent. The gain of this controller is 250.

A gain of 250 is much too high for most processes. If gains of this magnitude are used, the control valve will continue to move from one limit to the other. In this case, it would be necessary to reduce controller gain. The inherent gain of the baffle/nozzle detector would be too great. Recall that in the discussion of detector properties, high gain was a problem. Mechanisms are employed to reduce detector gain to more usable levels.

SUMMARY

Controllers are somewhat more complicated than the instruments discussed thus far. This assumption is based on the output of the controller being dependent upon the inputs. Secondly, the output of some controllers responds to input signals that are time dependent.

A controller has setpoint and measured variable signals applied to a comparison mechanism. The output applied to a detector is a comparison of the two inputs. A feedback mechanism connected to the output is returned to the detector mechanism for corrections. If the controller is working properly and the process is under control, the measurement signal will equal the setpoint signal. The output of the comparison mechanism generates an error signal and applies it to the detector. This signal applied to the feedback mechanism is used to correct the output signal accordingly.

The amount of output that a controller will change depends on the design of the controller. Gain, which is the ratio of output over input, is a common controller operation. As a general rule, controller gain must be held to a minimum in order to reduce extreme output variations.

ACTIVITIES

CONTROLLER INVESTIGATION

1. Refer to the operation manual of the instrument to be used in this activity.
2. Remove the housing or covering from the instrument.

3. Identify the function of the controller.
4. Locate where the input is supplied to the controller.
5. Note the location of the setpoint adjustment.
6. How is feedback achieved with this controller?
7. Does this controller have a lever mechanism?
8. How is detection achieved?
9. Explain the operation of this controller.
10. Reassemble the controller.

QUESTIONS

1. What is the difference between indicating and nonindicating controllers?
2. What are the components of a controller?
3. Explain the operation of the controller mechanism of Figure 19-7.
4. What is the comparison mechanism of a controller?
5. Explain how gain is achieved in a controller.

20
Control Theory

OBJECTIVES

Upon completion of this chapter, you will be able to:

- Distinguish between on/off, proportioning, integral, and derivative modes of control.
- List the basic functions of a controller.
- Define gain, sensitivity, and proportional band as it applies to the operation of a controller.
- Explain the response of an on/off controller.
- Explain the meaning of proportioning response.
- Describe the response of an integral controller.
- Explain the meaning of derivative response.

KEY TERMS

Derivative action. A control procedure that recognizes the rate or time at which a predetermined amount of change occurs in the operation of a system.

Deviation. The difference between instantaneous and setpoint values of a controlled variable.

Hyper reset. A term used by some manufacturers to describe the derivative response.

Index. A guide or pointer that is used to find something or a value on a scale.

Integral response. A control that provides an output that is related to the time integral of the difference between the setpoint and the process variable.

Mode of operation. A method, procedure, or type of control.

Preact. A term used by some manufacturers to describe the derivative response.

Proportional band. A range of values that a process variable operates within when a controller is active.

Proportional response.   A mode of operation in which the output is a linearly related function to the difference between a setpoint value and the process variable being controlled by a system.

Rate action.   Another name for the derivative mode of operation.

Reset action.   Another name for the integral mode of operation.

Sensitivity.   The ratio of change between the output and input of an instrument or the least amount of input that will produce an output.

Throttling.   An action that refers to the compression, reduction, or control of flow through a system.

Throttling band.   A narrow range or band of operation through which the input must change to cause a change in output.

Two-position action.   A control procedure in which the final control element is moved from on to off or open to shut in the operation of a system.

INTRODUCTION

Basic understanding of automatic-control theory is important in working with pneumatic instrumentation. This knowledge can help a person analyze instrument operation as well as understand why certain instruments are selected for certain control functions. In this chapter, some of the common control responses to achieve automatic control will be discussed.

CONTROLLER BASICS

Most industrial processes such as flow, temperature, level, or pressure can be controlled at or near a selected reference value called the setpoint. A controller is the instrument that is primarily used to achieve this function. The controller looks at the signal that represents an actual process variable value. It then compares this value with a setpoint value. Finally, it acts to minimize the difference between the two signals. A simplified block diagram of a pneumatic controller is shown in Figure 20-1.

The basic functions of a controller include:

1.   Receiving a signal to measure the value of the variable being controlled.
2.   Comparing that value to a reference value.
3.   Determining error in the value.
4.   Providing a controlled output.

These basic functions apply in general to all controllers.

MODES OF CONTROL

Pneumatic controllers can provide four different kinds of control. All of these controls can be used in automatic process applications including on-off or two-position, propor-

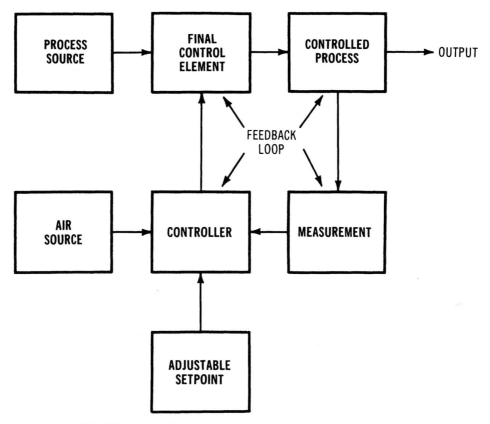

FIGURE 20-1   Simplified block diagram of a controller system.

tional control, integral response, and the derivative response. These basic forms of control are primarily achieved by changing the feedback components of the controller, as shown in Figure 20-1.

TWO-POSITION, OR ON/OFF ACTION

At times, the two-position, or on/off mode of control is considered as a control approximation. This mode of operation is not particularly accurate and does not, in this application, employ the feedback loop used by other modes of control. Figure 20-2 shows an on/off pneumatic controller of the throttling type.

The nozzle detector of the on/off controller is often described as a high-gain mechanism. The relay of this instrument has a gain of at least 3 or more, meaning that a small change in input error signal will cause a large output change. The action of this controller in response to a gradually increasing change in input is shown in Figure 20-3. Input is changed to output by a strip called the throttling band. This band of control is usually 1 percent to 2 percent of the total input range.

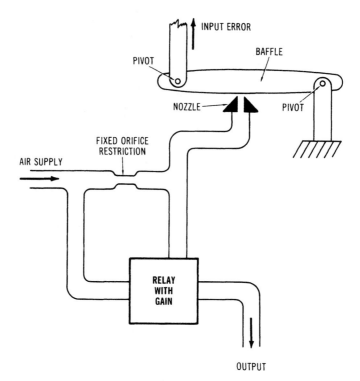

FIGURE 20-2   A throttling type of on/off controller.

In practice, on/off controllers are used to change control operations. Continuous use of the on/off switch indicates that the controller gain is set too high. Too low a gain setting will cause lengthy operating periods. Two-position control in pneumatic systems is also called a high-gain mode of operation because of the gain function.

PROPORTIONING RESPONSE ████████████████

Suppose now that the temperature of a tank of water is controlled by a steam heater. Water flows through a coil placed inside of a tank. Steam is forced inside of the tank and steam surrounds the coil. Heat from the steam passes through the coil walls and into the water. The steam entering the tank is controlled by a valve. The temperature of the water in the coil is controlled by tank steam. To measure water temperature, a strip-chart temperature recorder is used. This recorder has a range from 0° to 100° Celcius, or C. The water temperature is to be held at 50°C.

Assume that operation begins when the steam and water valves are half open, or at the 50 percent position. The water temperature will be 50°C. Further assume that the temperature of the water increases to 60°C, or a 10 percent change. This change calls for the steam valve to be closed to the 40 percent open position, which represents a 10 per-

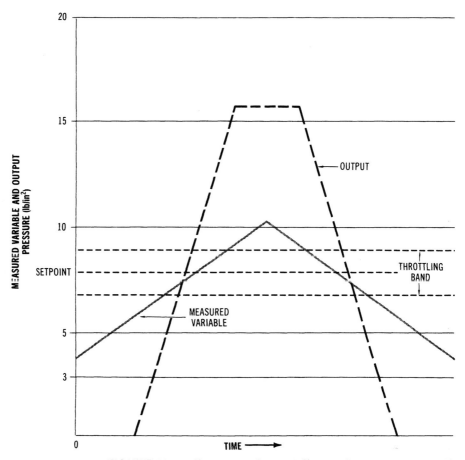

FIGURE 20-3 Response of an on/off controller.

cent reduction in steam. With the steam reduced by 10 percent, the temperature of the water in the coil will drop by 10 percent. This change in steam is needed to maintain the water at 50°C.

Assume now that the temperature of the water increases to 70°C, which represents a 20 percent increase in water temperature. To correct this increase, the steam valve opening is closed to its 30 percent position, which represents a 20 percent reduction in steam temperature. With the steam reduced by 20 percent the temperature of the water in the coil will drop by 20 percent. This reduction in steam is needed to return the water to 50°C.

When the preceeding adjustments are made, they represent a proportional change in temperature. A 10 percent change in water temperature is controlled by a 10 percent change in steam valve control. In other words, the amount of change in valve position is in proportion to the amount of the change in temperature. This is called the proportioning response.

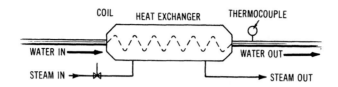

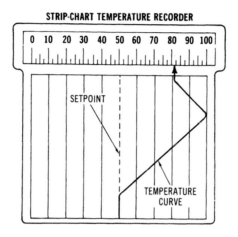

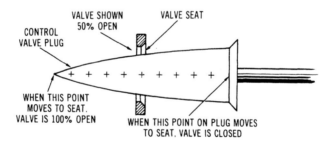

FIGURE 20-4   Elements of a temperature control system.

In the case just described, the ratio of temperature change to valve change is 1 to 1. So if the temperature reading were zero, the valve would be wide open. If the temperature were 100°C, the valve would be closed. A 100 percent change in temperature would result in a 100 percent change in valve position. These ideas might be represented as shown in Figure 20-4.

Proportional Band

In the case just discussed, the conditions were that the valve moved 100 percent to change the temperature an equal amount, which is never the case in an actual situation.

In practice, the valve may move only part of its travel. In this case, say that 50 percent of travel is needed to change the temperature 100 percent. As a result, the ratio would be 100/50, or 2 to 1.

In another application, the valve may move 100 percent of its travel and the temperature might move 50 percent. This ratio of temperature to valve change would be 50/100 or 1 to 2.

In the first example, the ratio is found to be 1 to 1. In this case, the controller can be said to have a proportional band of 100 percent. In the case where the ratio is 2 to 1, the band is said to be 200 percent. When the ratio is 1 to 2, the band is described as being a 50 percent proportional band. Proportional band is described as the ratio of measurement change to valve position change. This is also seen as a change in controller output taken as a percent.

In the previous discussion, a controller was described as a device that can be adjusted to perform different functions. An adjustment can be made to change the amount that the value will move for a given measurement. Adjustments to the throttling range and the overall sensitivity of the mechanism may also be altered. Controllers with these adjustments are called adjustable proportional-band instruments. Adjustable gain mechanisms have been described in previous chapters. The need for such an adjustment is called *matching the controller to the process.* This means that different processes require different valve movements for the measurement changes. The movements required depend on a variety of considerations. Among these are the size of the equipment, the kind of material being processed plus the rate at which the material will be processed. In addition, there is the size of the piping, desired pressures, and temperatures to consider. In general, there are as many different valve responses as there are chemical processes.

Gain, Sensitivity, Proportional Band

These three terms are used to define the proportional response. Previously, gain was described as the ratio of output to input. Gain is defined as the ratio of output pressure change to measurement change. Gain has been used throughout this book. Technically, gain is a dimensionless number. That is, the units used to express output are the same used to express input. Therefore, they cancel each other. Gain is usually applied to pneumatically coupled controllers. In this particular case, the ratio would be psi of output to psi input. The units would then cancel out leaving a dimensionless number. For example, if the transmitted measurement pressure changed 1-psi, then the output pressure changed 3-psi. The gain would be 3 to 1 or 3.

Sensitivity is defined as the ratio of output pressure to change in measurement. This is where measurement is expressed as indicator travel. For example, assume that a pen moves 2-in on its chart. This causes the output to change 1-psi. The sensitivity is 1-psi per 2-in, or 0.5-psi per 1-in.

To relate sensitivity to gain, the following expression applies:

$$\text{sensitivity} = \text{12-psi} \times \text{gain/chart width}$$

To relate proportional band to gain, use the following:

$$\text{proportional band} = 1/\text{gain} \times 100 \text{ percent}$$

To relate proportional band to sensitivity, use the following:

$$\text{PB}/100 = 12\text{-psi}/S \times \text{chart width}$$

where,

S is the sensitivity

PB is the proportional band.

Proportional band was defined as the ratio of measurement change to output change. This is expressed as a percent. For example, suppose that a change in measurement relative to the setpoint is 10 percent of the chart. The controller output changes 1.2-psi, then the proportional band is 100 percent. Why? Because 1.2-psi is 10 percent of output and the ratio is:

$$10 \text{ percent}/10 \text{ percent} = 1$$

and $1 \times 100 = 100$ percent proportional band.

Suppose the output changed 12-psi for a 10 percent change in measurement setpoint deviation. A change of 12-psi represents a 100 percent change in output.

$$10/100 = 0.1$$

and 0.1×100 percent $= 10$ percent proportional band

Suppose now, that for a measurement setpoint deviation equal to 100 percent of the chart, the output changed 60-psi or 50 percent.

$$100/50 = 2$$

and 2×100 percent $= 200$ percent proportional band.

Regardless of the terminology being used, it is important that you should be able to recognize that these terms all decribe the same thing: the relationship between the input and the output.

INTEGRAL RESPONSE

Using the process described earlier, suppose that the temperature increased from 50°C to 60°C as shown in Figure 20-5. If someone were operating the process, the valve would be readjusted. Closing it would bring the temperature to the setpoint. After waiting several seconds, it might be observed that the temperature had dropped. Although the tem-

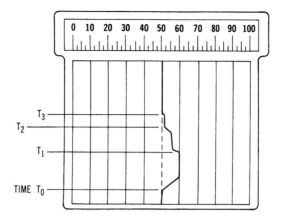

FIGURE 20-5   Effect of experimental adjustments in process control system.

perature may have dropped, it may not yet be on the setpoint. If this were the case, the valve would need to be closed further and the results observed. Finally, after a number of valve adjustments, the temperature would equal the setpoint. In Figure 20-5 the valve adjustments would be T_1, T_2, and T_3 which is where the temperature coincides with the setpoint and no further adjustments are required.

The valve is readjusted whenever the temperature is away from the setpoint. When the temperature is too high, the valve is closed a small amount. When the temperature is too low, the valve is opened a little.

Remember, the first temperature change that appeared on the chart occurred at T_o. There was no change in valve position. There must have been some change in the process however. Some changes that could cause this temperature rise are changes in the steam supply. There could also have been changes in the water throughput. The steam pressure might have increased. Even though the valve position would remain the same, more steam would flow through because of the increased pressure.

Suppose the water flow through the heater was reduced. The temperature of remaining flow would then increase because the same steam feed heats a reduced amount of water. On the other hand, if the water usage were increased, then the temperature would fall. If the steam pressure dropped, the temperature would fall. So the temperature can stay the same without valve adjustments. But process conditions must also remain the same. In practice, process conditions never stay exactly the same.

If a process were controlled with a proportioning controller, the valve would automatically compensate for a temperature change. While this is true, is it sufficient to bring the pen back to the index?

Suppose the requirements for hot water have been increased. The effect would be a reduction in the water temperature to about 40°C. The valve would open due to proportioning response and, as a result, the temperature might return to 50°C. If it should, the valve would be returned to 50 percent opening. But remember, the 50 percent opening would no longer be sufficient to heat the new water load. It was just sufficient to hold the original water load at 50°C. Therefore, it is impossible to heat the increased load with the opening needed to heat the original water load to 50°C. It must be concluded that if the load changes, the measurement cannot return to the setpoint unless:

1.   The load on the system returns to the original condition.
2.   The index is raised, causing the valve to open an amount sufficient to heat the additional water to 50 percent. The valve opening then shifts the physical setpoint index mechanically so that it coincides with the setpoint.

When these two conditions are met the controller is mechanically aligned for balance conditions that suit the new load. Such a readjustment is called manual reset.

For a continually changing process, a manual reset may not be sufficient. The utility of the reset mechanism will now be discussed.

If measurement is away from the setpoint, the control valve must be continually repositioned. This repositioning continues until the measurement returns to the setpoint. The proportional response is not adequate to return the pen all the way back to the setpoint. The basic reset mechanism will show that unless the error signal is zero, the controller mechanism will cause the valve to drive fully open or closed. The driving will stop only when the measurement and setpoint are together. If the automatic reset mechanism is used, the controller responds proportionally. This is due to the proportioning part of the mechanism. It also responds whenever the measurement is away from the setpoint. The magnitude of this reset response is determined by:

1.   How far the measurement is away from the setpoint.
2.   How long the measurement is away from the setpoint.

The magnitude of the reset response is determined by the setpoint and the measurement deviation. These two values are multiplied by the time that the deviation exists.

This relationship will now be analyzed mathematically. The magnitude of the reset response is related to the area under the measurement curve. The position of the setpoint is the base line or deviation multiplied by time. This is the reason for referring to this as integral response. One definition of integration is the determination of the area under a curve.

Remember that the reset mechanism uses a restriction which is adjustable. The restriction establishes the rate at which a corrective action will take place for a given measurement/setpoint deviation. If the restriction is wide open, the rate of correction is very fast and causes the control valve to be driven alternately from fully open to fully closed. If the restriction is closed, no reset response will be obtained.

There are two problems with integral response. These are a continual source of trouble. For example, it is impossible to have gain and integral-response values that will cause the measurement to overshoot the setpoint. The direction of reset response will change. The output will reverse only when the measurement crosses the setpoint. In the preceding example, the temperature was swiftly approaching 50°C from a starting point of 60°C. The integral action would continue to close the valve until the measurement was 50°C.

A second problem arises when a process is shut down. If the process is shut down there is no way for the changing valve to bring the measurement back to the setpoint.

Nevertheless, the controller continues to move the valve. If the shutdown is long enough, the valve will be driven to either fully open or fully closed, depending on the reset-restriction setting. It may take as much as several hours for the valve to be returned to its normal operating position. It is for this reason that a process using a proportional-plus-reset controller must be started manually. Proportional integral controllers are arranged to cut out the integral response for start-up on batch-type operations.

Integral Response Summary

1. The integral response is a function of the distance between the setpoint and measurement multiplied by the time that each of these differences exist. This refers to the area under the measurement curve.

2. The effect of integral response is to cause the valve to move until the measurement and setpoint are together, regardless of the actual output pressure.

3. The integral response will reverse its effect on the valve only when the measurement crosses the setpoint.

4. During shutdowns, a proportioning-plus-reset controller causes the valve to drive fully open or closed depending on the reset adjustment and the duration of the shutdown. If this drive is excessive, the process must be put on manual control or a special batch-type controller must be used.

DERIVATIVE RESPONSE

The last type of control is called derivative response. This type of control is also called hyper reset and preact by some manufacturers. Others may call it rate action. In this discussion it is called derivative response. Derivative is a procedure that recognizes the rate of time at which a predetermined change occurs in the operation of a system.

Using the process described earlier, suppose that a pair of temperature curves as shown in Figure 20-6 has been obtained. Examination of the curves shows a 10° C increase in temperature. The proportioning response for curve 1 and curve 2 are identical.

The area under curve 1 is a bit less than the area under curve 2. The total reset response is somewhat less for curve 1 than for curve 2. Curve 1 represents a process rising above the setpoint more rapidly than the one represented by curve 2. If it is to be brought back into control, a larger change in valve position must take place. Curve 2 represents a process rising above the setpoint at a much slower rate. This means that a less rapid change in the valve position would be required to control the temperature.

It would be desirable to have a control response that recognizes the rate at which the temperature changes from the setpoint. It might also be useful to have a control response that would recognize the slope of the curve. Mathematically, the slope of the curve can be defined as the time derivative. This is the reason for calling this control action a derivative response.

The rate of response of derivative control is established by the derivative restriction. Derivative control is used to give the valve a boost in the right direction so that it will respond quickly to a rapid change in temperature and to flatten it out. Derivative

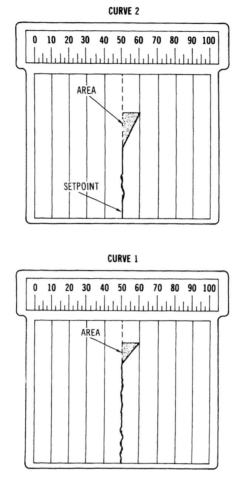

FIGURE 20-6   Temperature curves showing different rates of loss of control.

control also causes the control valve to overtravel for a short period of time. The valve overtravels much in the same way that an operator would close down the valve a few extra turns for a couple of minutes. Whenever the curve is flat there is no derivative response, regardless of the actual position of pen on the chart. There is no derivative response because derivative control reacts only to the slope of the curve. A flat curve has a zero slope. The derivative response is therefore zero.

Derivative control responds to the slope of the curve. It might be said that derivative control can anticipate what the measurement will be at some future time. Though this view is not strictly valid, it might be helpful in developing some feel for derivative control. Remember, if the slope is zero, there is no derivative response. This means that the measurement curve follows the setpoint.

COMBINED THREE-MODE RESPONSE ▬▬▬▬▬▬▬▬▬▬

Three control responses have been considered in this discussion thus far. The method employed was to assume a certain type of measurement/setpoint deviation, then propose a type of valve correction required to bring the measurement to the setpoint. Curves were used to demonstrate the need for the three types of response. To summarize, consider the random curve of Figure 20-7. This curve could represent any variable or measurement.

A very small portion of the curve can be described by specifying three things:

1. How far is the portion of the curve from the setpoint?
2. What is the time duration of this portion of the curve?
3. What is the slope of this portion of the curve?

A technician should be able to recognize these three conditions. A control adjustment is then made in light of the total situation as the curves are observed. Control mechanisms should be able to recognize the conditions listed as the following:

1. How far is the portion of the curve from the setpoint? It has been shown that the proportioning mechanism does this part of the job.
2. Specifications 1 and 2 multiplied represent the area under the portion of the curve. The integral response mechanism responds to these specifications.
3. What is the slope of the curve? The derivative mechanism responds to the slope.

Therefore, the total response of the control mechanism consists of the sum of the three individual responses of proportioning, integral, and derivative acting simultaneously. This combined function is called proportional, integral, and derivative response. It is identified by the abbreviation PID.

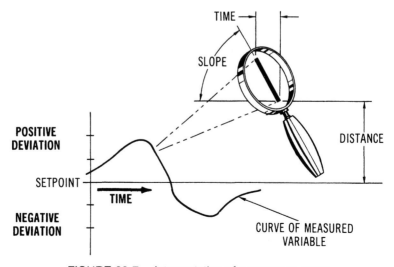

FIGURE 20-7   Interpretation of a response curve.

SUMMARY

Automatic control theory applies to instruments that implement some type of system control. Automatic control responses or modes of operation are of the proportioning, integral, and derivative types.

Proportioning response refers to adjustments that are made according to corresponding value changes of the same proportion. A 10 percent increase in temperature would result in a 10 percent reduction in control valve position to maintain the temperature at a desired level. Gain and sensitivity influence the proportioning response and its band.

Integral response is a function of the difference between setpoint values and measured values. Integral response also refers to time over which this difference exists. In effect, this type of response causes measured and setpoint values to shift or deviate until they come together.

The derivative mode responds to changes. It also has the ability to anticipate or estimate what measurements will occur based on present data. This mode of control recognizes the rate at which changes occur.

ACTIVITIES

CONTROLLER IDENTIFICATION ▬▬▬▬▬▬▬▬▬▬▬▬▬▬▬

In this activity, an operating procedure will be presented and you will be asked to identify the type of controller that will accomplish the procedure.

1. Assume that a controller has the final control valve half open and the amount of flow through the system coincides with the setpoint value of the system. If the flow decreases by 10 percent, the control valve opens by 10 percent to increase the flow to the setpoint value. What kind of controller is involved in this operation?

2. Assume now that a temperature controller has a setpoint value of 100°C and the system is off, and the temperature remains at this value. After a short period of time, the temperature drops to 98°C. The heat source automatically turns on, causing the temperature to increase in value to 102°C and is then turned off. The heat source remains off until the temperature drops to 98°C, at which point the heat source is again turned on to increase the temperature. What kind of controller is involved in this operation and what is the band of control?

3. Assume now that a temperature control system has the setpoint value adjusted to produce a temperature of 50°C. When the temperature rises to 60°C the heat source valve is closed somewhat in an effort to bring down the temperature. After a short period of time, the temperature drops to a value of 55°C but has not returned to the setpoint value. The heat source valve is closed further while we observe the result for a period of time. The heat source valve is closed even more while we observe the results. This process continues until the temperature of the system finally equals the setpoint value. What control operation is involved in this system?

4. Assume now that we have a temperature control system that has its setpoint adjusted to a value of 100°C. When the temperature rises to 110°C, the heat source valve is

turned down very quickly to compensate for the rise in temperature. This action tends to reduce the time of the temperature rise and flatten the response of the operation. This type of controller tends to respond to the slope of the temperature change curve and has the ability to anticipate changes in value. What control procedure is involved in this system?

5. Make a graph that shows the operation of each controller just described.

QUESTIONS

1. What is meant by the term mode of control?
2. Explain the operation of a throttling type of on/off controller.
3. What is the throttling band of an on/off controller?
4. What is the proportioning response of a controller?
5. Define the term sensitivity as applied to a controller.
6. What is the integral response of a controller?
7. What is the derivative response of a controller?
8. What is a three-mode controller?

21
Controller Functions and Mechanisms

OBJECTIVES

Upon completion of this chapter, you will be able to:

- Explain how gain is achieved in the operation of a controller.
- Identify different gain mechanisms.
- Explain the operation of a reset controller.
- Explain how a proportional-plus-reset controller operates.
- Describe how the derivative mode of operation is used in controllers and transmitters.

KEY TERMS

Adjustable gain mechanism.   An arrangement of components that make it possible to change the ratio of output to input in the operation of a controller.

Adjustable restriction.   A device that limits the pressure drop or resistance in a line by altering or changing the cross-sectional flow area.

Automatic reset controller.   An instrument that produces a self-generated output which is related to the time difference between the instantaneous and setpoint values of a process variable.

Closed-loop.   A group of components that measures a process variable and compares it with the setpoint value. If a difference in value occurs, a corrective error signal is developed and applied to the final control element to alter its operation.

Comparison mechanism. A device consisting of two bellows that are attached to a measurement input and a setpoint input that determines the similarity of the two signals.

Derivative restriction. A pressure limiting device that is placed in the proportioning feedback line of a controller.

Derivative capacity. A bellows assembly that has a storage function that adds pressure to the output of the system for regulation.

Equilibrium. A state of balance between opposing forces or actions.

Open-loop. A system in which no comparison is made between the actual value and setpoint value of input signals.

Parallelogram mechanism. An assembly of components that are bonded together by a parallel lever configuration.

Pneumatic gain mechanism. An assembly of components that can automatically change the ratio of output to input air pressure.

Positive feedback. A mechanism that returns part of a signal and adds it to the original to produce regeneration.

Proportional-only controller. A controller whose output is a percentage of the difference between the setpoint and the process variable being measured.

Proportional-plus-reset. A controller that combines proportional and reset functions in a single unit.

Repeat. An additional change in output equal to the proportional response.

Repeats per unit of time. The number of proportional responses that occur within a given time.

Reset-only controller An instrument whose output is the time difference between the instantaneous and setpoint values of a process variable.

Three-way valve. A multiple orifice flow control device with three flow openings.

Variable capacity. A bellows assembly that is added to a line to regulate pressure by altering its storage capability.

INTRODUCTION ▬▬▬▬▬

The functions of a controller are determined by the instrument's different modes of control. A controller can be made to achieve a single function only, as in the design of a proportional-only controller. Controllers can also be made to achieve multi-functions, as proportional-plus-reset and proportional-plus-integral-plus-derivative controllers are made to do. Thus, the function, or mode of control, of a control instrument is largely determined by its application.

In general, the internal construction of a controller is called a mechanism. In order to understand how an instrument responds when signal changes occur, controller functions and mechanisms, particularly the gain mechanism, will be discussed in this chapter.

GAIN MECHANISMS

Elementary controllers which have no internal feedback were discussed in a previous chapter. It was determined that an elementary controller has a very high gain. As a result, an extremely small change in measurement input causes a full range of valve travel. This particular fact was determined because the input had to change only 0.004-in to fully cover the nozzle. Previously, it was shown that high-gain controllers have a limited use and that there is a need to reduce controller gain. In this section, the method used to reduce controller gain will be discussed.

Components

The components of a low-gain controller are:

1. A comparison mechanism.
2. A detector mechanism.
3. A feedback mechanism composed of a bellows/spring combination.

Arrangements

A low gain controller is arranged in the same manner as in an elementary controller, except for the feedback mechanism. This mechanism is in a position to move the control lever as the nozzle backpressure changes. The feedback mechanism and its exact operation will be discussed in this chapter.

Operation

Remember that it is necessary to move the baffle only 0.002-in to cause a full change in output. Because of the lever lengths involved, the measurement must move only 0.004-in to cause a full change in output. Figure 21-1 diagrams these concepts.

Figure 21-2 shows the control lever, a nozzle, a detector mechanism, and a bellows/spring combination connected to the nozzle backpressure. Suppose the input, or error signal, is moved 0.004-in. The output would change from 3- to 15-psi. This change in output, acting on the bellows/spring, would move the lower end of the input lever.

Assume that the baffle is positioned before the nozzle so that there is a 3-psi output pressure. A change in input of 0.004-in will momentarily cover the nozzle. Theoretically, it should obtain a 15-psi output. But, as the nozzle becomes covered, the backpressure increases. This increase in backpressure moves the baffle away from the nozzle. The feedback movement need only move 0.004-in to return the baffle back to its position before the input is changed. In other words, the feedback motion balances the input motion.

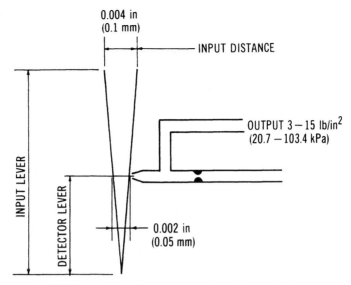

FIGURE 21-1   Baffle movement for full output change.

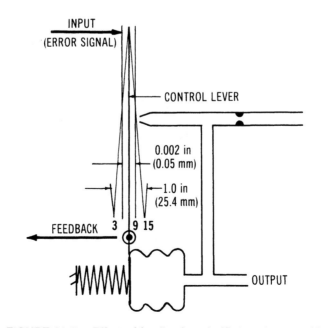

FIGURE 21-2   Effect of feedback on baffle/nozzle assembly.

Suppose that the bellows/spring were such that it would deflect a full inch. This would change the pressure from 3- to 15-psi, meaning that the input would also change. If the input is changed 1-in and obtains a 1-in change in output, the gain of the controller is 1. By introducing the feedback bellows, it is possible to change controller gain from 1 to 250.

DETECTOR RANGE AND GAIN

Suppose now that the input moves 1-in towards the nozzle. If it does, the output will change from 3- to 15-psi. As it changes, the feedback bellows moves 1-in. The baffle position next to the nozzle will be the same as it was before the input changed 1-in. If the baffle were back in the same position, the output would have to go back to 3-psi. If the output were 3-psi, the bellows/spring could not have moved its 1-in. When the output changes from 3 to 15-psi, the baffle/nozzle clearance must change from 0.002-in to zero. Then there must be a difference between the input and the feedback motions. This difference is equal to twice the 0.002-in detector range, or nozzle clearance, and is shown in Figure 21-3. In other words, if the input moves 1-in, the feedback motion will equal 1-in minus 0.004-in. So, instead of putting a numerical value in balance, it is almost balanced or within the range of the detector.

Detector range is important because self-balancing or feedback instruments do not return precisely to their original positions. The exactness of balance is limited by the

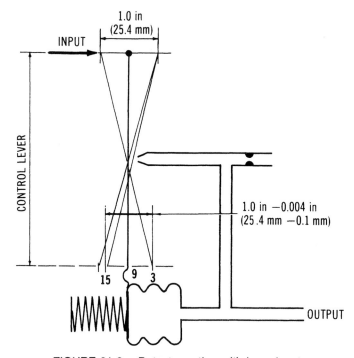

FIGURE 21-3   Detector action with large inputs.

range of the detector. A detector using 0.002-in has been used as the detector range up until this point. Remember that 0.002-in only shows the range for a particular instrument. Actual detector ranges vary between instruments.

To better understand the concept of detector range, examine the operation of the controller. Suppose the input were to change 1-in. Suppose also that the input changes very slowly. This example is shown in Figure 21-3. The error signals are moved to the right and the nozzle backpressure builds up. Changes in the error signal and feedback signal occur at the same time. When the nozzle is covered, it changes the error signal. The feedback signal then uncovers the nozzle. As the error signal continues, the output eventually becomes 15-psi and the nozzle must be covered. In other words, the nozzle clearance must change from 0.002-in to zero. Therefore, the feedback must be slightly less than the input motion. The feedback motion is actually 0.004-in less than the input motion.

In the discusssion of mechanisms, it has been assumed that all mechanisms were geometrically similar; that is, the nozzle is halfway between the error signal input and the feedback signal. Consequently, the gain is 1. If the detector range is neglected, the output is the same as the input. Controllers with a gain of 1 are as limited as controllers with a gain of 250. A controller should be arranged so that it is possible to change the gain from 250 to 1.

Some controller adjustable gain mechanisms are the same as those used in transmitters and positioners. In addition to those already discussed, two other types of adjustable gain mechanisms will be considered. There are some alternative terms concerning gain. Among these are sensitivity and proportional band. These expressions will be discussed in the chapter on controller theory. However, mechanisms are the same, no matter what label is used to describe them.

ADJUSTABLE GAIN MECHANISMS

A bellows/spring component makes it possible for a controller to have a gain of almost 1. A controller with a fixed gain of 1 is of limited use. A controller needs to be arranged so that it can obtain gain anywhere from 250 and 1. Two mechanisms that adjust controller gain will now be discussed.

Components

An adjustable gain mechanism can be one of four different types:

1. Parallel lever.
2. Angle-gain mechanism.
3. Parallelogram mechanism.
4. Pneumatic-gain mechanism.

Arrangements

The gain mechanisms are located relative to the comparison, the feedback, and the detector mechanisms, as shown in Figure 21-4. By adjusting these mechanisms the

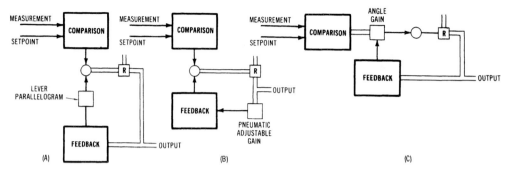

FIGURE 21-4   Some methods for obtaining adjustable gain. (A) Parallelogram mechanism; (B) Pneumatic-gain mechanism; (C) Angle-gain mechanism.

ratio of output to input is changed. If no feedback motion is fed back to the detector, the gain will be 250. If all of the feedback motion is fed back to the detector, the gain will approach 1.

PARALLEL-LEVER MECHANISM OPERATION

Figure 21-5 shows a parallel-lever adjustable-gain mechanism. Compare this mechanism with Figure 21-2. Notice that a second lever has been introduced. This second lever drives the baffle through a contactor. Suppose the contactor is raised so that it is in line with the feedback-lever pivot point. Also suppose that the error signal changes so that the baffle moves towards the nozzle. The nozzle backpressure will build up and the feedback

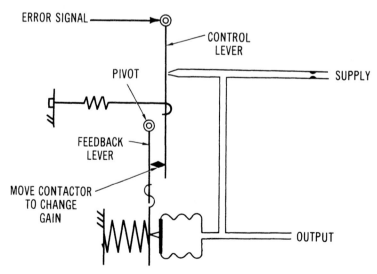

FIGURE 21-5   Parallel-lever adjustable gain mechanism.

will move. Since the contactor lines up with the pivot, the feedback movement will not be felt by the baffle. For all practical purposes, the adjustable gain mechanism has been taken out of the circuit. Therefore, the contoller has a high gain. The gain depends on the nozzle location and length of the lever.

Now consider what happens when the contactor is located near the feedback bellows/ spring. Again, assume an error signal that moves the baffle toward the nozzle. The nozzle backpressure will build up, and the bellows will move to the left. This permits the baffle to move away from the nozzle. If the contactor is at its low point, then all the feedback motion will be fed back to the baffle. For this contact position the gain of the controller is low, perhaps 1. By changing the contactor position from one extreme to the other, controller gain can be altered. Gains of 250 to 1 are common.

The lever gain mechanism of a controller is the equivalent of those used in positioners. By adjusting the lever lengths, it is possible to adjust the ratio of output to input, also known as adjusting the gain. The parallel-lever mechanism is applied to the controllers. The ratio of feedback motion to error signal is adjusted. The error signal is a result of a difference in two pressures. The output of a controller is pressure. It can be described as the ratio of output to input pressure. It is important to note that the dimensions used to describe input and output are of no consequence because the dimensions cancel each other.

Suppose the input is expressed in inches and the output is expressed in pounds. It would be incorrect to talk of controller gain in these terms. It then becomes necessary to use alternate expressions. Among these are sensitivity and proportional band. These terms will be fully discussed in chapter 22.

ANGLE-GAIN MECHANISM ▰▰▰▰▰▰▰▰▰▰▰▰▰▰▰▰▰▰▰▰▰▰▰▰

Now consider the angle-gain mechanism. Figure 21-6 is a diagram of a controller with an angle-gain mechanism. A bellows comparison mechanism and a proportional-only feedback mechanism is shown in Figure 21-6. Arranged between these two mechanisms is the angle-gain mechanism. Recall that the angle-gain mechanism consists of a baffle/nozzle detector mounted on a disk so that it can be rotated relative to various signals. In positioners, the two signals involved were measurement and actuator position. In controllers, the two signals are the error signal and the feedback signal. The operation of a gain mechanism is the same for positioners and controllers. The angle-gain mechanism changes the relationship of the feedback and error signals. In general, it changes the ratio of output to input.

Suppose now that the baffle is rotated relative to the nozzle. This movement will form an angle of 0° to the feedback signals. Figure 21-7 shows the baffle in a 0° position. Follow the operation of the controller for this baffle position.

If the error signal increases, the baffle will move to cover the nozzle. The error signal required is 0.002-in. Since the nozzle is covered, the backpressure will increase to 15-psi. This increase will result in an upward feedback motion. But, since the baffle is parallel to this feedback motion, the position will not be changed by the feedback motion. Consequently, as far as the baffle/nozzle is concerned, there is no feedback motion. For

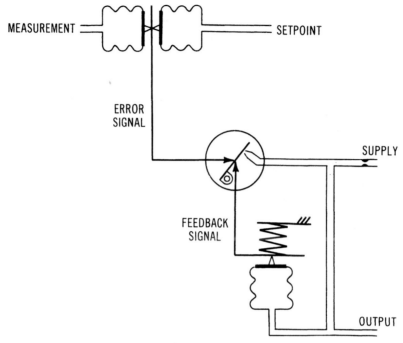

FIGURE 21-6   Angle-gain mechanism.

a 0° baffle position, the feedback is taken out. The controller has a gain of 250. When the baffle is rotated, a 0° angle is formed between it and the error signal. Figure 21-8 shows this configuration.

Suppose the error signal increases. The baffle is parallel to the error signal. The signal can travel the full range without a change in the backpressure and output. For a 0° baffle/error-signal relationship, no functional controller is needed because it is impossible for the error signal to cause a change in controller output.

Assume now that there is a small angle approaching 0° between the two signals.

An increase in error signal will move the baffle toward the nozzle, which will cause the nozzle backpressure to build up. Now almost all of the feedback motion gets fed back to the detector because the baffle is near 90° relative to the feedback. This is a large contribution of the feedback motion in relation to the error-signal motion. The ratio of output to input is a number of approaching zero. An example would be 0 divided by 1, which represents the feedback motion. If the output were zero, the gain would be zero. Since the output only approaches zero, the gain of the controller approaches zero.

Remember that when the baffle was at 45° relative to the feedback and error signals, the gain was 1. When the baffle was approaching 0 near the error signal, the gain approached zero. In other words, it is possible to get gains less than 1 in addition to gains from 1 to 250.

It is suggested that the reader review the discussion of the angle-gain mechanism in Chapter 16. The action of the instrument may be changed by rotating the angle-gain

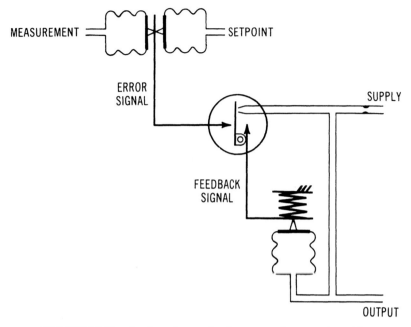

FIGURE 21-7   Angle-gain mechanism—maximum gain setting.

mechanism to a new quadrant. The controller action can be reversed by rotating the angle-gain mechanism. The baffle is then in an opposite quadrant.

The parallelogram and pneumatic mechanisms will be discussed in following chapters. All adjustable gain mechanisms have the same function—that is, to change the ratio of the output to the input. They accomplish this by changing the amount of feedback signal fed to the detector. The angle mechanism changes the error and the feedback signals.

Notice when the baffle is rotated, the amounts of error and feedback signals are changed. As the baffle is moved, the error signal applied to the baffle will decrease. The feedback signal, at the same time, increases. The consequence of these actions rests in the problem of aligning controllers, thus, it is important to understand how adjustable gain mechanisms are used to change the ratio of controller output to input.

In earlier chapters, discussion was limited to controllers having a proportional response. Such controllers are described as proportional-only controllers. In the next section of this chapter, controllers having automatic reset, or mode reponse, will be discussed.

THE RESET FUNCTION

The reset function is sometimes called integral control. This function has an output whose rate of change is proportional to an error signal applied to its input. When there is a large signal change, the output changes rapidly to correct the error. As the error signal gets smaller, the output changes more slowly. This action is purposeful to minimize

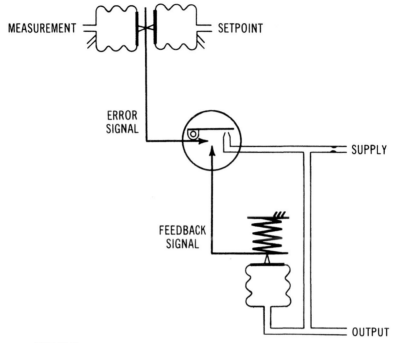

FIGURE 21-8   Angle-gain mechanism—minimum gain setting.

for over correction and reduce an offset error. As long as there is an error, the output of this control operation will continue to change. Once the error is driven to zero, the output change also goes to zero, which means that the controller has a form of inertia. It tends to hold the output which was necessary to eliminate the error signal applied to its input. Controller manufacturers call this action reset, automatic reset, or integral control.

Components

The components of a reset controller are the same as those used in the proportional-only controller. The difference between the controllers lies in the arrangements of the feedback component. The components of a reset controller are:

1. A comparison mechanism.
2. A detector mechanism.
3. A feedback mechanism consisting of a bellows/spring.
4. An adjustable restriction.

Arrangements

Figure 21-9 is a diagram of a controller with automatic reset feedback. Compare this diagram with the diagram of the proportional-only controller. Notice in the reset type that

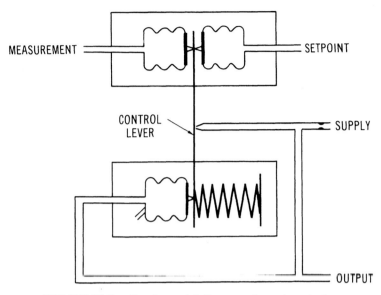

FIGURE 21-9   Fundamental diagram of reset controller.

the bellows/spring is reversed. The bellows is at the left of the lever in the reset type. It is located to the right of the lever in the proportional type.

The reset controller has some extremely interesting properties. On the surface, a controller arranged like the one shown in Figure 21-9 appears useless. The task at hand is to understand its operation. Next, it is important to recognize the characteristics of the mechanism.

Operation

Assume that the controller is balanced. The method used to get it into balance will be postponed for now. Suppose that in a balanced condition, there are 9-psi in the measurement and setpoint bellows. The baffle is positioned in front of the nozzle so that there is also a 9-psi backpressure. Backpressure is then applied to the feedback bellows.

Assume that there is an increase in measurement. As a result, the baffle will move toward the nozzle. The nozzle backpressure will build up, and there will be an increase in pressure in the feedback bellows. This increase in pressure will cause the bellows to expand. As a result, the baffle will move toward the nozzle. The nozzle backpressure will increase further. The baffle will move further toward the the nozzle. As the baffle moves, the backpressure will continue to increase. The bellows will continue to expand. The baffle will then keep moving toward the nozzle. When will this chain of events stop? Not until the backpressure equals the pressure.

Suppose the balance is upset between the baffle, nozzle, nozzle backpressure, and bellows/spring. The mechansim will drive until the the output air pressure goes to its limit. This action has been likened to a cat chasing its tail. The faster the cat runs, the faster its tail goes.

Observe how the mechanism in Figure 21-9 operates when the measurement signal is reduced. Again, the controller starts in equilibrium. First, the measurement signal is reduced. The baffle will move away from the nozzle and the nozzle backpressure will decrease. The pressure decrease in the bellows will cause the spring to move the baffle further away from the nozzle. As a reult, the nozzle backpressure will decrease further. Since the pressure is reduced, the spring will continue to move the baffle away. This chain of events will continue until the nozzle backpressure drops to zero. Again, notice how the equilibrium has been upset. There is no balance between the error measurement signal and the feedback motion. In fact, the feedback motion drives the instrument to complete unbalance. This type of feedback is called positive feedback because it adds to the original signal.

Remember the operation of the controller. It was assumed to be in balance. How was the controller brought into equilibrium? It is important to understand how an integral controller can be brought into equilibrium.

To start, apply a 9-psi pressure to the setpoint and the measurement bellows. This same pressure is applied into these two bellows. The upper end of the controller lever will assume a fixed position when this occurs. Position the nozzle from the baffle so that no backpressure can build up. Without pressure on the feedback bellows, it will collapse. The spring will then be fully extended. There will also be no pressure in the feedback bellows. Let these be the starting conditions when the nozzle is moved toward the baffle.

Now assume that the nozzle is mounted on a lever. When the lever is moved, the nozzle can also be moved relative to the baffle. Figure 21-10 shows a lever-mounted nozzle arrangement. Start with the nozzle far enough away from the baffle that there is no backpressure and so there is no pressure in the feedback bellows. Slowly move the nozzle toward the baffle and watch the output. Continue to move the nozzle even closer to the

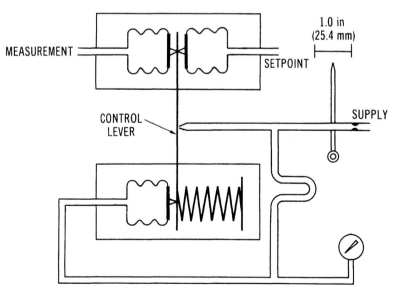

FIGURE 21-10   Reset controller with moveable nozzle.

baffle. The nozzle is close enough to the baffle to cause the nozzle backpressure to start to build up. As it begins to build, move the nozzle away from the baffle. If this is not done, the backpressure begins to build again. As it acts on the feedback bellows, it moves the baffle further toward the nozzle. A chain has begun that will cause the output to drive to 15-psi. The nozzle is then moved away from the baffle at a slower speed. It is a slower speed than that at which the bellows is driving the nozzle toward the baffle. In this way, the backpressure will increase slowly, and it can be seen when the bellows and nozzle movements coincide. At this point, the nozzle backpressure will stop building. Therefore, the bellows will cease expanding and the nozzle will also stop. Now the controller is in equilibrium.

CONTROLLER EQUILIBRIUM

This brings up an interesting question. Where is the equilibrium point? In the previous discussion, equilibrium occurred when the nozzle moved at the same rate the feedback bellows was moving the baffle. When the rates are the same, the clearance is fixed. Equilibrium is established whenever the baffle/nozzle clearance is such that there is a back-pressure—any backpressure.

The description of how an equilibrium point can be established should help to explain integral action. Notice how the baffle tends to "chase after" the nozzle. The nozzle is constructed so that as it moves, so does the baffle. This action attempts to maintain a fixed clearance.

Consider a different way of bringing the controller into equilibrium. Figure 21-11 is the same mechanism shown in Figure 21-10. The only difference is that now the output

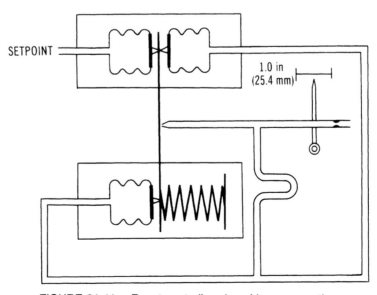

FIGURE 21-11   Reset controller, closed-loop connection.

does not stop at the pressure gauge. It is instead fed back to the measurement bellows. Therefore, as the baffle/nozzle clearance changes, the measurement input signal will also change. A controller connected in this fashion is said to be closed-loop connected. Consider how a closed-loop connected controller will operate.

Do not be confused that the measurement and setpoint bellows have been interchanged. All that has happened is that the action of the controller has been changed, that is, from direct to reverse. If the controller is to balance when closed-loop connected, it must be reverse acting.

If the setpoint pressure is zero, chances are that the nozzle will be completely uncovered. So, the pressure and the feedback pressure will also be zero. Since the measurement bellows is connected to the controller, the pressure on the measurement bellows will be zero. In any case, assume that this is the condition of the controller. Now increase the setpoint pressure.

The nozzle backpressure will increase. As the backpressure increases, pressure within the feedback bellows will also increase. This, then, drives the baffle closer to the nozzle. This operation is similar to the automatic reset action discussed earlier. Notice that as the nozzle backpressure increases, the pressure to the measurement bellows increases. The measurement pressure causes the measurement bellows to move against the setpoint bellows. As the measurement bellows moves against the setpoint bellows, it causes the baffle to move away from the nozzle. Recall that the original change in setpoint signal caused the baffle to move toward the nozzle. By simply closing the controller loop, a balancing action between the setpoint and measurement signals has been obtained. The feedback signal still tends to unbalance the controller.

The question concerning "the precise moment of equilibrium" has not been answered. In this regard, first find out if the controller will come into balance. It will then be decided at what point the controller will come into balance. If the setpoint pressure is increased, the baffle tends to cover the nozzle. This increase drives the baffle closer to the nozzle. The output pressure will continue to increase. The increase in output pressure is also applied to the measurement bellows. The measurement bellows is such as to position the baffle within 0.002-in of the nozzle. Why is this? Remember earlier discussions when the nozzle was moved by hand. It was shown that the controller could be brought into equilibrium by the rate of nozzle and baffle movement or whenever the baffle was within 0.002-in of the nozzle. The baffle is automatically brought within the 0.002-in separation because the output pressure continues to drive. The driving will not quit until the measurement bellows of the baffle is within that distance.

RESET OFFSET ▬▬▬▬▬▬▬▬▬▬▬▬▬▬▬▬▬

To drive the baffle within 0.002-in of the nozzle, a difference between setpoint and measurement pressures of 4-psi is needed. If the setpoint pressure is 9-psi, then the measurement pressure would 13-psi. The fact that there is a difference in signals is all right when talking about bringing controllers into balance. In talking about controller mechanisms, it was said that the measurement should equal the setpoint signal. But, in this instrument, the measurement is 4-psi greater than the setpoint signals. A difference of this kind is called reset offset. How can it be eliminated?

Suppose, since measurement pressure is too high, the nozzle is moved away from the baffle. The nozzle backpressure will decrease. The output and the measurement pressures will also decrease since they are the same. Moving the nozzle away from the baffle will decrease pressure on the feedback and measurement bellows. This pressure decrease will cause the baffle to move away from the nozzle. As the baffle moves away, the measurement pressure will decrease, which will cause the baffle to move toward the nozzle. Therefore, since the decrease in the measurement causes this movement, a balance between the two are reached.

The equilibrium point again must be within the 0.002-in baffle/nozzle clearance. All that was done to the nozzle when it was moved was to change its position relative to the feedback. Changing the nozzle position made it possible to bring the controller into equilibrium at a new measurement pressure. The position of the nozzle can keep on making changes until the measurement pressure equals the setpoint pressure. The controller has now been brought into equilibrium. Now that a balanced condition has been obtained, plug off the controller output. Next, connect the setpoint signal source to the measurement and setpoint bellows. When this is done, the controller is open-looped connected.

The adjusting of the nozzle relative to the feedback mechanism constitutes one of the controller alignments. Unless the nozzle is near the feedback mechanism, there will be a difference between the measurement and setpoint signals. This adjustment will be called feedback-mechanism alignment.

Look again at an open-loop connected integral controller. For the nozzle to move at the same rate as the feedback bellows, the baffle/nozzle clearance must be fixed. The controller would then come into equilibrium. The equilibrium point could occur at any baffle/nozzle clearance within the range of the baffle/nozzle. There is no need to move the nozzle at the same rate of the baffle and then stop. Now, look at the controller operation from a slightly different viewpoint. First, put a small needle valve in the line to the feedback bellows. If this adjustable restriction is closed down, the pressure in the feedback bellows will change slowly. Suppose that the pressures in the setpoint and measurement bellows are the same. This condition can easily be obtained by applying equal pressures to both bellows. With these conditions, the error signal now is zero. Again move the nozzle toward the baffle until there is an increase in backpressure. This increase in backpressure will slowly be applied to the feedback bellows because of the restricion. Just as slowly, the nozzle backpressure will increase because of the "cat-chasing-it's-tail" action. The feedback pressure continues to increase until it reaches the supply pressure.

Depending on the restriction setting, the supply pressure will equal the output in either a few seconds or several hours. The restriction will be set so that there is ample time to watch this action. As the output pressure drifts toward the supply pressure, small adjustments will be made. The adjustment causes the nozzle to move away from the baffle. After each adjustment, look at the output pressure to see if it drifts. If it does, make additional adjustments to the nozzle. If the output starts to drift in the opposite direction, the nozzle has been moved too far. At this point, it is on the other side of its 0.002-in clearance. If the output stops drifting, the nozzle has then been properly aligned to the feedback mechanism. This adjustment serves the same purpose as adjusting the nozzle with the controller closed-loop connected until the measurement and setpoint pressures are equal.

PROPORTIONAL-PLUS-AUTOMATIC-RESET ▬▬▬▬▬

Automatic reset action was described from the point of view of a mechanism. It was shown that the feedback elements were arranged relative to the baffle so that any error signal would cause the output to drive from one extreme to the other. It was also shown that the controller can be balanced at any output between 3- and 15-psi. Balancing is achieved by simply positioning the baffle relative to the nozzle. At equilibrium, the baffle is within 0.002-in from the nozzle. Consequently, the controller can be balanced at any output pressure. This is an important aspect of the automatic reset controller.

The open-loop method of connecting the controller has been discussed. Controller action using open- or closed-loop results in the automatic reset hookup being brought into equilibrium. The output of a controller not balanced and open-loop connected drives to supply or zero pressure. This depends on whether the error signal is plus or minus. The controller output which is closed-loop connected and not balanced is different from the measurement signal. The difference is called reset offset.

In the proportional-only controller, the output is always proportional to the error signal. The ratio of output to input depends on the linkage. In most cases, this linkage is adjustable, making possible a full range of gains anywhere between 250 to nearly zero. In discussing proportional controllers, the feedback element was arranged so that the feedback motion was subtracted from the error-signal motion. This was described as negative feedback. Negative feedback is opposite to the automatic reset controller. In the automatic reset controller, the feedback motion adds to the error signal. This type of feedback is called positive feedback.

Components

The components of a proportional-plus-reset controller are:

1. A comparison mechanism.
2. A detector mechanism.
3. A feedback mechanism with both negative and positive bellows/spring feedback elements.

Arrangements

The basic proportional-plus-reset controller is shown in Figure 21-12. The comparison mechanism is arranged at one end of the lever. At the opposite end of the lever are the two feedback bellows/springs. Mounted between the comparison and the feedback mechanism is the detector. Notice that there is a restriction in the line to the reset bellows. To simplify the diagram, the two feedback springs are not shown. It is necessary for the

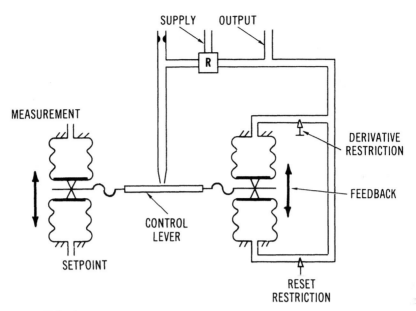

FIGURE 21-12 Elementary proportional-plus-reset controller.

actual mechanism to have either the springs or an equivalent. In some controllers, the spring properties of the bellows are used.

Operation

Controller operation begins with the same pressure on both comparison bellows and being aligned. This condition means that there is zero error signal to the detector. The baffle positions to the nozzle so there is 11-psi of backpressure. The same 11-psi is also in the proportional and reset bellows. In this case, the reset restriction is closed down, but not off. The error signal moves to cover the nozzle.

The output pressure will normally build and proportioning action will take place. The increase, in time, will pass through the reset restriction and load up the feedback bellows. This loading will move the baffle closer to the nozzle, causing a change in backpressure. These movements will continue until the backpressure reaches the supply pressure. In other words, normal automatic resetting will occur. So, the operation of the proportional-plus-reset controller is different from the proportional-only and reset-only controllers. Further, the proportioning and resetting actions are separate in time only. For shorter times, the controller acts like a proportional-only controller. For longer times, the controller takes on the aspects of a reset-only controller. The actual times depend on the setting of the automatic reset restriction.

Notice how the controller operates with the reset restriction wide open. Assume the same conditions exist as before, plus an error signal that moves the baffle towards the nozzle.

There will be an increase in backpressure that will be fed to the proportioning bellows. But, since there is no restriction in the reset-bellows line, the same increase in backpressure will be fed also to the reset bellows. The reset-bellows pressure changes at the same rate as the proportioning-bellows pressure. These two pressures cancel each other out. In other words, there will be no feedback motion. Since there is no feedback motion, the controller acts like the high-gain controller discussed earlier. Remember that in a high-gain controller, small changes will cause the output to range from 3- to 15-psi. This control action frequently is described as on/off. Regardless of the fact that such controllers are called on/off, they are narrow or high-band controllers.

Until now, cases where the reset restriction is almost closed or all the way open have been considered. But, what happens if the reset restriction is fully closed? Again assume that the controller is in equilibrium and that the error signal moves the baffle toward the nozzle.

Normal proportional action will happen and the output will increase. Since the reset restriction is closed, the change in output will not be felt by the reset bellows. Therefore, it is impossible to get any automatic reset action. For all practical purposes, by closing the reset restriction, reset has been eliminated. The controller becomes the proportional-only type.

The three cases discussed summarize the limitations of the proportional-plus-reset controller. Notice how it was possible to change the controller from a high-gain proportional controller to a proportional-plus-reset controller. As the restriction was changed from fully open to closed, the controller became a low-gain proportional-only type. For intermediate restrictions, the time it took for the resetting action to swamp out, the proportioning action was changed. The more the restriction closed, the longer the reset time. Finally, notice that there had to be an error signal before there was any controller action at all.

In the discussion of actual mechanisms, it was found that they can be related component-for-component to the fundamental types. These were discussed in the previous two chapters.

Reset time is stated as "repeats" per unit of time or vice versa. Two examples are 3-repeats per minute and 10-minutes per repeat. Three minutes per repeat is the same as one-third repeat per minute.

A repeat is an additional change in output equal to the proportional response. The proportional response is that change in output which occurs in zero time, the magnitude of which is proportional to the error signal times the controller gain.

PROPORTIONAL-PLUS-RESET CONTROLLER

A proportional-plus-reset controller is a mechanism where the two controller modes are combined. Each mode individually contributes to the output signal. The contribution varies with time, proportioning acting first and reset performing second. An important aspect of the proportional-plus-reset controller is that it can be in equilibrium at any output pressure. Unless the error signal is zero, the output will drive from one limit or the other. A controller of this type with adjustable gain will now be studied.

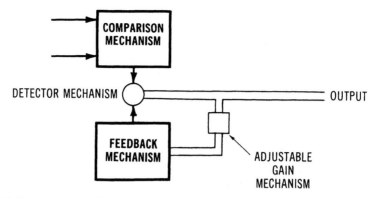

FIGURE 21-13   Block diagram of reset controller with adjustable gain.

Components

A proportional-plus-reset controller with adjustable gain consists of:

1.   A comparison mechanism.
2.   A detector mechanism.
3.   A feedback mechanism.
4.   An adjustable gain mechanism.

Arrangements

Figure 21-13 is a block diagram showing the major components of a proportional-plus-reset controller. In most respects, it is identical to the controller block diagram of Chapter 19 (Figure 19-6). Notice, however, that the adjustable gain mechanism is located between the detector and feedback mechanism. In the block diagram of Figure 19-6, there is no adjustable gain mechanism. The two input signals are air pressures applied to opposing bellows. These two bellows drive one end, stated as "repeats" per unit of time or vice versa. Two examples are 3-repeats per minute and 10-min per repeat. Three minutes per repeat is the same as one-third repeat per minute.

A repeat is an additional change in output equal to the proportional response. The proportional response is that change in output which occurs in essentially zero time, the magnitude of which is proportional to the error signal times the controller gain.

PROPORTIONAL-PLUS-RESET CONTROLLER ▬▬▬▬▬▬▬

The proportional-plus-reset controller combines the two controller modes. Each mode contributes to the mechanism's output signal. The contribution varies with time, the proportioning bellows acts first and the reset acts next. The important aspects of the proportional-plus-reset controller are that it can be in equilibrium at any output pressure. Unless the error signal is zero, the output will drive from one limit to the other.

Components

1.   A comparison mechanism.
2.   A detector mechanism.
3.   A feedback mechanism.
4.   An adjustable gain mechanism.

Arrangements

Figure 21-13 is a block diagram showing the major components of a proportional-plus-reset controller. In most respects, it is identical with the controller block diagram of Figure 19-6. Notice, however, that the adjustable gain mechanism is located between the detector and feedback mechanism. In the block diagram of Figure 19-6, there is no adjustable gain mechanism. The two input signals are air pressures applied to opposing bellows. These two bellows drive one end of the control lever. At the other end of the lever are the feedback components, which also are two opposing bellows. The detector is the baffle/nozzle/relay combination. The output of the relay is the output of the controller. The output is fed into the adjustable gain mechanism. This mechanism has not been discussed. For the time being, consider it to be a fixed gain mechanism. Further into the chapter, discussion will center around the construction and operation of this mechanism.

Figure 21-14 is a component diagram of the controller. Compare this with Figure 21-12. Notice that, except for the gain mechanism, the two units are identical. The con-

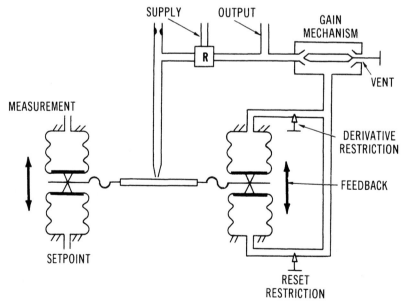

FIGURE 21-14   Diagram of control mechanism in a Multitrol controller. (*Courtesy of Fisher Control International, Inc.*)

troller of Figure 21-14 is arranged with the control lever positioned horizontally. A photograph of the internal mechanism of the controller is shown in Figure 21-15.

Operation

Operation of the Fisher controller is essentially the same as that of other instruments. These controllers are open-loop connected. Now look at the closed-loop connected controller. Recall that the term closed-loop means that the output is connected to the measurement bellows. Assume that the controller has a zero air supply. Also assume there is a 9-psi setpoint signal applied to the setpoint bellows. Observe what happens when the air supply is applied.

Since there has been no air supply, the pressure in the measurement, proportioning, and reset bellows is zero. Immediately before the air supply is turned on, there is a large error signal, causing the baffle to be positioned against the nozzle and permitting backpressure to build. The increase in backpressure drives the relay, causing the output air pressure to build. The output air pressure connected to the measurement bellows produces a measurement force that opposes the setpoint force and reduces the error signal.

An increase in output also acts on the proportional bellows. The action tends to move the baffle away from the nozzle. The input will quickly jump to that value which will automatically balance the setpoint signal. This occurs when the input is applied to the measurement and proportioning bellows.

Now look at the reset bellows. The increase in pressure when turning on the air supply will slowly pass through the reset restriction. It will then enter the reset bellows. The bellows will return the feedback end of the lever to a position when there was no air supply. This happens because having the same pressure in the proportioning and reset

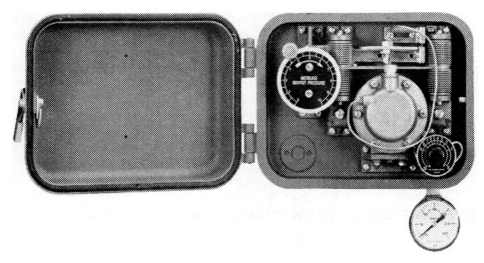

FIGURE 21-15   Photograph of a Multitrol controller. (*Courtesy of Fisher Control International, Inc.*)

bellows is the same as having zero pressure in these bellows. Therefore, the feedback end of the lever will have assumed the same position that it had when the controller had no supply.

ALIGNMENT ▰▰▰▰▰▰▰▰▰▰▰▰▰▰▰▰▰▰▰▰▰▰▰▰

If the pressures in the measurement and setpoint bellows become equal, the comparison end of the lever will assume a predetermined position. Here, a setpoint signal of 9-psi was applied. The measurement must also be equal to 9-psi because the error signal is going to be zero. The controller output will drive whatever output is necessary to position the baffle within 0.002-in of the nozzle. The controller could not care less what the measurement signal is and whether or not it equals the setpoint signals. All the controller can do is drive the measurement until the baffle is within 0.002-in of the nozzle.

There is a need for a controller to be balanced with the measurement and setpoint equal. The nozzle must be positioned to within 0.002-in of the instrument center line. The baffle falls on this center line. The two sets of bellows loaded with the same pressure cause the controller lever to fall on the center line.

Assume now that the air supply is turned on. If the measurement pressure does not equal the setpoint pressure, the nozzle is moved until they equalize. When these conditions are met, the controller is said to be aligned. A test of controller alignment is that the measurement equals the setpoint in all levels. One of the tasks associated with controllers is to make adjustments be able to pass the alignment test.

Look at how the components are lined up on the center line of the controller. The total line is determined when two points are established. If there are any more than two points, a straight line cannot be expected. When these two points are established, the third point is automatically predetermined.

The controller has three points, all of which fall on the center line. They are the feedback, the detector, and the comparison mechanism points. The feedback point is fixed, after a length of time. The nozzle is fixed relative to the baffle by its position on the instrument chassis. The only unfixed point is the comparison mechanism point. Therefore, the comparison mechanism must drive so that its point falls on the line established by the other two. For example, the same pressure in both of the comparison mechanism bellows is needed. But, the resulting bellows position has to be accepted. The nozzle, instead, becomes the third point that must be put on the center line.

If this action is understood, there is no problem in understanding any automatic-reset controller. The Fisher controller is relatively easy to grasp because it is symmetrical about a center line. In following chapters, controllers that are not symmetrical will be discussed.

PNEUMATIC GAIN MECHANISM ▰▰▰▰▰▰▰▰▰▰▰▰▰▰▰▰▰▰

In earlier work, mechanical-gain mechanisms of the parallel-lever and angle-mechanism types were discussed. Some manufacturers use what is called a pneumatic gain mechanism. A pneumatic gain mechanism is essentially a three-way valve. This mechanism is arranged so that as it is adjusted, a vent or port closes as the inlet port opens. The valve

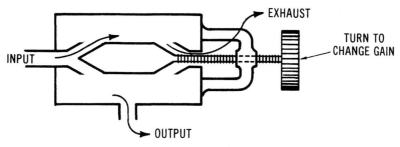

FIGURE 21-16   Pneumatic gain mechanism.

can also work in the opposite way. Between these two ports is an output connection. Figure 21-16 shows a pneumatic gain mechanism. Figure 21-14 shows the same mechanism installed in the controller circuit.

Recall that the function of an adjustable gain mechanism makes it possible to change the ratio of output to input. The output of the pneumatic gain mechanism is the pressure between the two ports. The input to the gain mechanism is the relay output.

If a mechanism can change the ratio of output to input, it is an adjustable gain mechanism. Determine if the mechanisms shown in Figure 21-14 and 21-15 will change the ratio.

Assume that the plug is positioned so that it closes the exhaust port. For that plug position, all of the relay output will pass by the plug and out into the feedback bellows. For this plug position, the output equals the input and the gain is 1. Turning the adjustment to slightly open the exhaust port will close down the input port. Part of the input signal will then be released into the atmosphere. As a consequence, the output pressure will be less than the input pressure. In other words, the ratio of output to input will be changed. Even though the mechanism is essentially a three-way valve, by adjusting relative port openings, the ratio of output to input can be changed. Therefore, such a mechanism is an adjustable gain mechanism.

A pnematic gain mechanism is an important type of mechanism and is widely used in the "stack type" force balance controller. However, this type of controller will not be discussed at any length in this discussion.

DERIVATIVE MODE CONTROLLERS ▰▰▰▰▰▰

In the first part of this chapter, proportional control and reset control were discussed. The two were then combined to form a proportional-plus-reset controller. A proportional-plus-reset controller manufactured by Fisher Controls Company was then discussed in detail. In this controller, there was a restriction in the line to the proportioning bellows. In the previous discussion, it was assumed that the restriction was wide open and could be neglected. The response of this restriction to derivative control is very important and will be discussed at this time. As is so often the case, the derivative controller to be discussed is described by various terms. The term derivative shall be used to describe this controller. Some manufacturers use the term rate action to describe this condition of operation. Derivative mode is a mathematical description of the controller's function.

Components

Only one component is needed in addition to the components of a proportional or proportional-plus-reset controller. This will obtain a proportional-plus-derivative or a proportional-plus-reset-plus-derivative controller. That component is an adjustable restriction. A variable capacity is added to stabilize the output.

Arrangements

The adjustable restriction or the variable capacity is placed in line to the proportional bellows. Figure 21-17 is a diagram of a proportional-plus-derivative controller. The only difference between this diagram and the diagram of the proportional-only controller is the presence of the restriction in the line to the proportioning bellows.

Operation

For example, assume that the reset restriction is wide open. The operation of the proportional-plus-derivative controller is identical with the operation of any proportional-only controller. But, if the derivative restriction is closed, this eliminates the proportioning feedback motion. Therefore, the controller will act like a high-gain controller. The restriction is adjusted from open to closed. The characteristics of the controller have been changed. Now, the controller has no feedback. On the surface it may appear that the gain is being changed.

Assume that the controller has the derivative restriction closed down but not shut off. Also assume that an error signal moves the baffle toward the nozzle. The nozzle back-pressure increases and so will the output. The derivative restriction is almost closed. The

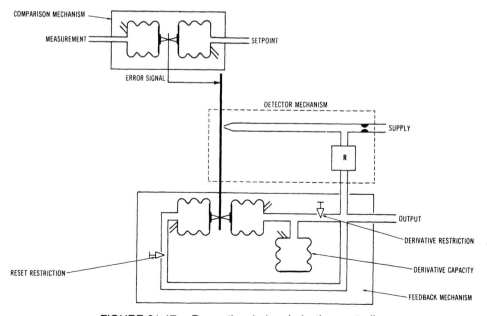

FIGURE 21-17 Proportional-plus-derivative controller.

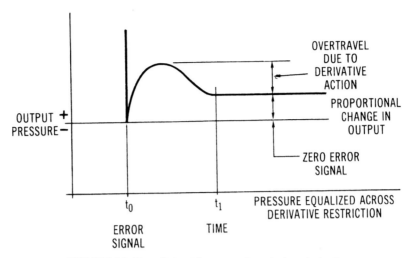

FIGURE 21-18   Output in proportional-plus-derivative.

increase in output will not be immediately felt by the proportioning bellows. For a while, the controller will act like a high-gain type. But, the increase in nozzle backpressure will work past the restriction and load up the proportioning bellows. This will permit the bellows to operate in a normal way. The pressure increase in the proportioning bellows will move the baffle away from the nozzle. This will decrease the output.

Following a longer wait, the pressure within the proportioning bellows and the output pressure will equalize across the derivative restriction. When this happens, the output pressure will be directly proportional to the error signal. Since there is no difference in pressure across the restriction, the circuit acts as if the derivative restriction were not there.

The derivative restriction "holds back" the proportional feedback motion, causing the output to change to an amount greater than if the restriction was not there. For the amount of time the derivative restriction is present, controller gain is not increased.

As was the case with automatic reset, the effect of the derivative is time-dependent. Given enough time, the derivative has no effect. In Figure 21-18, a plot of output pressure versus time is shown. Notice that the output pressure on a change of error signal inceases almost immediately. This is beyond the time that proportional-only would cause. But, with time, the output reduces to an amount that would be obtained with a proportional-only controller.

DERIVATIVE IN TRANSMITTERS

The use of derivative action is not limited to controllers. Certain transmitters, especially temperature transmitters, use derivative action. In those cases, a restriction is placed in the line to the feedback bellows. Figure 21-19 is a diagram of a temperature transmitter with derivative action. Operation of a derivative transmitter is similar to the operation of a controller.

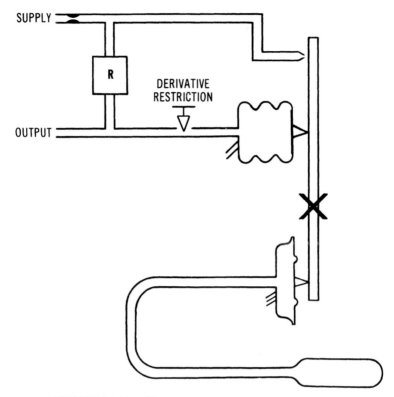

FIGURE 21-19   Temperature transmitter with derivative.

Suppose that the measurement changes. In this case, the derivative restriction slows the rate at which the balancing bellows can act. The balancing motion is then delayed in time. The output of the transmitter is then greater than the change in measurement. Given time, the transmitter will come to balance at an output pressure that balances the new measurement signal. Temporarily, the output travels an amount greater than expected when compared to the input.

The reason for arranging the transmitter so that the output overtravels the input will be discussed later. For now, it enables the controller to handle rapid process changes more quickly.

SUMMARY

When a controller has a high level of gain, a small change in the measurement input will result in full travel of the output. In general, controller gain must be reduced to permit the controller to have a more effective range of operation.

By incorporating a bellows/spring component, it is possible to obtain an adjustable gain. Mechanisms of this type are used to adjust gain from 250 to 1. Two methods of controlling gain are the parallel-lever type and the angle-gain mechanism.

The output of an automatic reset controller will continue to drive until the baffle comes within 0.002-in clearance of the nozzle. Unless and until the baffle is within the 0.002-in clearance, the output will continue to change. If there is a plus error signal, the controller will drive in one direction until the measurement crosses the setpoint. The action will continue in the reverse direction. If there is a minus signal, the controller will drive in the opposite direction. When there is a zero error signal, the controller output will be constant. This means that such a controller can come to equilibrium at any output pressure. The output will drive in one direction until the measurement crosses the setpoint. When it does, the drive will reverse direction.

If the controller is aligned, the output will drive until the measurement equals the setpoint.

The operation of the proportional-plus-reset controller is different from the proportional-only and reset-only controllers. The proportioning-plus-reset controller combines the operational features of both controllers in a single instrument.

In the operation of the proportional-plus-reset controller, it is possible to change the controller from a high-gain proportional controller to a proportional-plus-reset controller. A low gain proportional-only controller has the restriction changed from fully open to fully closed. For intermediate restrictions, the time it takes for the resetting action to swamp out the proportioning action was changed. The more closed the restriction, the longer the reset time. It should have been evident that there had to be an error signal generated before there could be any controller action at all.

The nozzle derivative-mode controller temporarily prevents normal proportioning or balancing action. This is done by introducing a restriction in the line to the proportioning or balancing bellows. By adjusting the restriction, the time that it will take before normal proportioning occurs can be changed. The effect of this is to cause the output to overtravel.

ACTIVITIES

CONTROLLER MECHANISMS ▬▬▬▬▬▬▬▬▬▬▬▬▬▬▬▬▬▬

1. Refer to the operational manual of the instrument to be used in this activity.
2. Remove the housing or covering from the instrument.
3. Locate the gain mechanism.
4. Identify the type of gain mechanism used.
5. Explain the operation of the mechanism.
6. Identify the components of the gain mechanism.
7. If possible, apply air to the instrument while monitoring the output.
8. Adjust the setpoint to a usable value and then change the value of the measurement input.

9. Is there a physical change in the operation of the mechanism when there is a variation in the measurement value?
10. Describe any physical change that takes place in the gain mechanism.
11. Reassemble the instrument.

RESET FUNCTION

1. Refer to the operational manual of the instrument used in this activity.
2. Remove the housing or outside covering from the instrument.
3. Locate the reset mechanism of the instrument.
4. What are the components of the reset mechanism?
5. Identify the measurement and setpoint inputs of the reset mechanism.
6. Locate the control lever and nozzle of the mechanism.
7. If possible, apply air to the input and monitor the input and output pressure of the instrument.
8. Adjust the setpoint to a usable value that will produce equilibrium.
9. Change the value of the setpoint while observing the output pressure.
10. Describe the response of the instrument when a change in equilibrium occurs.
11. Reassemble the instrument.

QUESTIONS

1. What are the components of a low-gain controller?
2. What are some of the common types of adjustable gain mechanisms?
3. What are the components of a reset controller?
4. What does controller equilibrium describe?
5. What is reset offset?
6. Explain the operation of the elementary proportional-plus-reset controller of Fig. 21-12.
7. What is meant by the term derivative mode of control?
8. What is the time function of derivative control?

22
Controllers

OBJECTIVES

Upon completion of this chapter, you will be able to:

- Explain the operation of different pneumatic controllers.
- Identify the fundamental parts of a controller.
- Translate a fundamental diagram into the actual components of a controller.
- Identify the basic operating features of a controller.

KEY TERMS

Adjustable sensitivity.   A variable detector mechanism that alters the ratio of output to input signal control.

Air-to-close.   A condition indicating that when air is applied to a control element, it will close.

Bellows in a can.   A self-contained bellows mechanism housed in a metal enclosure.

Capsular air chamber.   A small, compact receptacle that holds or stores air for system operation.

Comparison bellows.   A bellows assembly that distinguishes between the measurement and setpoint signals applied to it.

Eccentric.   A mechanical connection or device that is offset from a conventional center.

Error-signal.   A change in air pressure that is the result of a difference between measurement and setpoint values in the operation of a controller.

Onstream time.   The time that a final control element or device is placed in operation by connecting it to the source of air supplying the system.

Pen assembly.   A mechanism for scribing a mark on a chart or paper.

Sensitivity.   A ratio showing a change of output to a change of input. The least amount of input signal that is capable of producing an output.

INTRODUCTION ▬▬▬▬

In the preceding chapters controller fundamentals, operational theory, functions, and internal mechanisms were discussed. Some general concepts about the operation of a controller and its importance to industry were developed. A controller is the decision-making element of an automated system. It tells the final control element what to do and when to do a specific operation. A pneumatic controller performs its functions through the control of air from a central source.

There is a wide variety of pneumatic controllers currently being used in industry. These instruments all have the capability of exercising some form of automatic control over a process variable. In general, there is a great deal of similarity in pneumatic controllers. They can respond as on-off devices, have proportional-only control, and be equipped to perform two or three modes of operation such as proportional-plus-reset or proportional-plus-reset-plus-derivative control. In spite of their similarities, there are many differences in pneumatic controllers. These include type of linkage employed, the geometry of the components, and the construction of the comparison mechanism. Many controllers are of the motion-balance type. They employ some type of angle-gain mechanism. These instruments are available in two housing styles, such as the miniature plug-in unit or a large case design that houses the entire assembly.

In this chapter, the operation and unique features of several controllers shall be considered. The discussion will start with a complete controller and relate it to a fundamental diagram. This approach is an attempt to emphasize the relationship of controller parts and subassemblies to one other. It is not intended to single out specific controllers in order to point out their good and bad features. It is, however, intended to show some of the unique features of a controller that distinguish one instrument from other.

FOXBORO CONTROLLERS ▬▬▬▬

The Foxboro Model 40 controller is a motion-balance instrument. It uses a linkage comparison mechanism, an angle gain mechanism, and a baffle/nozzle relay combination as the detector. This particular series of instruments is available in single operational modes and combined assemblies. Combined units have two or three modes of control. This discussion centers around an instrument that achieves only one of these modes of control. It will be identified in the following discussion.

Arrangements and Components

Figure 22-1 is a photograph of the Foxboro Model 40 controller. Let us first attempt to determine whether this controller is a proportional-only, a proportional-plus-reset, or a proportional-plus-reset-plus-derivative controller. If it is a proportional-only controller, it

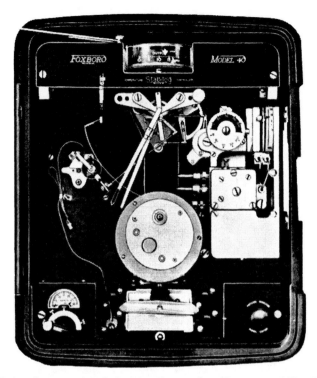

FIGURE 22-1   A proportional-plus-reset controller. (*Courtesy of The Foxboro Co.*)

will not have adjustable restrictions in the instrument. The feedback mechanism consists of a bellows/spring. A *proportional-plus-automatic-reset* controller will have one adjustable restriction and two bellows in the feedback mechanism. If it is a proportional-plus-reset-plus-derivative controller, there will be two adjustable restrictions and two bellows in the feedback mechanism.

Having this information is sufficient to answer the question that will establish the number of modes. The location is determined by the elements comprising the comparison, detector, and feedback mechanisms. In addition, it must be determined if an adjustable gain mechanism is used. If one exists, what kind is it?

An examination of Figure 22-1 shows two scales in front of the instrument in addition to the scales marked supply and output. One of the scales is circular. The other is a strip gauge. These scales make it appear as if there were two adjustments, indicating that the controller has an adjustable-gain mechanism and a reset restriction.

If the controller has reset, the feedback mechanism should consist of two opposing bellows. Figure 22-2 is a photograph that shows the feedback mechanism. Notice that there are two bellows. Therefore, it can be concluded that the subject instrument is a proportional-plus-integral-plus-derivative controller. Also notice in Figure 22-2 that the gain mechanism is an angle-type gain mechanism. It is an angle-type gain mechanism because the gain dial is rotated. The nozzle and the baffle rotate about an axis. This

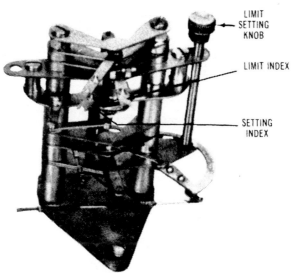

LIMIT
SETTING
KNOB

LIMIT INDEX

SETTING
INDEX

FIGURE 22-2   Feedback mechanism of a controller. (*Courtesy of The Foxboro Co.*)

conclusion is supported by the fact that the error-signal link is at right angles to the feed-back motion. This configuration implies it is an angle-type gain mechanism.

The error signal and the feedback signal coverage on an adjustable gain mechanism have been located. The error signal originates at the comparison mechanism. If the error link is followed, it is directed to what might be a comparison mechanism. The mechanism certainly is not made up of bellows. Clearly, it consists of linkage. To determine if it is a comparison mechanism, the linkage mechanism must meet the test of the comparison mechanism. Recall that the purpose of the comparison mechanism compares the measurement to the setpoint signals. If there is no difference, the output of the comparison mechanism is zero. In other words, there is a zero error signal. If the measurement is below the setpoint, there is a minus error signal. Increasing or decreasing the setpoint the same amount should result in the same error signal.

A study of the linkage of the comparison mechanism in the Foxboro Model 40 controller will show that it is a type-2 motion-balance linkage. Recall that the type-2 motion-balance linkage consists both of fixed and floating levers. On all motion-balance linkages, there are two inputs. The output is taken from the floating lever. Any controller using a linkage-type comparison mechanism, the two inputs are the measurement and the setpoint signals.

In motion-balance linkage, there will not be any change in output if the two signals are moved at the same time. Yet, if the same value of measurement or setpoint signals do change but are still the same, the output still will not change. If the measurement signal increases, the error signal will have one polarity. But if the measurement signal decreases, the error signal will have the opposite polarity. Figure 22-3 is a series of diagrams showing the relationship of the actual to the fundamental mechanisms.

The examination of the controller has shown that the comparison mechanism is a linkage-type mechanism. The feedback mechanism is a two-bellows mechanism having

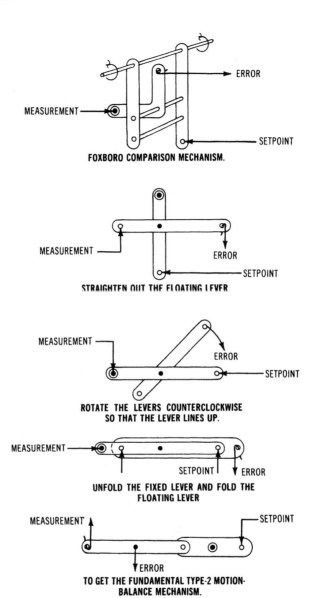

MEASUREMENT

ERROR

SETPOINT

FOXBORO COMPARISON MECHANISM.

MEASUREMENT

ERROR

SETPOINT

STRAIGHTEN OUT THE FLOATING LEVER

MEASUREMENT

ERROR

SETPOINT

**ROTATE THE LEVERS COUNTERCLOCKWISE
SO THAT THE LEVER LINES UP.**

MEASUREMENT

SETPOINT

ERROR

**UNFOLD THE FIXED LEVER AND FOLD THE
FLOATING LEVER**

MEASUREMENT

SETPOINT

ERROR

**TO GET THE FUNDAMENTAL TYPE-2 MOTION-
BALANCE MECHANISM.**

FIGURE 22-3 Relationship of fundamental to actual detector controller mechanism. (*Courtesy of The Foxboro Company*)

both positive and negative feedback. This means that it also has reset and proportional modes of operation. The gain mechanism is the angle-type gain mechanism. The detector is the baffle/nozzle and relay combination. The relay is mounted under the proportional-band dial.

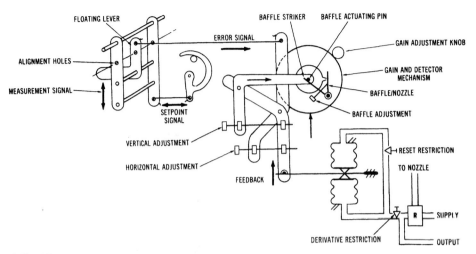

FIGURE 22-4   Essential elements and adjustments of a controller. (*Courtesy of The Fox-boro Company*)

Figure 22-4 shows all of the essential elements and adjustments. Some of these elements have been distorted, particularly the ones around the gain mechanism. Functionally, the diagram is the same as the actual mechanism.

Operation

Any increase in the setpoint signal will raise one end of the floating lever to the left. Now the error signal is also to the left. The error signal positions one end of the control lever. The other end of the control lever is positioned by the feedback signal. The control lever consists of three levers. These are pinned and locked together. The essential lever, or where positioning takes place, is a straight line between the error and feedback signals.

The leftward movement of the upper end of the control lever moves the baffle striker to the left. This permits the baffle to cover the nozzle. The nozzle backpressure increases and the relay output pressure increases; this increase in output first acts on the proportional bellows. It causes an upward movement of the feedback bellows. It also causes an upward feedback signal, which lifts the control lever vertically. At the same time, the signal lifts the baffle actuating pin. The striker forms an angle with this vertical movement. The baffle striker is permitted to move to the right, uncovering the nozzle. Remember that the original error signal caused the baffle to cover the nozzle.

The actuating pin, or the baffle, now comes to balance and causes a baffle/nozzle clearance that results in an output pressure. Since it acts through the feedback mechanism, the error signal is balanced. But this change in output also acts on the reset bellows and causes the feedback mechanism to return the feedback signal to another position. This is the position it was in before the setpoint signal was changed. As a result of these interactions, reset drifting will continue in an effort to maintain a feedback signal. The reset bellows will undo this earlier effort.

The original error signal was in the direction to cause the baffle to cover the nozzle. Also, since the nozzle is direct-acting, the output will drive due to reset action. Remember that output driving in a reset controller can be prevented by reducing the error signal to zero.

The error signal can be reduced to zero only if the measurement moves to the setpoint. Assume now that the output of the controller is connected to a control valve in a process. As the valve moves, the measurement will change. This will continue to happen until it reaches the setpoint. When it reaches the setpoint, the output of the controller will stabilize the control valve. The valve will then reach a fixed position, and the measurement will be in line with the setpoint.

CONTROLLER ALIGNMENT

If a measurement is to line up with the setpoint value, the controller must be aligned correctly. Alignment means that the comparison, feedback, detector, and gain mechanisms must all interact properly.

Before considering the total alignment of a controller, each individual mechanism must be aligned. Having done this, the individual mechanisms are then aligned as a group. When the measurement and the setpoint are together, the output will not drift. This means that it will be possible to rotate the gain dial without changing the output.

Comparison Mechanism Alignment

Comparison mechanism alignment is accomplished by inserting an alignment pin through the holes provided in the mechanism linkage. Having done this, assume that the setpoint and measurement indexes do not agree. The micrometer on the setpoint index is then rotated, which permits the measurement index to agree with the setpoint index.

Consider now the feedback mechanism. When the pressure is zero, the pressure in both bellows is the same. This causes the bellows to assume a fixed position. Since the pressures in the individual bellows cancel out each other, it makes no difference what the pressure level is. The bellows will return to the same fixed point when the pressure equalizes. As a consequence, the point at which the feedback signal is applied to the control lever is fixed. There are no adjustments available to change this point. This point becomes one of the fixed points of the three necessary to establish the alignment center line of the controller and leaves the two other points to consider.

When the comparison mechanism is aligned with the alignment pin, it fixes a second point because the error signal length is not adjustable. Now two points of the necessary three have been fixed. If the controller is to be brought into alignment, the only remaining point is the detector position. The nozzle position of the detector is fixed, but the baffle is adjustable. The adjustment of the baffle makes it possible to align the detector relative to the comparison and feedback mechanisms. The remaining mechanism needing alignment is the gain mechanism.

Gain Mechanism Alignment

Alignment of the gain mechanism determines if it is possible to change the gain when there is a zero error signal. Alignment is done without having the output change with the gain adjustment. If there is no error signal, this mechanism should be able to change the gain without changing the output. This means that zero times any number is zero. It is possible to change the gain of an angle-type gain mechanism without the output changing. This adjustment occurs only if a gain dial change does not alter the baffle/nozzle clearance. If it does, the output must also change.

Baffle/nozzle clearance will not change if the baffle actuating pin is centered on the axis of the gain mechanism. If it is on the axis of rotation, the baffle striker will ride around the pin. The baffle will remain in a fixed position relative to the nozzle.

To test for gain mechanism alignment, rotate the gain mechanism. Maintain a zero error signal and look at the output. If the output changes, the actuating pin is not centered. If the pin is centered, it must be free to move both vertically and horizontally. Examine the diagram. Notice that as the vertical adjustment is rotated, it moves one end of a two-sided 90° lever horizontally. The other side of the lever that carries the actuating pin moves vertically and constitutes the vertical adjustment.

Consider the horizontal adjustment. Remember to disconnect the error-signal link. Also, hold the feedback end of the lever in a fixed position. Change the horizontal screw. As it is turned, the error-signal end of the control lever will move horizontally. Now reconnect the error-signal link.

Recognize that the comparison end of the error-signal end of the control lever is fixed. The feedback end of the control lever is also fixed. When the horizontal adjustment is made, the middle of the controller lever tends to bulge. The bulge can either be to the right or the left, depending on which way the horizontal adjustment screw is turned. As it bulges, it moves the actuating pin horizontally. Adjusting these vertical/horizontal screws will position the baffle-actuating pin and permit it to be centered on the axis of rotation of the angle-gain mechanism.

Alignment Principles

As is the case with instrument adjustments, there is a certain amount of interaction between the adjustments which can be complicated. In addition, another complication arises because the detector mechanism is used to determine if the gain mechanism is aligned. The very thing used to determine if the gain mechanism is aligned may not itself be aligned. This not an uncommon problem in alignment. It is not limited to the alignment of the gain and the detector mechanisms. Comparable problems occur when aligning feedback and comparison mechanisms.

The problem is solved by alternately correcting the alignment of each of the two interacting mechanisms. In other words, corrections are alternately made to the baffle and to the gain mechanisms. A good rule to remember is that one adjustment can adjust for one condition only; the adjustment cannot be used to achieve two alignments.

SPECIAL ASPECTS ▰▰▰▰▰▰▰▰▰▰▰▰▰▰▰▰

The Foxboro Model 40 controller differs from other controllers in certain areas. First, notice that the derivative restriction is in series with the reset restriction. This means that a change in backpressure must pass through the derivative restriction before it can pass through the reset restriction. An instrument so arranged is described as a series controller. It differs from the parallel controllers previously diagrammed.

Overall, there is no essential difference between a series and a parallel controller. The difference arises in the frequency response of the controllers.

A second aspect involves the measurement micrometer. It is possible to change the measurement micrometer without introducing errors in the controller alignment due to the location of the micrometer adjustment. It is arranged so that changes are made on the measuring side of the comparison mechanism rather than on the display side.

THE FOXBORO 58 CONTROLLER

The Foxboro Model 58 controller is a pneumatically connnected instrument. It is classified as a force-balance controller. Remember that transmitters are classified as motion-balance or force-balance instruments. Force-balance instruments can be either the moment-balance type or true force-balance type. Motion-balance instruments can be either linear motion-balance or angle motion-balance. The same kind of categorizing is possible with controllers.

True force-balance controllers are constructed so that the forces directly oppose each other. There are no levers involved. This type of controller is frequently referred to as a stack controller.

Although it was not mentioned before, the instruments previously discussed were angle-type motion-balance controllers. They were motion-balance because the feedback and error signal do not directly load each other. They are the angle-motion type because the motions act at distances from a pivot. In other words, levers are involved.

Components

The Foxboro Model 58 controller consists of the typical controller components. That is, a comparison, a feedback, a detector, and adjustable gain mechanisms.

Arrangements

The comparison mechanism consists of two bellows applied to a plate. The feedback mechanism also consists of two bellows and is applied to the same plate. This plate is a two-dimensional control lever. The plate is located below a reference plate. It is essentially part of the instrument chassis. It serves as a reference for all the controller components. Mounted in the reference plate is the nozzle. The nozzle is arranged to look at the position of the force plate control lever. This is driven by the four bellows. That part of the force plate under the nozzle is the baffle. Between these two plates is a bar-type fulcrum. This fulcrum can be rotated 90°. At one extreme, it centers on the feedback bellows. At the other, it centers on the comparison bellows.

As the fulcrum rotates, it changes from the centered feedback position to the centered comparison-bellows position. The relative contribution of the feedback signal to the error signal is changed. Figure 22-5 is a simplified diagram of the controller showing the arrangement of the components. Figure 22-6 diagrams the gain mechanism. Figure 22-7 is a photograph of the controller mechanism.

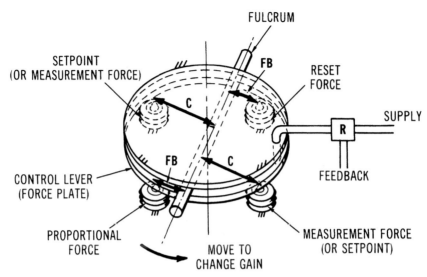

FIGURE 22-5   Simplified diagram of a controller mechanism. (*Courtesy of The Foxboro Company*)

The gain mechanism used in this controller is a parallel-lever gain mechanism. Do not let the two-dimensional construction of the lever system hide its characteristic. Notice that as the fulcrum is moved, the lever lengths of the two pairs of bellows relative to the pivot are changed. When the bellows remain fixed, the lever lengths over which they act change as the fulcrum is rotated.

Operation

Suppose that the controller is in equilbrium and the measurement signal increases. An increase in measurement signal will lift the force plate toward the nozzle. The nozzle backpressure will increase. The output will increase, and the proportional force will increase. Since the proportional bellows is on the opposite side of the fulcrum, it tends to move the baffle away from the nozzle. This causes the feedback lever to be shorter than the comparison lever. There must be a greater pressure change on the proportioning bellows than on the measuring bellows if the forces are to balance. An important point to recognize is that the feedback force directly opposes the measurement force. Since the lever lengths are different, the air pressure will be different.

Now examine the controller, with the fulcrum positioned so that the feedback lever and the comparison lever lengths are the same. They will be the same when the fulcrum forms a 45° angle relative to the bellows center lines.

Assume that there is an increase in measurement. The increase in measurement will move the baffle closer to the nozzle and cause an increase in the output. This, acting on the proportioning bellows, will result in a force opposing the change in measurement force. It tends to move the baffle away from the nozzle. The change in measurement force

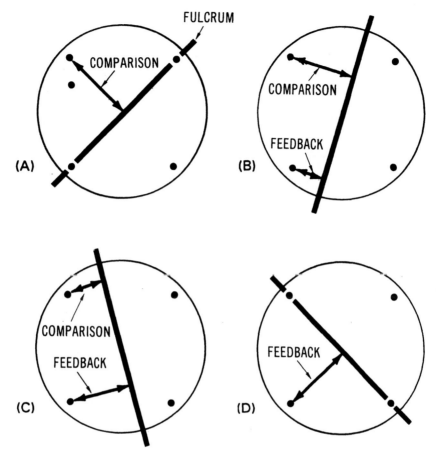

FIGURE 22-6 Different settings of gain mechanism. (A) Maximum comparison; minimum feedback. (B) High comparison-to-feedback ratio; (C) Low comparison-to-feedback ratio; (D) Minimum comparison; maximum feedback.

moves it toward the nozzle. The output will change until the forces on the force plate come into equilibrium.

Suppose that the controller is connected open-loop. Because of the peculiarities in the control loop, the measurement is not changed as the controller output is changed. In other words, suppose there is a difference between the measurement and setpoint pressures, hence, a difference in force. This difference in force is an error signal.

When the measurement changes, the output changes, which changes the proportioning bellows at the feedback mechanism. It tends to restore the baffle to its position prior to the change in measurement. However, this same change in output is also applied to the reset bellows. Depending on the reset restriction, the pressure in the reset bellows will equal the pressure in the proportioning bellows.

FIGURE 22-7   A controller mechanism. (*Courtesy of The Foxboro Co.*)

When this happens, the feedback-force signal will go to zero. There will be no coun-
terbalancing force to the error-signal force. Therefore, the output will continue to change
in an effort to balance out the error signal and can be recognized as normal reset action.

As a consequence, the controller output will drive to one limit or another unless the
error signal is eliminated. The error signal can be eliminated only if the measurement
pressure becomes the same as the setpoint pressure.

Since the gain mechanism is unique, it will be worthwhile to study it further. Figure
22-6 is a diagram showing several positions of the fulcrum relative to the bellows. If the
fulcrum lines up with the feedback bellows, as in (A), there can be no feedback signal.
This is because the force plate cannot be moved. The controller has an extremely high-
gain. If the fulcrum is rotated, it lines up with the comparison mechanism, as in (B).
There can be a large difference between the measurement and setpoint pressures. None of
this difference in pressure will be detected by the nozzle because the measurement and
setpoint bellows cannot move the force plate. Assume now that the fulcrum is rotated
between these two limits, as shown in (C) and (D). The comparison lever length becomes
shorter as the feedback lever becomes longer. Hence, the amount of feedback signal required to balance a given error signal changes. This, of course, represents a change in
gain of the controller.

Alignment

There are two alignment adjustments. These consist of springs that load the force plate. One spring is on the center line of the comparison mechanism bellows. The other is located on the center line of the feedback bellows.

The alignment test for the Model 58 controller is identical to that of any other adjustable-gain, pneumatic-reset controller. That is, does the measurement agree with the setpoint for all lever or setpoint pressures? This controller should be aligned closed-loop connected.

It will be possible to change the gain without the output changing. This occurs only if the force plate is parallel to the reference plate. If it is to be made parallel, two adjustments are required. If the comparison is to be aligned, the effective area of the bellows must not change with the internal pressure. Any small difference in an effective area can be corrected by making adjustments to the spring.

In order to align the gain mechanism, the two plates are made parallel by adjusting the springs. It will not be possible to also align the comparison mechanism by adjusting the same springs. It is not possible to correct the two faults with only one alignment. Either the gain alignment or the comparison-mechanism alignment must suffer. If both mechanisms are not in alignment, trouble is indicated in the bellows or in the gain mechanism.

It is important to recognize that a misaligned controller indicates faulty components. This is true regardless of the controller. A controller, once aligned, should not go out of alignment. Depending on the individual controller design and manufacturing policies, there may up to twelve alignment adjustments. The lack of alignment adjustments in no way changes the definition of whether or not the controller is aligned. Therefore, the number of adjustments does not change the test that is made to determine alignment. If a controller is not aligned, several adjustments should straighten out the problem. But, keep in mind that individual components may also cause the problem. These problems should be located and corrected. There is no need to continually realign controllers any more than there is reason to continually recalibrate measurements.

HONEYWELL CONTROLLER ▬▬▬▬▬▬▬▬▬▬▬▬▬▬▬▬▬▬▬▬▬

The Honeywell Air-O-Line controller is a motion-balance instrument that uses a linkage comparison mechanism. Note that the feedback mechanism, though functionally equivalent to other mechanisms, is ingeniously arranged and is quite unique.

Components and Arrangements

Figure 22-8 is a pictorial diagram of the Honeywell Air-O-Line controller. The derivative unit of this controller is made up of a resistance and a variable capacity. These are placed between the relay output and the proportioning bellows. If the derivative restriction becomes wide open, the controller responds only as a proportioning-plus-reset controller.

The controller is a proportional-plus-reset instrument, meaning that it has several mechanisms. These include a comparison unit, error-detection, and a two-bellows

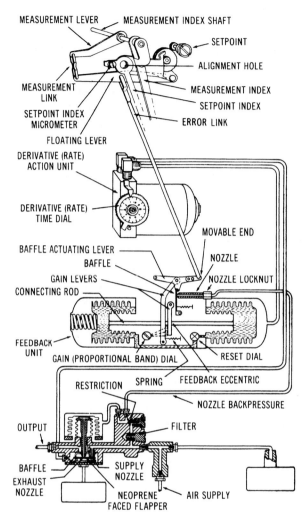

FIGURE 22-8   Pictorial diagram of the Air-O-Line controller. (*Courtesy of Honeywell, Inc.*)

feedback mechanism. It also has a restriction in the line to the reset bellows. Since it is an adjustable gain controller, an adjustable gain mechanism must be included.

Several different mechanisms which are aligned with each other will be discussed. The comparison mechanism will be examined first.

Alignment

The comparison mechanism is a linkage-type mechanism different from the bellows-type mechanism. As in the Foxboro controller, it is important to first determine the type of linkage in the unit.

Remember that there are two input signals and one output signal from a comparison mechanism. The output signal is the error signal and is taken off the floating lever. The two input signals are the measurement and the setpoint. If the mechanism consists of three levers, it is a type-3 mechanism. Two-levers make a type-2, and 1 lever is called a type-1. Study the pictorial diagram of Figure 22-8 and the simplified diagram of Figure 22-9. Locate the floating lever. Observe that one end of the floating lever is supported by a link and a lever. The supporting lever is driven by a shaft. The shaft rotates as the measurement changes. Study the other end of the floating lever. Notice a series of gears. Further notice that as the gears rotate, a lever is caused to rotate. This lever is attached to the setpoint end of the floating lever. Since it has three levers, it can be concluded that it is a type-3 linkage comparison mechanism.

How is this mechanism aligned? It is aligned if a zero error signal equals the measurement index. Recall that in aligning any controller, one mechanism is selected as the reference and all others are related to it. In the Foxboro Model 40 and the Air-O-Line unit, alignment centers around the comparison mechanism. The comparison mechanism is attached to the instrument chassis. This is accomplished by inserting an alignment pin through an alignment hole in the instrument chassis. The setpoint or the measurement is changed so that the alignment pin is able to enter the hole vacated by the error-signal link. If the measurement and setpoint indexes do not agree, the setpoint micrometer is adjusted to bring them into agreement. The comparison mechanism is now aligned and the error link can be reconnected.

Now look at the adjustable gain mechanism. Determine what kind of gain mechanism is used and then consider its alignment. Notice that it consists of two levers. Notice also that as the gain dial is rotated, a contactor between the two levers is moved. Therefore, it can concluded that the gain mechanism is a parallel-lever gain mechanism. Remember that the parallel-lever gain mechanism has only one input signal applied. There is only one output signal developed by the mechanism. In other words, as the gain dial is changed, the amount of output is varied. The input stays fixed. Examine the diagram to see that there is one input and one output to the mechanism. When the dial is changed, the ratio of output to input is also changed. The input to the gain mechanism is the feedback signal. The output is the lever movement detected by the nozzle.

Now consider instrument alignment. A requirement for any gain mechanism alignment is that it be possible to change the gain with no change in output, that is, provided that the error signal is zero. There should be no change in output as the contactor is moved up or down if the levers are parallel. Alignment is the process of making the levers parallel.

The levers are driven by an eccentric pin mounted on a rod. The rod connects the feedback bellows. As the nozzle backpressure changes, the relay output will change and the feedback will drive. The drive of the feedback mechanism acting on the gain mechanism moves the levers, causing them to move either to the left or to the right. Any lever movement will be detected by the baffle/nozzle. The movement will then be reflected by a change in nozzle backpressure. Adjustments will need to be made so that there will be no lever movement as the gain is changed. Either adjust the nozzle or the actuating pin on the feedback-mechanism connecting rod.

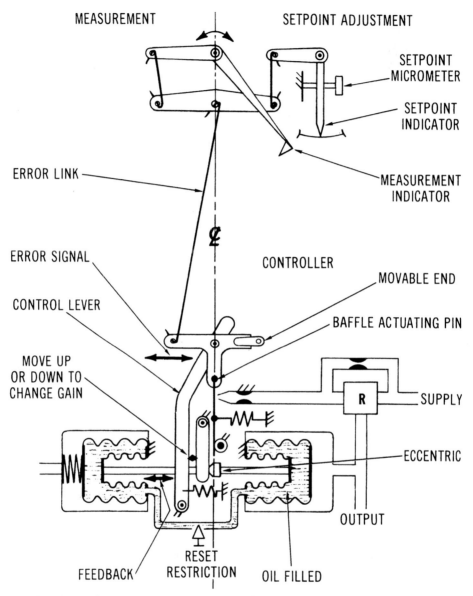

FIGURE 22-9   Simplified diagram of the Air-O-Line controller. (*Courtesy of Honeywell, Inc.*)

The nozzle can be adjusted by turning it on its threaded base. The feedback point can be adjusted by rotating the eccentric actuating pin mounted on the rod.

Remember that an automatic-reset mechanism drives to bring the baffle or nozzle within the baffle/nozzle clearance. The feedback mechanism will search out the position it must assume to bring the baffle to the nozzle. If an attempt is made to align this mechanism, the eccentric is adjusted. If the eccentric is adjusted, it momentarily repositions the lever. The baffle moves in front of the nozzle. As far as the mechanism is concerned, this adjustment is correct. But, by changing the eccentric, the baffle has been moved. In this case, the nozzle backpressure changes. This change in backpressure acts on the feedback mechanism. It then drives the mechanism so it returns the baffle to its position before the adjustment. Because the baffle has the same position, the parallel levers must have the same position.

As the eccentric is moved to shift the levers to the left, the feedback mechanism drives to move the whole eccentric. The levers are also back to the same position before changing the eccentric. It must then be concluded that adjusting the eccentric will not align the gain.

Suppose the nozzle is moved away from the baffle. The reset mechanism will then drive to bring the baffle back to the nozzle. To do this, it must move to the right permitting the baffle to drive within the baffle/nozzle clearance. Since the top of the controller lever has moved to the right, the incline has obviously changed. The bottom of the controller lever is fixed. And, through the contactor, the second lever has been also changed. Since one end of it is also fixed, the two levers are parallel. A conclusion can be drawn that by adjusting the nozzle, gain mechanism levers can become parallel. Therefore, through nozzle adjustment, the gain mechanism can be aligned.

The last mechanism needing alignment is the feedback mechansim. Two points on the controller center line have been established. These are the comparison and the error-detector mechanisms. Recall that the feedback mechanism will drive so that its takeoff point falls on the controller center line. For alignment, it should be positioned so that its takeoff point is on the center line. It should touch this position when there is the same pressure in both bellows. The takeoff point can be adjusted by rotating the eccentric on the connecting rod. Examine the controller output. If the output drifts, then the feedback is attempting to bring its takeoff point to the center line. Make adjustments with the eccentric until the drift stops. When the drift stops, it shows that the takeoff point is on the controller center line.

The center line of the Air-O-Line controller is a straight line originating at the alignment hole in the chassis. The centerline originates at the nozzle tip and ends at the feedback eccentric. The gain mechanism levers will be parallel to the center line.

To reverse the action of the controller, the error link is simply moved to the other end of the baffle actuating lever. This changes the polarity of the error signal by the detector; therefore, it changes the polarity of the controller output.

SPECIAL FEATURE ████████████████████

Honeywell uses a nonbleed relay in its controller. The nonbleed relay was discussed earlier. The feedback mechanism is one of the components that sets itself apart from other controllers. Unlike other controllers, the relay output does not pass through the reset restriction. Oil flows through the reset restriction. This oil is confined within a compartment. The inside of the compartment is composed of a proportioning bellows, a seal bellows, and a reset bellows. The compartment also has a line connecting these bellows systems.

When the reset restriction is closed, it is easy to recognize that the fluid cannot compress. It must be concluded that the connecting rod will follow the proportioning bellows when the nozzle backpressure positions the proportioning bellows. The spring opposing the proportioning bellows is mounted against the reset bellows. Since the fluid is noncompressible, as the proportional bellows is loaded, it is opposed by the spring. The whole assembly then moves to the left. This movement compresses the spring as the relay output increases. Now open the reset restriction. If a pressure difference within the two compartments exists, oil will flow through the restriction in an effort to equalize the two pressures. The normal reset drift will then set in.

COMPONENTS

Figure 22-10 shows a photograph of the Air-O-Line unit removed from the controller. This component houses the oil-filled bellows assembly and control lever. It can easily be removed from the controller. Note again the internal structure of the assembly shown in Figure 22-8 and 22-9.

An interior view of an Air-O-Line controller with the chart removed is shown in Figure 22-11. This photograph shows the location of the Air-O-Line unit. It also shows the supply line, control gauges, and the chart indicting mechanism. The electric chart drive assembly of the unit is located in the center of the controller along with the indicating hands of the recorder controlled by the linkage levers.

TAYLOR INSTRUMENT CONTROLLERS

A.B.B. Kent Taylor makes a controller that is widely used in process control applications today. The Taylor Model 120R instrument is a combination unit. It records process information on a chart and controls process variables. This instrument achieves proportioning-plus-reset control by using a linkage-type of comparison mechanism. This instrument is quite similar to many of the other instruments already discussed. Its major difference is in the gain mechanism. Figure 22-12 shows a view of this controller with the

FIGURE 22-10   Control unit of an Air-O-Line controller. (*Courtesy of Honeywell, Inc.*)

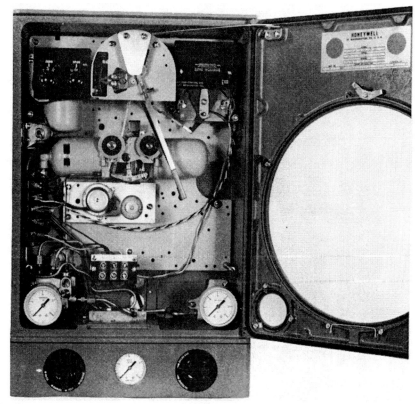

FIGURE 22-11   Interior view of an assembled Air-O-Line controller with the recording chart removed. (*Courtesy of Honeywell, Inc.*)

front cover door opened. Note particularly the location of the control mechanism, pen assembly, and setpoint control.

Figure 22-13 shows an internal view of the Taylor controller. This picture is a partial cutaway of the cover plate and recording chart. This view shows the bellows assembly, control mechanism, relay valve assembly, connecting link, Bourdon element, and gauges. This particular instrument is set up as a temperature controller for a filled system. A capillary tube is attached to the Bourdon element through a connector at the bottom of the case. The same controller may be used to record and control pressure, flow, level, and other process variables with some minor modification.

Components

1. A comparison mechanism.
2. A detector mechanism.
3. A feedback mechanism.
4. An adjustable gain mechanism.

FIGURE 22-12   Fulscope controller with cover door open. (*Courtesy of ABB Kent Taylor*)

The comparison mechanism is a linkage type. The detector mechanism is a baffle/nozzle/relay detector. The feedback mechanism is a bellows in a can. Functionally, it is the equivalent of two opposing bellows mechanisms. If the derivative function is furnished, it is placed in the line to the proportioning bellows.

Arrangements

The comparison mechanism is mounted so that its major center line coincides with the axis of the measurement index. The detector is arranged so that its nozzle portion looks directly at the floating lever of the comparison mechanism. An examination of the mechanism will show that there is no error link. Nevertheless, any error in position of the floating lever is detected by the nozzle. The error link is very short. The baffle portion of the detector mounts directly on the floating lever rather than being driven by the float-

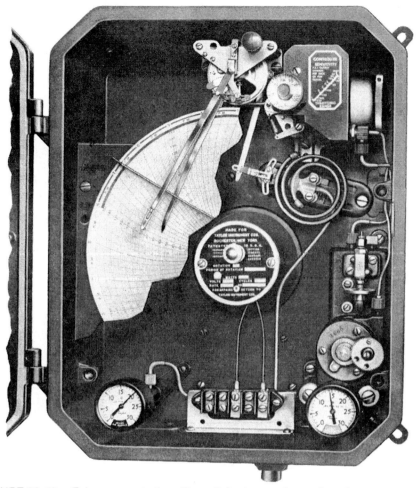

FIGURE 22-13   Fulscope controller with partial cutaway of recording chart and cover late. (*Courtesy of ABB Kent Taylor*)

ing lever by a link. The feedback mechanism is mounted off to one side. The output of the adjustable gain mechanism positions the nozzle, as shown in Figure 22-14.

Look now at the comparison mechanism more closely. Figure 22-15 is a simplified diagram of the Taylor controller. Notice that one end of the floating lever pivots on a geared disk. The disk can be rotated. As it rotates, one end of the floating lever is raised and lowered. This is the setpoint input signal. Driving the other end of the floating lever is an actuating pin mounted on a lever. This lever is positioned by the measurement signal. It forms the second input to the comparison mechanism. The comparison can be considered to be a type-2 or a type-3 motion-balance linkage. If the disk is considered to be a lever, it is a type-3 comparison mechanism.

FIGURE 22-14   Two views of the Fulscope controller unit. (*Courtesy of ABB Kent Taylor*)

For the feedback mechanism, this controller uses the bellows-in-the-can construction. Recall that this is the functional equivalent of two opposing bellows. A problem with all bellows-in-the-can mechanisms is how to take the bellows position out of the can. Taylor accomplishes this by introducing a seal bellows within the feedback bellows, which makes it possible to connect a link to the movable end of the bellows assembly. An intoduction of one seal bellows introduces a problem in itself. If the bellows is to be brought into equilibrium, a different pressure is going to be needed. The pressure inside must be different than what is required outside. A solution to this problem is the introduction of a second seal bellows so that the effective areas of the inside and outside of the bellows will be the same. The bellows will come to an equilibrium with the same pressure on each side.

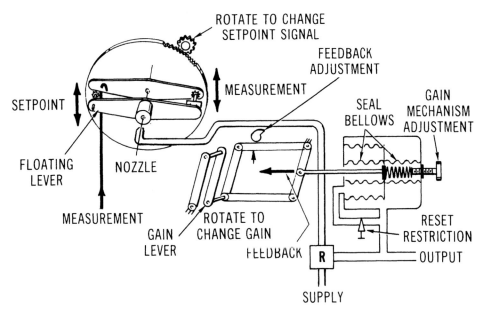

FIGURE 22-15   Simplified diagram of the Fulscope controller. (*Courtesy of ABB Kent Taylor*)

A proportioning spring is mounted within the second sealed bellows. Attached to the spring is an adjustment. As this adjustment is changed, the loading on the spring changes. This, in turn, changes the initial position of the bellows.

Operation

Basically, operation of the Taylor controller is very similar to that of other controllers. A difference is in the location of individual components and the action that occurs due to their placement. The pictorial diagram of Figure 22-16 will be used in the following explanation. Reference numbers are used to direct attention to the specific components being discussed.

The Taylor controller has adjustable sensitivity and automatic reset action for a steam-heating process. The final control element (42) is an air-to-close type of valve used to alter steam which controls the temperature of a process. A capillary tube (1) is used as the controller sensing element.

A 20-psi air supply enters the controller and is registered on the supply pressure gauge (2). It is applied to the air relay valve (31). In the relay valve, the air supply divides into two paths. A small amount of air passes through an orifice (32) and feeds the nozzle (15). It then moves through the nozzle air line (29) and the chamber (37). The major flow of air is applied to the control valve (42) and the bellows (25) of the adjustable sensitivity unit. Air flow to the back the bellows (27) is unrestricted. Air to the front bellows (26) must pass through a needle valve (39). This provides some restriction.

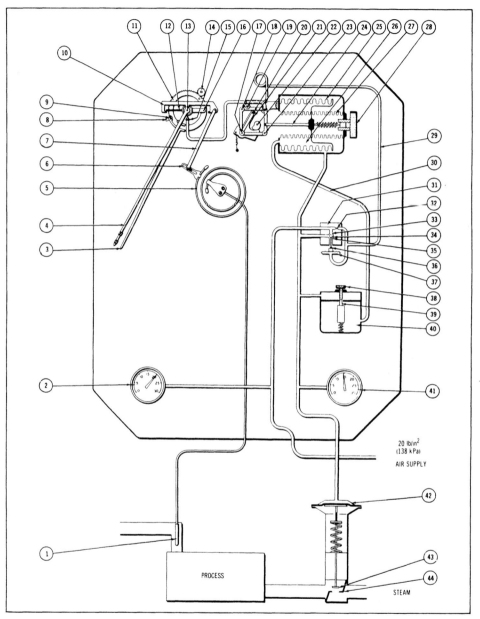

FIGURE 22-16   Pictorial diagram of the Fulscope controller. (*Courtesy of ABB Kent Taylor*)

Let us assume that the pen (4) and the pointer (3) are both together at the control point initially. In this condition, the output pressure is in a state of equilibrium. It is adjusted to approximately 10-psi. The baffle (12) and nozzle (15) then change position. The backpressure of the capsular air chamber (37) of relay valve (31) throttles the air supply to the diaphragm motor (42). Both bellows (27) and (26) of the adjustable sensitivity unit move to a value of 10-psi.

When the sensor signals an increase in temperature, it causes the pen (4) to move upward and away from the pointer (3). The movement of the pen also causes the baffle (12) to pull away from the nozzle (15). An increased clearance between the nozzle and baffle causes a reduction in the backpressure of the capsular air chamber (37). The change in backpressure causes the ball (34) of the relay valve (31) to shift downward. This action, in turn, increases the output pressure of the control valve (42) and bellows (25) of the sensitivity unit. As a result, the control valve (42) closes, which ultimately reduces system heat.

The flow of air to the bellows (25) is primarily unrestricted. Any increase in pressure applied to the bellows will force the connecting rod (24) to move to the left. This action causes the nozzle (15) to follow the baffle (12) which, in turn, limits the clearance between the nozzle and baffle. Any change in this clearance will cause a corresponding increase in output pressure to some new value. The pressure at the front of the bellows (26) will begin to increase because of the differential pressure appearing across the needle valve (39). As the pressure in the front of the bellows (26) increases, the nozzle is pulled away from the baffle. This increases the output pressure. The output pressure continues to increase. The difference across the needle valve (39) is maintained until the temperature of the process cools down. This is the value that will bring the pen back to the control point. Equilibrium is again established and the baffle (12) has returned to its original position above the nozzle (15). There is no longer a pressure difference across the needle valve (39) because both bellows elements are at the same new pressure.

When the pivot point (20) of the arm (21) is directly under pivot point (18), there is no nozzle follow-up action. The sensitivity of the controller is at its highest value. The sensitivity of this unit can be changed. It is lowered when the pivot point is moved in a clockwise direction by turning knob (22) in a counterclockwise direction.

When the reset rate knob (38) is turned counterclockwise, it forces needle (39) downward. This, in turn, increases the annular opening. This action increases the reset rate of the instrument. It responds more rapidly to a change in the pressure of bellows (26).

Parallelogram Gain Mechanism

The parallelogram gain mechanism is a one-input, one-output mechanism. In this respect it is comparable to the parallel-lever mechanism. Fundamentally, the parallelogram is a four-bar linkage arranged in a parallelogram, as shown in Figure 22-17. One point of the linkage is fixed to the instrument chassis. The opposite point is supported by a lever that can be rotated relative to the parallelogram. The lever is lined up with the vertical edge of the parallelogram. There will be no output from the parallelogram even though there is a change in input. Try picturing an input movement. As the input moves, three lower

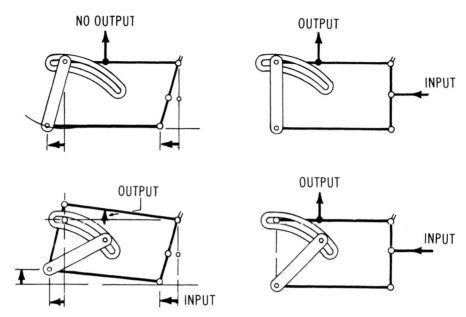

FIGURE 22-17   Action of parallelogram gain mechanism.

levers of the parallelogram will swing on the fourth lever. There will be no change in the position of the fourth lever. No change in output will occur.

The feedback signal moves to the left. The bottom lever of the parallelogram will also move toward the left. The left end of the lever is supported by the adjustable gain lever. The direction of its travel is determined by the travel of the gain lever. The gain lever forms a 45° angle. Left movement applied to the gain lever will cause it not only to swing to the left, but also to swing up. This upward movement raises the parallelogram. The raising of the parallelogram is the output.

As the angle of the gain lever increases, there is more parallelogram upward movement for a given input. As the angle of the gain lever is reduced to zero, the output for a given input is reduced. In fact, at zero angle, the output, regardless of the input, is zero.

Alignment

The overall problem of aligning this controller is the same as with all other controllers. Several mechanisms must each be aligned within themselves and then to each other. The basic Fulscope alignment reference is the comparison mechanism. However, this alignment was accomplished by the use of an alignment pin that was inserted through the chassis. The comparison mechanism was adjusted until the alignment pin could be entered into a reference hole in the floating lever. In the Taylor controller, the baffle/nozzle is used to determine if the floating lever is centered. The problem, however, is how to determine whether the nozzle itself is properly aligned.

Remember that this same kind of problem was encountered in the alignment of the Foxboro Model 40 controller. The method to use when aligning two components is to alternately make corrections on each of them. Corrections are done in steps, bringing them into alignment. This is called a half-and-half adjustment.

The comparison mechanism is aligned by alternately reversing the action of the mechanism. Alignment starts by examining the output, making corrections to the setpoint micrometer, and then to the nozzle position. The action of this controller is reversed by rotating the floating lever. The setpoint end of the lever is moved to the opposing side by rotating the large geared disk. The disk supports the pivot on which the floating lever is mounted.

For the gain mechanism to be aligned, it must be possible to rotate the mechanism's gain lever without a change in output. This is done when the measurement and setpoint are given. The mechanism will be aligned if the movable end of the parallelogram is centered on the axis of ratation of the gain lever. If it is centered on the axis, changing the angle will not change the position of the movable end of the parallelogram. The angle gain lever is not adjustable. The position of the parallelogram needs to be horizontal to cause the movable end of the parallelogram to line up with the axis of rotation. Only one adjustment is needed. This is accomplished by an adjustment that loads the spring. Adjustments are made until it is possible to rotate the gain knob without a change in the output.

The gain mechanism and the comparison mechanism have been aligned. Now the feedback mechanism must be aligned. If it is aligned, the output will hold constant if the error signal is zero. The output will hold constant providing the nozzle is within its clearance relative to the baffle. If it is not, the feedback mechanism will drive in an effort to bring it within its clearance. If the output drives, nozzle position relative to the feedback mechanism must change.

An eccentric is furnished that changes the nozzle position relative to the feedback position. This adjustment is comparable to the eccentric on the connecting rod of the Air-O-Line controller. Unlike the Honeywell unit, the eccentric of the Taylor controller is on the nozzle side of the gain mechanism. In the Honeywell instrument, the eccentric was on the feedback side of the mechanism. Adjustment of this eccentric will result in a new baffle/nozzle position for a given feedback mechanism position, thus stopping the reset drift.

There is a practical problem in using this adjustment. It is difficult to precisely adjust the eccentric. It is extremely difficult to stop drift by using the eccentric adjustment alone. The solution is to use the eccentric for coarse adjustment and to make minor adjustments using the gain mechanim. Using the gain mechanism adjustment to correct the feedback alignment introduces small errors in the gain mechanism alignment. As a practical matter, these errors in gain mechanism alignment are not serious.

Discussion of the alignment procedure for the Taylor Fulscope controller should serve to reinforce some basic alignment principles. For a detailed alignment procedure, refer to the manufacturer's operation manual.

FISCHER & PORTER CONTROLLERS ■■■■■■■■■■■■■■■■■■■■■

A number of different controllers have been discussed up to this point. Nearly every controller mentioned has used some type of linkage comparison mechanism. This mechanism simply compares the measurement and setpoint inputs through a linkage assembly and develops a corresponding output.

The Fischer & Porter Model 53PR controller employs a mechanism that uses two opposing bellows to perform the comparing functions. Units of this type are commonly called pneumatically connected controllers. A controller of this type can be located some distance from other instruments. Individual instruments may be located at any point as long as the necessary tubing is run to each instrument. These instruments are frequently called plug-in or miniature controllers. These controllers may or may not be plug-in units. In general, controllers are all about the same size. A common trend in controller technology leans heavily toward the design of miniature assemblies.

A disadvantage of the pneumatically connected controller is the resulting increase in maintenance cost. The increased cost is offset in part by an increased onstream time. To this potential increase of onstream time, the instruments must be calibrated to a standard. The controller must be aligned for all input/output levels and all gain settings. If it is not so aligned, it is not possible to take advantage of the plug-in feature without extensive retuning.

Components

The components of the Fischer & Porter Model 53R controller are the usual components found in most controllers. The functioning of this controller is no different than most. However, there is one additional component that makes this controller unique. That additional component is a second restriction. The restriction is located between the relay output and the nozzle-backpressure systems. As the relay output pressure changes, the amount of air going to the nozzle changes. Because of this change, the amount of baffle/nozzle clearance needed to change the output from 3- to 15-psi is reduced. Fischer & Porter call this restriction "regenerative-feedback." The presence of the regenerative-feedback affects the operation of the controller.

Arrangements

The feedback mechanism consists of two sets of bellows placed in a can. The two bellows are connected by a rod, as shown in Figure 22-18. Pivoting on this platform is a control lever. The feedback motion is in line with the bellows. As the bellows connecting rod moves, it carries the platform and the control lever.

The error signal is at a right angle to the feedback mechanism. When the error signal changes, the control lever rotates. On the control lever is a flapper or baffle positioning ball, which positions the baffle in front of the nozzle.

The comparison mechanism consists of two opposed bellows. The output of the comparison bellows is a displacement that drives the error link.

The detector is a baffle/nozzle nonbleed relay combination. The relay has the modification that was mentioned earlier. The baffle is bent 90°, forming a right angle. Baffles, up to this point, formed a straight angle. The baffle has this particular angle because of the introduction of the nozzle.

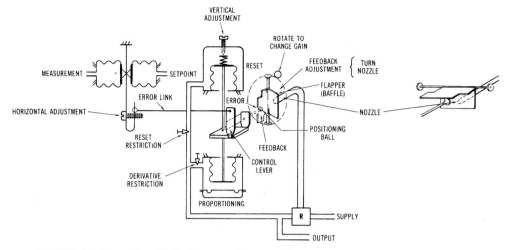

FIGURE 22-18   Simplified diagram of a controller mechanism. (*Courtesy of Fischer & Porter Company*)

Operation

First, look at Figure 22-18. Suppose the measurement signal increases. The error link will move to the right. This movements rotates the control lever. As the lever rotates, the baffle positioning ball moves to the right, permitting the baffle to cover the nozzle. The nozzle backpressure will build, as will the relay output. The relay output is applied to the proportioning bellows, lifting the control lever. The action against the relay output will later be applied to the relay bellows. Depending on the angle of the feedback, its movement will be applied to the baffle. This will cause the baffle to move away from the nozzle. Remember that the measurement signal caused the baffle to move towards the nozzle.

The controller output will drive to the pressure that brings the feedback signal into equilibrium. The feedback movement, or pressure, needed for error-signal movement will depend on the gain setting.

Later, the pressure change in relay output will pass by the reset restriction. This action loads the reset bellows. When the pressures in the reset and proportioning bellows are equal, there will not be any feedback signals. If there is an error signal, the controller output will drive to a certain limit, which will be recognized as the normal resetting action.

Suppose the derivative restriction is closed down. Suppose further that it closes before the change in relay output is applied to the proportioning bellows. As a result, the feedback action will slow down and the controller output will "overtravel." That is, the output will not change, or, the output will not change any more than if the normal proportioning response were not slowed down.

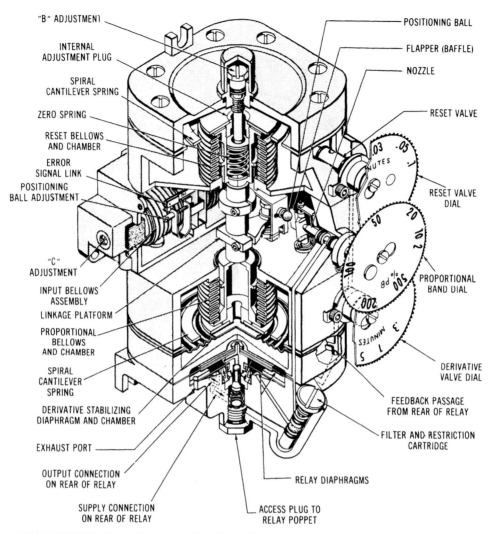

FIGURE 22-19   Partial cutaway drawing of a controller. (*Courtesy of Fischer & Porter Company*)

Figure 22-19 shows a partial cutaway drawing of the Model 53PR controller. Study this drawing and the mechanism diagram to see the relationship of the components.

Alignment

The alignment consists of relating the comparison, the feedback, the gain, and the detector mechanisms to each other so that, for all levels of output and for all gain settings, the output will not drift. This is especially true for the controller that is open-loop connected. If the controller is closed-loop connected, there will be no offset between the setpoint and measurement output. This is the case for all measurements and all gain settings.

Consider the gain mechanism alignment. If this mechanism has a zero error signal, it is possible to rotate the gain. The gain can be moved from low to high without the output changing. This will be possible only if the positioning ball is centered on the gain disk. The nozzle is also centered on the disk. To align the gain mechanism, adjustments are needed. The adjustments move the positioning ball both up and down.

Since the nozzle is adjustable, it must be aligned. An aligned nozzle helps in the alignment of the gain mechanism. Since these two alignments interact, the entire assembly must be aligned. When this alignment occurs, there is a problem with the half-and-half adjustment of the open-loop connection.

First try to remedy the situation by using the horizontal adjustment. Reverse the action and correct for half of any change in output by adjusting the nozzle. These adjustments are made until the output does not change. But with this repetition, the positioning ball is centered on the horizontal centerline. Simultaneously, the nozzle is brought into alignment with the feedback mechanism. Rotate the baffle to a 90° angle to the connecting rod. By correcting the vertical adjustment, the positioning ball can be centered. The ball is centered on the vertical centerline of the gain mechanism.

Consider now the comparison mechanism alignment. Recall that if the measurement and setpoint signals are equal, there is no error signal regardless of the signal level. This means that when the pressure in both bellows changes from 3- to 15-psi, the output does not change. If it does change, the comparison mechanism is not properly aligned.

Previously, it was stated that error signals occur at one pressure level when there is no error signal at a different pressure level. This happens because the effective area of the bellows changes with internal pressure. Usually, bellows are manufactured and matched by the manufacturer to avoid this problem.

In the Fischer & Porter controller, there are adjustments for aligning the comparison bellows. Alignment is accomplished by changing the lengths of the two bellows. The procedure for aligning the controller when it is open-looped connected has been discussed. The procedure can also be aligned for a closed-loop connection.

SUMMARY

The Foxboro Model 40 instrument is a motion-balance controller. It has a linkage comparison mechanism, an angle-gain mechanism, and a baffle/nozzle/relay combination as the detector. This controller can be used to achieve single mode, two mode, or three mode control.

The Foxboro Model 58 instrument is pneumatically connected and is of the force-balance type. Applied error and feedback signals cause a forcing action, with the difference in the two signals driving the baffle. Controllers of this type employ a feedback mechanism, a detector, and an adjustable gain mechanism.

The Honeywell Air-O-Line controller is a motion-balance instrument that uses a linkage comparison mechanism in its operation. The feedback mechanism is functionally similar to units used in other controllers. This instrument has an unusual physical arrangement of its components.

The Air-O-Line unit is a proportional-plus-reset controller. It employs a comparison mechanism, and error-detecting mechanism, and a two-bellows feedback mechanism with a restriction in the line of the reset bellows. The developed output is an error signal that is taken from a floating lever. With three levers involved in the assembly, the unit is classified as a type-3 linkage comparison mechanism.

A special feature of the Air-O-Line controller is the incorporation of a nonbleed relay. In this unit, the relay output does not pass through the reset restriction. Oil flow is confined within a compartment composed of the proportioning bellows and a seal bellows, a reset bellows and a seal bellows, and a line connecting the two bellows systems. If there is a difference in pressure within the two bellows compartments, oil will flow through the restriction in an effort to equalize the two pressures.

The Taylor proportioning-plus-reset controller is a motion-balance instrument. It uses a linkage-type comparison mechanism. The major difference between this unit and others is the gain mechanism. Gain is achieved by a parallelogram structure that has one input and one output. Operation is the same as other controllers, except that the components are arranged in a unique structure.

The Fischer & Porter controller is a motion-balance unit. It is similar in many respects to other instruments. This controller uses a bellows-type of comparison mechanism. Units of this type are called pneumatically connected controllers. A pneumatically connected controller permits individual pieces of equipment to be placed at remote locations. Regenerative feedback in this type of controller happens when the relay output pressure changes and the amount of air furnished to the nozzle also changes.

Alignment of the Fischer & Porter controller consists of relating the comparison mechanism, the feedback mechanism, the gain mechanism, and the detector mechanism to each other. Physical alignment is achieved by adjusting or changing the length of the two bellows.

ACTIVITIES

CONTROLLER ANALYSIS

1. Refer to the operational manual of the specific controller to be used in this activity.
2. Remove the housing or open the front door of the instrument.
3. Identify the model and manufacturer of the instrument used in this activity.
4. Locate the comparison mechanism of the instrument.
5. What type of comparison mechanism is used?
6. Locate the gain mechanism of the instrument. Describe the gain mechanism.
7. Is the gain mechanism adjustable?
8. Locate the baffle/nozzle assembly. Describe the construction of this mechanism.
9. How is feedback achieved in this instrument?
10. Describe the mode of operation achieved by this controller.
11. Does this controller have an alignment procedure?
12. Reassemble the instrument.

QUESTIONS

1. What are some of the features to look for in identifying the mode of operation of a controller?
2. If a controller has two levers in its mechanism, what is the lever type classification?
3. What are the distinguishing features of an angle-gain mechanism?
4. What controllers of this chapter are motion-balance instruments?
5. What are the identifying features of a linkage comparison mechanism?
6. What are the identifying features of a bellows comparison mechanism?

23
Universal Controllers

OBJECTIVES

Upon completion of this chapter, you will be able to:

- Explain what is meant by the term universal controller.
- Identify some of the common inputs applied to a universal controller.
- Recognize the output of a universal controller.
- Explain the manufacturing philosophy behind the development of the universal controller.

KEY TERMS

Ball gauge. A check valve assembly that permits flow in only one direction.

Bump pressure. A sudden forceful blow, impact, or jolt that is caused by a surge or change in pressure to a system.

Header. A length of pipe or vessel, to which two or more pipe lines are joined, that carries air or fluid from a common source to various points of use.

Manual regulator. A pressure control alteration that is adjusted by hand.

Modular contruction. An instrument construction procedure which employs a group of components mounted on an independent board or assembly that permits flexibility and variety of its use.

Universal controller. An instrument that has a wide range of applications and can be adapted to fit a number of specific controller functions.

INTRODUCTION

For years, pneumatic controllers have been specialized instruments designed for specific industrial applications. Each controller generally called for a unique application. Each of these controllers contained components designed specifically for that instrument. This obviously called for specialized personnel to maintain each type of controller. A large inventory of replacement parts was also needed.

A recent trend in manufacturing philosophy has led to the design of a universal type of controller. This controller serves as a common standard for a wide range of controller applications. Through the universal type of controller, operator training is simplified. Also, parts inventory is reduced and controller maintenance problems surround a single instrument. The Taylor 440R series of instruments discussed in this chapter are universal controllers that measure, indicate, and control process variables.

DESCRIPTION

The input to a universal controller can be absolute pressure, differential pressure and gauge pressure, temperature, or any volumetric load measured by a sensing element. Controller output is a standard 3- to 15-psi signal that is applied to a final control element. Modular construction simplifies maintenance. It also permits field modifications when control responses require changing. Figure 23-1 shows an exterior view of a Taylor 440R

FIGURE 23-1   Exterior view of a universal controller. (*Courtesy of ABB Kent Taylor*)

FIGURE 23-2   A typical control-room installation of universal controllers. (*Courtesy of ABB Kent Taylor*)

universal controller. The panel installation of Figure 23-2 shows four of these controllers in a typical control-room installation.

A typical pneumatic system application of the universal controller is shown in Figure 23-3. Air from a header is applied to the controller through the supply line. Controller operation alters the final control element according to the setpoint position. The Taylor 440R series is adaptable to field applications or control-room installations without any modifications.

Components

The components of universal controllers are similar in many respects to those used by controllers discussed earlier. A comparison, a feedback, a detector, and adjustable gain mechanisms are typical of an automatic reset controller.

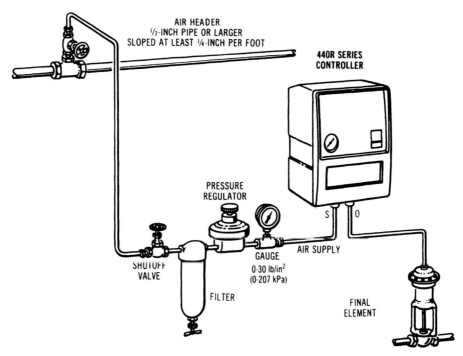

FIGURE 23-3   Typical pneumatic system application of the universal controller.

Arrangements

Figure 23-4 shows a Taylor 440R universal controller with the front cover removed. Notice the location of the Bourdon measuring element and the setpoint hand. Notice also the location of the process pointer, the pneumatic-set unit, the automatic/manual balance detector, and the gain dial. The gauge on the left of the instrument is used to indicate output pressure.

The 440R series of pneumatic controllers operate on the motion-balance principle. Motion from the pneumatic feedback unit balances motion produced by a process measuring element. In Figure 23-5, the scale plate of the controller is removed, which provides an inside view. This controller uses a low-pressure bellows measuring element rather than the Bourdon element of Figure 23-4. Note the location of the output relay, gain, setpointer, process pointer, process, and control links.

Operation

Simplified diagrams of the 440R, 441R, and 442R controllers are shown in Figures 23-6 through 23-8. These diagrams show controller operation for three different response modes. All of these controllers use the Bourdon spring measuring element of Figure 23-4. As noted, the Taylor 440R series is quite versatile. This series can easily be modified to achieve a variety of different control functions.

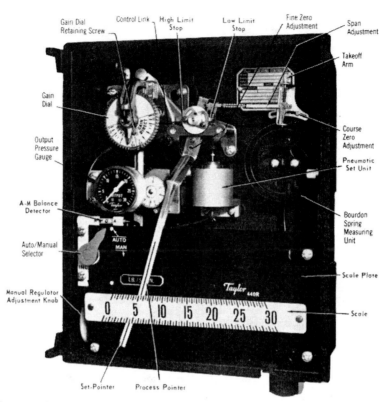

Gain Dial
Retaining Screw

Control Link

High Limit
Stop

Low Limit
Stop

Fine Zero
Adjustment

Span
Adjustment

Takeoff
Arm

Gain
Dial

Output
Pressure
Gauge

Course
Zero
Adjustment

Pneumatic
Set Unit

A-M Balance
Detector

Auto/Manual
Selector

AUTO
MAN

Bourdon
Spring
Measuring
Unit

Scale Plate

Manual Regulator
Adjustment Knob

Taylor

Scale

Set-Pointer   Process Pointer

FIGURE 23-4   A universal controller with the front cover removed. (*Courtesy of ABB Kent Taylor*)

In Figure 23-6, the temperature of the sensing element increases. The Bourdon element slightly uncoils. The process pointer then moves to the right and the baffle actuating pin moves to the left. Movement of the actuating pin lowers the baffle to decrease the gap between the nozzle and the baffle. As a result of this action, nozzle backpressure increases. The increase in backpressure in transmitted to chamber A of the output relay.

As the pressure in chamber A increases, the diaphragm assembly moves the relay stem downward. This closes the vent port and opens the air supply port of the relay. The output pressure is then increased. The output pressure continues to increase until it balances the downward force of the diaphragm assembly.

The 440R modification of the universal controller provides a fixed high-response mode of operation. In this condition of operation, gain is fixed at approximately 125. The controller simply turns the final control element on or off, depending on the sensed measured value and the setpoint value setting.

The 441R modification of the universal controller is shown in Figure 23-7. This mode of operation provides for proportional response control. The output pressure of the relay is fed to a follow-up bellows. This part of the controller is designed to raise the baffle actuating pin and baffle. Through this additional component, the original baffle/

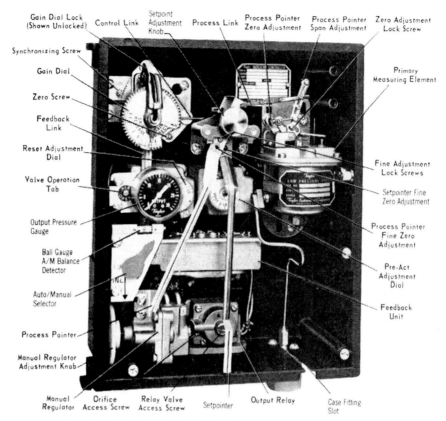

FIGURE 23-5   A universal controller with scale plate removed. (*Courtesy of ABB Kent Taylor*)

nozzle gap can again be balanced. The follow-up bellows assembly is part of the feedback unit. It is added to provide proportional response, which means that the output pressure of the relay will change along with the measured input and the setpoint value. The amount of output can be varied by rotating the gain dial. As the dial rotates, it moves the baffle/nozzle assembly around the baffle actuating pin. The resulting position of the pin determines the amount of motion the pin transmits to the baffle. This is due to process measurement changes and feedback action. When the gain is decreased, the baffle becomes less sensitive to process measurement changes. The baffle is more sensitive to feedback.

A modification of the 442R universal controller is shown in Figure 23-8. It permits both proportional and reset responses. The feedback unit is equipped with an additional bellows. This bellows provides the reset response.

The reset response of the 442R automatically returns the process variable to the setpoint after a change in load. This is accomplished by opposing the action of the follow-up bellows with the reset bellows. The output pressure is fed to the reset bellows through a needle valve. The reset needle valve permits regulation of output pressure values, which

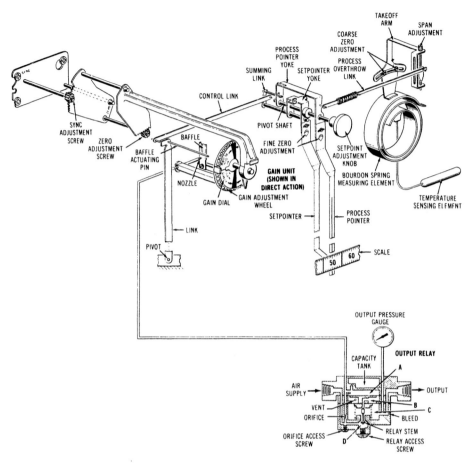

FIGURE 23-6   Schematic diagram of the 440R controller with fixed high response. (*Courtesy of ABB Kent Taylor*)

determines the time factor of the reset action. A modification of this type is quite important; it is used in applications where the output is subjected to prolonged changes.

A modification of the 442R permits the universal controller to be changed to a proportional-plus-reset-plus-preact response. The Taylor 444R is a controller of this type. The modification consists of placing a preact needle valve in the output pressure feedback line, located between the reset needle valve and the follow-up bellows. This modification is designed to reduce the offset caused by a process disturbance. It also reduces the recovery time following the disturbance. The preact needle valve permits the output pressure to be adjusted to compensate for any offset that occurs. Preact is used here in place of the term derivative response. The Taylor model 444R controller can therefore be called a proportional-integral-derivative, or PID instrument.

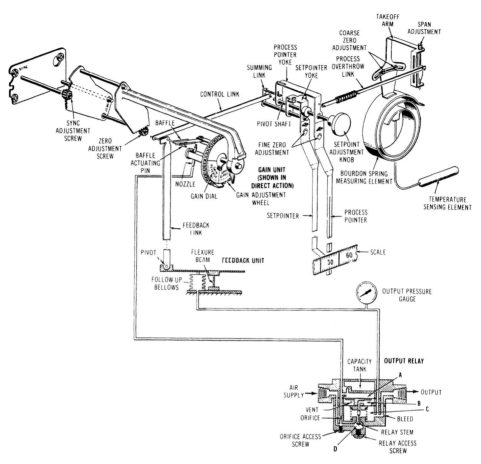

FIGURE 23-7   Schematic diagram of the 441R controller with proportional response. (*Courtesy of ABB Kent Taylor*)

The Gain Unit

The gain unit of the 441R controller modification is designed to maintain a minimum or maximum output pressure until the process pointer crosses a reverse trip point. Figure 23-9 shows a drawing of the gain unit. It is set for differential-gap response.

In the unit design, there are two trip points located an equal distance from the set-pointer. These two points are shown in Figure 23-10. The differential gap between the trip points can be set between 4 percent and 100 percent of the input span. When the process pointer reaches either trip-point location, the output pressure will reverse from maximum to minimum or from just the reverse. The trip points are set by turning the gain dial to the desired percent-gap setting. Controller operation can also be changed to direct or reverse action. This is done by positioning the gain dial to either of these positions.

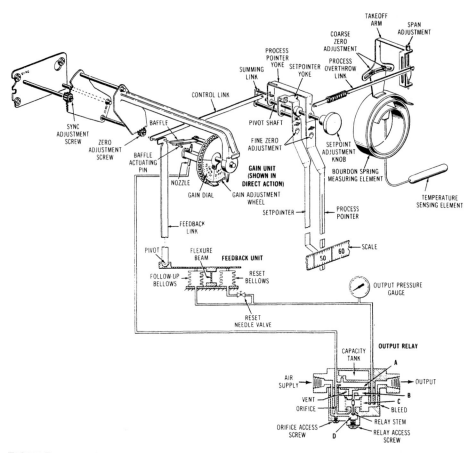

FIGURE 23-8   Schematic diagram of the 442R controller and reset responses. (*Courtesy of ABB Kent Taylor*)

When the controller is set for differential-gap operation, the baffle is located on the opposite side of the pin. Figure 23-9 shows the location of the nozzle, baffle, and baffle actuating pin. Compare the location of these components with the same ones shown in Figure 23-8. The process pointer control link moves the pin to decrease or increase the baffle/nozzle gap. The follow-up bellows then moves the baffle to cap or uncap the nozzle. The output pressure will go to minimum or maximum and remain at this pressure. This continues until the process pointer reaches the other trip point.

Automatic/Manual Operation

Automatic or manual operation of the universal controller is achieved by simply adding a switching unit. Figure 23-11 shows a view of this assembly. The unit also contains an automatic/manual, or A/M balance detector. Figure 23-12 shows a schematic diagram of the unit.

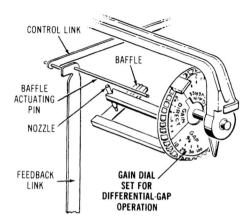

FIGURE 23-9   Schematic diagram of gain unit set for differential-gap response. (*Courtesy of ABB Kent Taylor*)

The manual regulator provides a pneumatic output signal through the A/M selector. The signal occurs when the switch is in the manual position. Turning the manual adjust knob downward causes the diaphragm and relay stem of the manual regulator to move to the right. An air supply from chamber C enters chamber B through a more open valve for an increased regulator output. Equilibrium is then established. Equilibrium occurs when the output pressure against the diaphragm in chamber B equals the force of the loading spring in chamber A.

Turning the manual adjust knob upward causes the diaphragm and relay stem to move to the left. This causes the air supply of chamber C to be throttled. The pressure in chamber B is sent to chamber A through an inner valve. The pressure is released into the

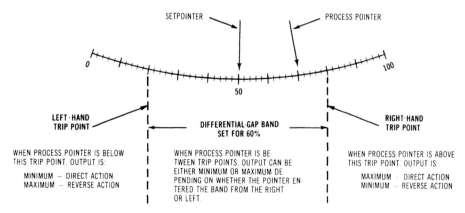

FIGURE 23-10   Differential gap operation. (*Courtesy of ABB Kent Taylor*)

FIGURE 23-11   Automatic/manual selector unit. (*Courtesy of ABB Kent Taylor*)

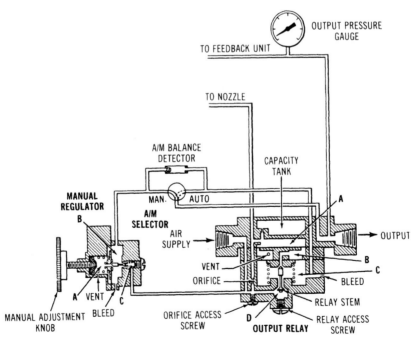

FIGURE 23-12   Schematic diagram of the automatic/manual selector assembly. (*Courtesy of ABB Kent Taylor*)

atmosphere through an open vent, which causes a decrease in output pressure. Equilibrium will occur when the decreased output pressure of chamber B balances the decreased spring force.

The A/M balance detector is a simple ball gauge. This detector has O-rings positioned at each end of the travel extremes. The ball is seated at the left side of the gauge, as shown in Figure 23-12. It indicates that the controller output exceeds the manual regulator output. When the controller and manual output pressures are equal to within 0.01-psi, the ball centers itself in the gauge. The A/M selector may then be rotated. This rotation occurs between automatic and manual control without a pressure bump.

Alignment

Alignment of the Taylor universal controller applies to four adjustments that are often necessary in order to assure accurate operation. This includes process-pointer calibration, pneumatic-setpointer calibration, controller alignment, and nozzle height adjustment. As a rule, modifications of the basic controller have some influence on these maintenance procedures.

Process-pointer calibration involves three general adjustments. These adjustments pertain to the process signal input and include zero, span, and linearity adjustments. The procedure is based on supplying 0 percent, 50 percent, and 100 percent input signals to the controller. If the instrument is being adjusted to respond to temperature, substitute 20 percent for 0 percent and 80 percent for 100 percent.

Zero adjustment is acheived by observing 0 percent input and the process pointer location. It should be at 0 percent, ±2 percent of full scale. If it is not, the process pointer fine zero is adjusted. If the pointer is off by more than 5 percent, the coarse zero on the takeoff arm is adjusted first. See Figs. 23-4 through 23-8.

Span adjustment is achieved when the applied input is 100 percent. The process pointer should be adjusted to 100 percent, ±2 percent of full scale. The process pointer span adjustment is located on the primary measuring element. Moving the process link pivot along the takeoff arm farther away from the center of the measuring element shortens the span. This causes more pointer action. As a rule, zero adjustment should be checked again after altering the span adjustment.

Linearity adjustment is achieved by observing 50 percent input and the process pointer location. It should be at 50 percent, ±2 percent. This adjustment is made by selecting another ball pivot on the right end of the process link. If the indication is low, the link is shortened. If the indication is high, the link is lengthened. Zero and span adjustments should be made again after altering the linearity.

Pneumatic-setpointer calibration is made by applying 0 percent, 50 percent, and 100 percent signals and observing the indication. Figure 23-13 shows the location of the span, fine zero, and coarse zero adjustments.

Controller alignment is used to position the baffle actuating pin. It is placed directly behind the center of the gain dial. Alignment occurs when the process pointer and setpointer are together anywhere along the scale. This alignment is important during start-up. It allows the gain dial setting to be changed without changing the operating point of the controller. This adjustment varies with different modifications of the controller.

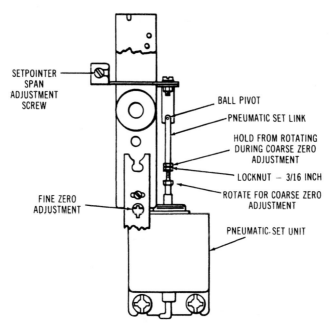

FIGURE 23-13   Span, fine zero, and coarse zero adjustments for the pneumatic set-pointer. (*Courtesy of ABB Kent Taylor*)

specific alignment procedure must follow the manufacturer's recommendations. These recommendations are in the operational manual for each specific controller.

Nozzle height adjustment is made when the controller cannot be aligned properly by the preceding adjustments. As a general rule, this adjustment is preset at the factory. Adjustment is usually not necessary unless nozzle replacements are made. Specific procedures for this adjustment must follow the manufacturer's recommendations.

SUMMARY

A recent trend in controller design has brought about the universal controller. This instrument has a wide range of applications. It can be adapted to fit a number of specific controller functions.

The input to a universal controller can be absolute pressure, gauge pressure, differential pressure, temperature, or any volumetric load measured by a sensing element. Output is a standard 3- to 15-psi signal that is applied to a final control element.

Components of the universal controller include a comparison mechanism, feedback, a detector, and adjustable gain.

The Taylor series of controllers operate on the motion-balance principle. Motion from the pneumatic feedback unit balances motion produced by the process measuring element. Operation applies to the fixed high-response mode, proportional response, and proportional-response-plus-reset.

Gain of the controller is designed to maintain a minimum or maximum output pressure. Gain continues until the process pointer crosses the reverse trip point. The trip points can be set between 4 percent and 100 percent of the input span.

Automatic/manual operation is an optional addition for the universal controller. Manual operation is achieved when the manual regulator develops a pneumatic output signal through the A/M selector switch. The automatic balance detector of the unit is a simple ball gauge. Switching between automatic and manual is achieved without a pressure bump.

Alignment applies to process-pointer calibration, pneumatic-setpointer calibration, controller alignment, and nozzle height adjustment. These adjustments are subject to some variation according to controller modifications.

ACTIVITIES

UNIVERSAL CONTROLLER CALIBRATION ■■■■■■■■■■■■

1. Connect a universal controller to an air supply, as shown in Fig. 23-3. Install a pressure gauge to monitor the supply pressure.
2. If an output pressure gauge is not included in the controller, attach one to the output to monitor the changes in output pressure.
3. Calibrate the process pointer of Fig. 23-4 as directed in the operational manual. Calibration should be based on 0 percent, 50 percent and 100 percent input to the controller.
4. Change the input air source to the pneumatic set connection and calibrate the set-pointer of Fig. 23-4.
5. Set the high and low limit stops to the upper extremes for unrestricted travel of the set-pointer.
6. Apply 3.0-psig (20-kPa) to the input. The set-pointer should read 0 percent ±2 of full scale. If it does not, adjust the set-pointer fine zero adjustment to obtain this value.
7. Apply 15-psig (100-kPa) for span adjustment. The set-pointer should read 100 percent ±2 of the full scale. If the indication in step 7 is low, rotate the span adjustment screw of Fig. 23-4 clockwise. If the indication is high, rotate the span adjustment screw counterclockwise.
8. Repeat the zero and span adjustments.
9. Apply 9-psig (60-kPa). The set-pointer should indicate 50 percent ±2 of full scale.
10. Turn off the air supply and disconnet the controller.

QUESTIONS

1. What is the philosophy behind the development of a universal controller?
2. What is the function of a universal controller?

3.   What are the components of a typical automatic reset controller?
4.   How is automatic/manual operation of a controller achieved?
5.   What is the operational principle of the universal controller discussed in this chapter?

24
Process Control

OBJECTIVES

Upon completion of this chapter, you will be able to:

- Explain what is meant by the term process control.
- Identify the elements of a process control system.
- Distinguish between open-loop and closed-loop control systems.
- Define the term automatic control.

KEY TERMS

Capacity.   The ability to hold, receive, or store, or a measure of content capability.

Control lag.   A delay in the response of a system caused by the control components.

Demand-side capacity.   The content capability of equipment or part of a system that uses energy or materials in its operation.

Dynamic variable.   The characteristic of a variable quantity that changes in some unique way or pattern.

Evaluation.   To examine, judge, or estimate the value or operational status of a system or the response of components in the system.

Inertia.   A property of matter which causes it to remain at reset or in uniform motion in the same straight line unless acted upon by some external force.

Lag.   A delay in the response of a function or operation.

Measurement.   A determination of the existence or magnitude of a variable.

Process dynamism.   A condition or theory that explains the interrelationship of an instrument or device to the process that uses it in an operation.

Supply-side capacity.   The content capability of the energy supplying source of a system.

Tolerance.   An allowable deviation from a standard value.

Transfer lag.   The time it takes to shift or change energy from one device or piece of material to another.

INTRODUCTION

In the study of instrumentation attention has focused on instrument mechanisms. These mechanisms are used to achieve some type of measurment and control. Process control deals with a specific application of instruments where operations are achieved automatically. The basic theory of process control is, however, well beyond the intended scope of this discussion. Nevertheless, a brief introduction to this type of system control might be very helpful.

Remember that the primary reason for automatic process-control instrumentation is increased production, improved quality, and reduced production-line costs. As a result, all process-control instrumentation must be evaluated according to the specific job that it must perform. In this regard, instrumentation often influences the process. The process to a far greater degree influences the instrumentation. It is, perhaps, valid to say that instrumentation does not have an identity of its own. Instrumentation at best can only reflect the process to which it is applied. Process dynamism must be understood before instrumentation can be judged. For example, a temperature recorder/controller may be completely unsatisfactory on one process. On a second process the same equipment may do an excellant job. An instrument is only "good" or "bad" in relation to what it can do.

CONTROL

The word control means many things to many people. Control as applied to electric motors usually refers to the equipment that starts and stops the motor. A control laboratory is a place where chemicals are tested with the objective of determining whether the chemicals meet specifications. The accounting department refers to certain practices as cost control. The warehouse personnel talk about inventory control. Each group has a control concept.

Industrial instrumentation personnel have their own definition of control. The operating people have their own particular definition. In spite of the diversity of applications, control has one basic concept. It is expressed as control as used by all people in the various activities. This concept is that control implies the ability to predict what will happen in the future. Because of this ability to alter factors, it can affect future events. There is control of an automobile if a prediction can be made of what will occur in the future. The extent of the control is limited timewise to the next few seconds. The degree of control is also limited by factors that cannot be predicted—for example, a blowout that might cause the automobile to go out of control.

Any job requires two pieces of information before control can be started. They all must perform some type measurement. For example, warehouse supervisors cannot control inventory unless they know how much stock is on hand. Chemists must take measurements to analyze materials. Accountants must know how much it costs to do certain jobs. All of these involve some type of measurement.

A second factor is absolutely necessary for control. This information is frequently lost or neglected. The second factor is a precise definition of the object being controlled. For example, control chemists may make precise and involved measurements that are turned into specific information on chemicals. Yet, it is impossible for them to make any decisions on quality control—that is, unless they are given a set of quality specifications. If chemists have these specifications, they can compare their measurements to the specifications. This permits them to exercise control by accepting or rejecting the chemicals examined.

In these two examples, it is necessary to measure what is being controlled. It is equally necessary to compare that measurement with what is needed. The control action must always be based on the difference between the two.

OPEN- AND CLOSED-LOOP CONTROL ■■■■■■■■■■

There are two basic types of control used in process applications. One is termed open-loop and the other is closed-loop. The meaning of each may be best conveyed by again using the analogy of the automobile driver. There are two ways to drive an automobile on a road. One way would be for the driver to look down the road and decide where the automobile is to go. This represents the measurement function. Secondly, the driver must decide where the automobile is and compare this with where it is to go, then decide how to adjust the steering wheel and throttle to get it there. This example is representative of the control function. Having done this, the driver could then travel to the destination blindfolded. Upon reaching the destination, the blindfold could be removed and the action repeated. This operation would be representative of a process exercising open-loop control.

The second way to drive an automobile would be for the driver to simultaneously determine where the automobile is to be on the road and then decide its actual position. This action is indicative of the measuring function. These positions are then compared. Next, the throttle and steering wheel are controlled. The vehicle will move to the desired location. In this type of operation, the driver continuously measures, compares, and controls the automobile. The sequence of events is continuous and at no time is this cycle of operation broken. The continuous loop of events including measurement, comparison, and control is closed. This type of operation is representative of a process exercising closed-loop control.

The open-loop method of controlling an automobile is perhaps used more often than one cares to admit. On a straight road with no side traffic, it might be possible for this method of control to actually work. Under these conditions of operation, the degree of success would be based on an ability to predict what will happen. This requires a control action. It includes such factors as the reaction of other automobiles, pedestrian movement, straight roads, tire and weather conditions, and mechanical failures. Since these conditions are not very predictable, open-loop driving in not at all practical. As a practical matter, our ability to predict events in driving is limited in time to only a very few seconds. It is also limited to those factors that are known. As a consequence, closed-loop control of an automobile is the only practical method of driving.

CONTROL IN A PLANT

Assume now that the ideas of the first part of this discussion are applied to an industrial plant. Some conclusions can be very swiftly made. These are:

1. Measurements that are not used for the comparison step, or do not result in control, need not be made.
2. Measurements that are in error are worse than no measurements at all.

The definition of control and its ability to alter a process determine what instruments will be used. This is also an indication of how well these must work. In general, the faster a process responds, the greater the difficulty in holding it in control.

A process is in control if we can predict what it will do in the future. It is possible to predict what will occur in the future. The equipment must respond to changes being made to the process when it goes out of control. In one of the examples it was shown that the location of an overflow pipe was all the equipment needed to maintain control of liquid level in a tank. Usually the control problems are much more complicated. Varying amounts of instrumentation and human attention are required. These combinations of human attention and instrumentation may be broken down into three groups, as shown in Figure 24-1. The three groups are:

1. All human attention, with very little instrumentation.
2. Human attention plus measuring instruments.
3. Measuring and control instruments with very little human supervision.

It is interesting to observe just how these three groupings affect control in the following control procedures:

1. The operator measures the temperature by feeling the pipe. This temperature is then compared with what the temperature should be. Depending on the comparison, control valves are adjusted.
2. The operator determines the temperature by examining an instrument that measures temperature. This is then compared with the desired temperature. Depending on the comparison, the valve is altered. Assume now that the operator continually monitors the instrument and changes the control valve setting. This action is actually affecting closed-loop control. If the operator periodically checks the chart and adjusts the valve, the adjustment is affecting open-loop control.

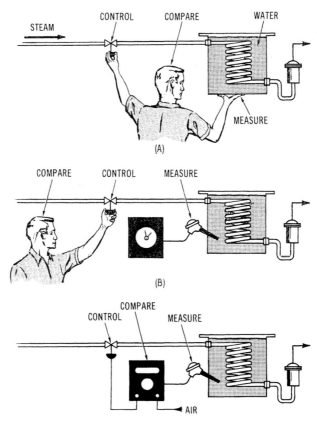

FIGURE 24-1   Three types of control. (A) Mostly human—very little instrumentation; (B) Human plus measuring instruments; (C) Measuring and control instruments with very little human supervision.

3.   The measurement function is made by the temperature sensor. This is applied to the controller. It compares the measurement with the setpoint and controls the valve accordingly. This is a continuous cycle of events without interruption. It is described as closed-loop control.

PROCESS-CONTROL ELEMENTS ▬▬▬▬▬▬

We will now study some of the essential elements of a process-control system. Process-control was developed to provide automatic regulation of manufacturing operations, which, in turn, permitted products to be developed more efficiently. Process control procedures are performed automatically. Little human intervention is needed in this operation. Automatic regulation is considered to be a primary function of the process-control system.

Dynamic Variables

Any physical value that can change or be changed by an outside influence is a dynamic variable. The word dynamic refers to something that is time dependent and results from some outside influence. A variable is simply something that has the capacity to change. In process-control systems, dynamic variables are regulated to achieve some type of manufacturing operation. Temperature, pressure and level are three examples of dynamic variables.

Regulation

The basic function of a process-control system is to adjust a dynamic variable to a desired value. The desired value is achieved by continually correcting the variable to keep it at that value. Regulation refers to this process. In this way, process-control is used to regulate the dynamic variable of a manufacturing operation.

Process

The flow of steam into the heat-exchanger of Figure 24-2 is a typical process. The process in this operation consists of adding heat energy to water. The steam coils, tank, pipes, and regulator valves are representative of the system. These are all needed together to accomplish the heating process. The process may involve a number of things that all relate to some manufacturing operation. A number of dynamic variables may be involved

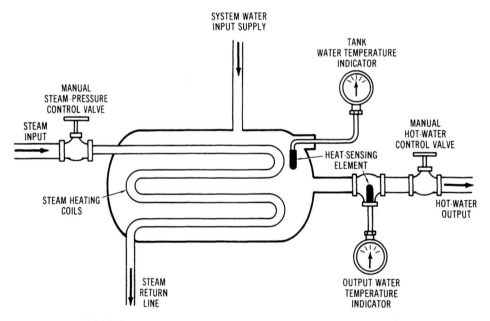

FIGURE 24-2   A manually operated heat-exchanger process.

in the process. It may be desirable to control all of the variables at the same time. Controls may apply to only one variable in some systems. In other systems, more than one control is needed.

In the heat exchanger, the temperature of the output water depends on several variables. These include hot-water output flow, temperature of the incoming water, and heat losses through the tank. Input energy to the system is determined by the rate of steam flow and the quality of the steam. When input energy or steam equals heat losses, the temperature of the output remains at a constant level, which represents a condition of balance. Balance will result in a change in hot-water output. The hot-water output is considered the primary process variable.

Measurement

When controlling a dynamic variable, there must be information about the variable being cotrolled. This information is found by measuring the operating value of the variable. The measurement function changes the process variable into a corresponding analogue value. Typically, sensors or transducers are used to achieve this function. For example, a sensor could be used to change a temperature value. A sensor was used in a filled system employing a capillary-tube sensor. Pressure would then be applied to the other elements of the process-control system. The final control element would be moved by pressure. It is this element that performs the control function.

Evaluation

After measurement of the process variable takes place, the information must then be examined. Examination determines what action should be taken. The controller of a process system is normally responsible for this function. Evaluation is performed by a human operator, by electronic signal processing, pneumatic signal processing, or with direct interaction from a computer. Human evaluation of process information is not associated with automatic control applications.

Controller evaluation of process information requires a measured signal. The signal is representative of the process variable and the desired setpoint value. These two inputs to the controller must be expressed in similar terms. The setpoint input is a manual adjustment that is set to a certain value. Evaluation by the controller consists of a comparison of the two input signals plus corrective action. If the setpoint and measured variable are equal, no corrective action is taken. If there is a difference in the setpoint and measured inputs, a corrective signal is initiated. When the variables are equal, no other signal is made. The measured and the setpoint inputs are held in close tolerances of one another through the controller.

Control Element

The control element of a process-control system is mainly responsible for altering the system output. It responds to the output of the controller and corrects the process variable so that the process variable compares with the setpoint input value. Typical control elements are valve positioners. These devices are used to alter the output of the system. In practice, the control element is called the final control element.

AUTOMATIC CONTROL ▬▬▬▬▬▬▬▬▬▬▬▬▬▬▬▬▬▬▬▬▬▬▬

Most industrial process-control applications perform efficiently when the value of the variable stays within certain limits. In practice, process variables are normally in a continous state of change. The system must respond to these changing values unless the instrument has reached a state of balance. The basic function of a process-control system deals with the manipulation of input-to-output energy. This energy is used to keep process variables within a desired operating range.

In practice, several factors influence the operation of an automatic process-control system. These factors may slow down the control process. The control may further be hindered from reaching a state of balance. System balance, process time lags, and capacity are important considerations in automatic process control.

System Balance

It is important to maintain a process in balance. Balance is achieved by matching the rate at which material is fed to the process to the rate at which the process uses material. Control valves must take a position that has the material being used up by the process. On the surface, this seems to be a simple job, however, exact measurements must be taken without any delays. The processes must be free of inertia and lags. The control valves must act without delay. If measurements could be achieved, system balance would be a simple job.

Lags

There can be delays with time and transfer in an instrument when the process changes. The process change must be substantial, in order for the instrument to place the valve in the correct position. The controller requires time to perform its job and even more time to get the controller output to the valve. When the valve does respond, output time is required for the fluid to move into the process. An additional interval is taken before the process equipment responds to a change in the feed rate. After these things are done, the process change is sensed, but by this time it may be too late. The new valve position may no longer be at the correct location for that particular process.

These time requirements are called *lags*. For example, the time it takes heat to flow through a tube wall is called transfer lag. The time it takes for the change to be recorded is called a measurement lag. The period it takes for a signal to the control valve to get material flowing through the valve to the process is called a distance-velocity lag.

Capacity

The basic function of control is to match valve movement with the movement of the process. Due to lags and instrument sensitivities, it is difficult to accurately match valve travel to the change of a process. Because of this difficulty, processes that can tolerate relatively large errors in valve positioning are the easiest to control.

Now consider the heating process shown in Figure 24-3. Suppose the hot-water storage capacity is quite large compared to the amount of water taken from the heater. If the

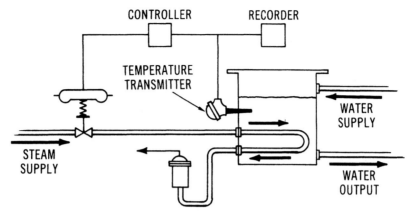

FIGURE 24-3   Hot-water control system.

capacity is, a large amount of steam will be required to change the temperature of the water. Further assume that the valve controlling the steam flow is fairly small. It might be possible to allow the valve to go wide open for a substantial time before the temperature changes. Alternately, it might use hot-water at a faster rate for a long time, with small changes in temperature. A large quantity of water acts as a "flywheel." The small steam supply affects the flywheel only if it acts over a long period of time. As a consequence, it can be concluded that this process is easy to control.

Suppose the process were one of holding a level in a tank. Assume the input to the tank is small. We are given the problem of holding the level constant by changing a valve opening. Changing a valve opening permits more or less fluid to flow from the tank. The valve is sufficiently large to discharge all the fluid that may enter the tank.

To effect good control, it is necessary that the out-flow be almost exactly equal to the in-flow. Any difference will result in a change in level. Suppose that the supply suddenly increases. This increase would mean that the control valve must suddenly open more. If it is slow to respond, the level will increase substantially. If the valve overtravels, the tank may run dry. In other words, the capacity of the tank is small compared to the capacity of the supply. In the heater example, the capacity of the steam supply is small and the heater capacity is large.

The process might be split in halves. One half is called the supply-side, and the other half, using materials or energy, is called the demand-side. The capacity of the equipment to use energy or material is called the demand-side capacity. The capacity of the input is called the supply-side capacity. These expressions are used in connection with the heater example and the tank-level example. It could be said that the heater example was one with a small supply-side capacity and a large demand-side capacity. The tank-level example was a process with a small demand-side capacity and a large supply-side capacity. Processes with large demand-side capacities are easier to control compared to processes with small demand-side capacities.

SUMMARY

Control is a general term that describes the ability of something to alter its normal course of operation. Process control deals with instrumentation that achieves control of numerous industrial manufacturing processes.

There are two major types of response that are of concern in process control. Open-loop control observes the situation, measures a process, and decides on what course of action needs to be taken through the judgement of an operator. Closed-loop control is an automatic process. A process measurement is taken and compared with a setpoint value. It then generates a correction signal that ultimately alters the process automatically.

To control a process automatically, the unit takes into account such things as material balance, time-lag consideration, and demand capacity.

ACTIVITIES

PROCESS CONTROL ANALYSIS ▬▬▬▬▬▬▬▬▬▬▬▬

1. Select an industrial process control system for analysis.
2. Identify the process being controlled by the system.
3. Identify the fundamental elements of the system.
4. Is the system open-loop or closed-loop?
5. If the system achieves control automatically, describe how balance is achieved.
6. How is measurement achieved by the system?
7. Does the system have any conditions that cause lag to occur?
8. Explain the operation of the system.

QUESTIONS

1. What is the difference between open-loop and closed-loop control?
2. What determines if a process control system is automated?
3. What is the meaning of dynamic variable?
4. Identify any automatic process control systems found in a residential building.
5. Why is measurement an important function in a process control system?
6. What is the function of the control element of an automated industrial process control system?

APPENDIX A

Pressure Conversion Chart

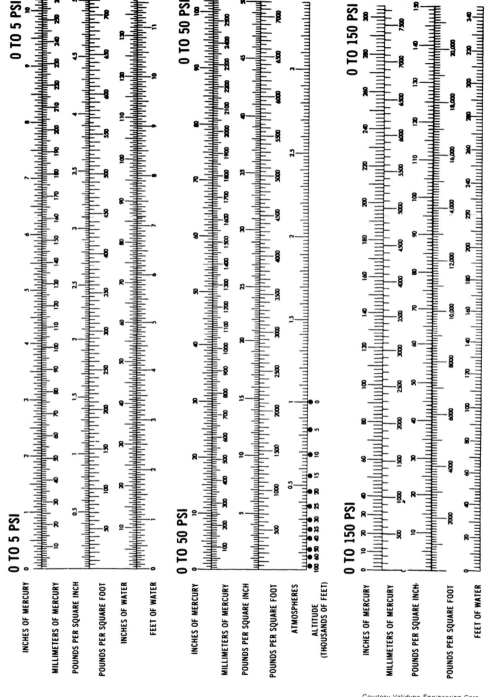

Courtesy Validyne Engineering Corp.

APPENDIX B

Comparison Chart for U.S. and Metric Units

U.S.		METRIC	METRIC		U.S.
1 inch	= 25.4	millimeters	1 millimeter	= 0.03937	inch
1 foot	= 0.3048	meter	1 meter	= 3.2808	feet
1 yard	= 0.9144	meter	1 meter	= 1.0936	yards
1 mile	= 1.609	kilometers	1 kilometer	= 0.6214	mile
1 square inch	= 6.4516	square centimeters	1 square centimeter	= 0.155	square inch
1 square foot	= 0.0929	square meter	1 square meter	= 10.7639	square feet
1 square yard	= 0.836	square meter	1 square meter	= 1.196	square yards
1 acre	= 0.4047	hectare	1 hectare	= 2.471	acres
1 cubic inch	= 16.387	cubic centimeters	1 cubic centimeter	= 0.061	cubic inch
1 cubic foot	= 0.028	cubic meter	1 cubic meter	= 35.3147	cubic feet
1 cubic yard	= 0.764	cubic meter	1 cubic meter	= 1.308	cubic yards
1 quart (liq)	= 0.946	liter	1 liter	= 1.0567	quarts (liq)
1 gallon	= 0.00378	cubic meter	1 cubic meter	= 264.172	gallons
1 ounce (avdp)	= 28.349	grams	1 gram	= 0.035	ounce (avdp)
1 pound (avdp)	= 0.4536	kilogram	1 kilogram	= 2.2046	pounds (avdp)
1 horsepower	= 0.7457	kilowatt	1 kilowatt	= 1.341	horsepower
1 lb/in^2 (psi)	= 6894.76	Pa (N/m^2)	1 Pa (N/m^2)	= 0.000145	lb/in^2 (psi)
1 lb/in^2 (psi)	= 0.07031	kg/cm^2	1 kg/cm^2	= 14.223	lb/in^2 (psi)

APPENDIX C

Fluid Power Symbols

Pneumatic Instruments

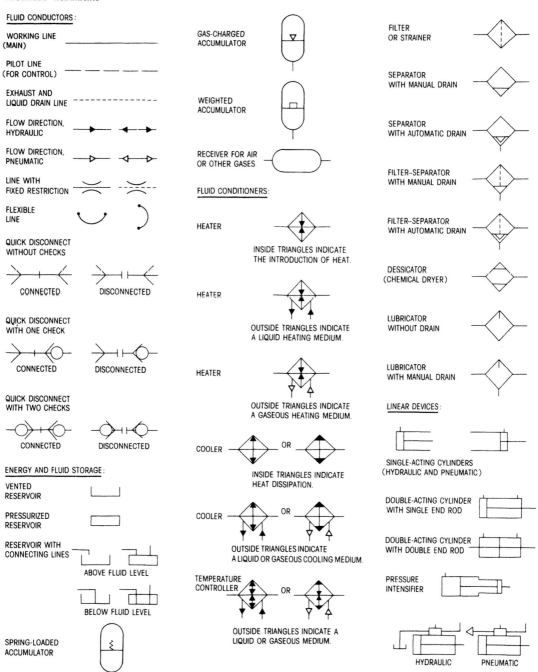

FLUID CONDUCTORS:

WORKING LINE
(MAIN)

PILOT LINE
(FOR CONTROL)

EXHAUST AND
LIQUID DRAIN LINE

FLOW DIRECTION,
HYDRAULIC

FLOW DIRECTION,
PNEUMATIC

LINE WITH
FIXED RESTRICTION

FLEXIBLE
LINE

QUICK DISCONNECT
WITHOUT CHECKS

CONNECTED   DISCONNECTED

QUICK DISCONNECT
WITH ONE CHECK

CONNECTED   DISCONNECTED

QUICK DISCONNECT
WITH TWO CHECKS

CONNECTED   DISCONNECTED

ENERGY AND FLUID STORAGE:

VENTED
RESERVOIR

PRESSURIZED
RESERVOIR

RESERVOIR WITH
CONNECTING LINES

ABOVE FLUID LEVEL

BELOW FLUID LEVEL

SPRING-LOADED
ACCUMULATOR

GAS-CHARGED
ACCUMULATOR

WEIGHTED
ACCUMULATOR

RECEIVER FOR AIR
OR OTHER GASES

FLUID CONDITIONERS:

HEATER

INSIDE TRIANGLES INDICATE
THE INTRODUCTION OF HEAT.

HEATER

OUTSIDE TRIANGLES INDICATE
A LIQUID HEATING MEDIUM.

HEATER

OUTSIDE TRIANGLES INDICATE
A GASEOUS HEATING MEDIUM.

COOLER   OR

INSIDE TRIANGLES INDICATE
HEAT DISSIPATION.

COOLER   OR

OUTSIDE TRIANGLES INDICATE
A LIQUID OR GASEOUS COOLING MEDIUM.

TEMPERATURE
CONTROLLER   OR

OUTSIDE TRIANGLES INDICATE A
LIQUID OR GASEOUS MEDIUM.

FILTER
OR STRAINER

SEPARATOR
WITH MANUAL DRAIN

SEPARATOR
WITH AUTOMATIC DRAIN

FILTER–SEPARATOR
WITH MANUAL DRAIN

FILTER–SEPARATOR
WITH AUTOMATIC DRAIN

DESSICATOR
(CHEMICAL DRYER)

LUBRICATOR
WITHOUT DRAIN

LUBRICATOR
WITH MANUAL DRAIN

LINEAR DEVICES:

SINGLE-ACTING CYLINDERS
(HYDRAULIC AND PNEUMATIC)

DOUBLE-ACTING CYLINDER
WITH SINGLE END ROD

DOUBLE-ACTING CYLINDER
WITH DOUBLE END ROD

PRESSURE
INTENSIFIER

HYDRAULIC   PNEUMATIC

SERVO POSITIONER

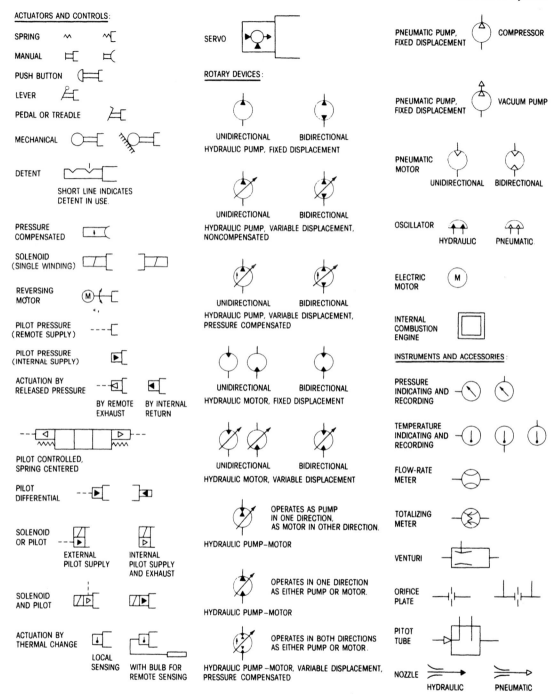

ACTUATORS AND CONTROLS:

SPRING

MANUAL

PUSH BUTTON

LEVER

PEDAL OR TREADLE

MECHANICAL

DETENT

SHORT LINE INDICATES
DETENT IN USE.

PRESSURE
COMPENSATED

SOLENOID
(SINGLE WINDING)

REVERSING
MOTOR

PILOT PRESSURE
(REMOTE SUPPLY)

PILOT PRESSURE
(INTERNAL SUPPLY)

ACTUATION BY
RELEASED PRESSURE

BY REMOTE   BY INTERNAL
EXHAUST   RETURN

PILOT CONTROLLED,
SPRING CENTERED

PILOT
DIFFERENTIAL

SOLENOID
OR PILOT

EXTERNAL   INTERNAL
PILOT SUPPLY   PILOT SUPPLY
                AND EXHAUST

SOLENOID
AND PILOT

ACTUATION BY
THERMAL CHANGE

LOCAL   WITH BULB FOR
SENSING   REMOTE SENSING

SERVO

ROTARY DEVICES:

UNIDIRECTIONAL   BIDIRECTIONAL
HYDRAULIC PUMP, FIXED DISPLACEMENT

UNIDIRECTIONAL   BIDIRECTIONAL
HYDRAULIC PUMP, VARIABLE DISPLACEMENT,
NONCOMPENSATED

UNIDIRECTIONAL   BIDIRECTIONAL
HYDRAULIC PUMP, VARIABLE DISPLACEMENT,
PRESSURE COMPENSATED

UNIDIRECTIONAL   BIDIRECTIONAL
HYDRAULIC MOTOR, FIXED DISPLACEMENT

UNIDIRECTIONAL   BIDIRECTIONAL
HYDRAULIC MOTOR, VARIABLE DISPLACEMENT

OPERATES AS PUMP
IN ONE DIRECTION,
AS MOTOR IN OTHER DIRECTION.

HYDRAULIC PUMP-MOTOR

OPERATES IN ONE DIRECTION
AS EITHER PUMP OR MOTOR.

HYDRAULIC PUMP-MOTOR

OPERATES IN BOTH DIRECTIONS
AS EITHER PUMP OR MOTOR.

HYDRAULIC PUMP-MOTOR, VARIABLE DISPLACEMENT,
PRESSURE COMPENSATED

PNEUMATIC PUMP,   COMPRESSOR
FIXED DISPLACEMENT

PNEUMATIC PUMP,   VACUUM PUMP
FIXED DISPLACEMENT

PNEUMATIC
MOTOR
          UNIDIRECTIONAL   BIDIRECTIONAL

OSCILLATOR
          HYDRAULIC   PNEUMATIC

ELECTRIC
MOTOR

INTERNAL
COMBUSTION
ENGINE

INSTRUMENTS AND ACCESSORIES:

PRESSURE
INDICATING AND
RECORDING

TEMPERATURE
INDICATING AND
RECORDING

FLOW-RATE
METER

TOTALIZING
METER

VENTURI

ORIFICE
PLATE

PITOT
TUBE

NOZZLE
          HYDRAULIC   PNEUMATIC

422

Pneumatic Instruments

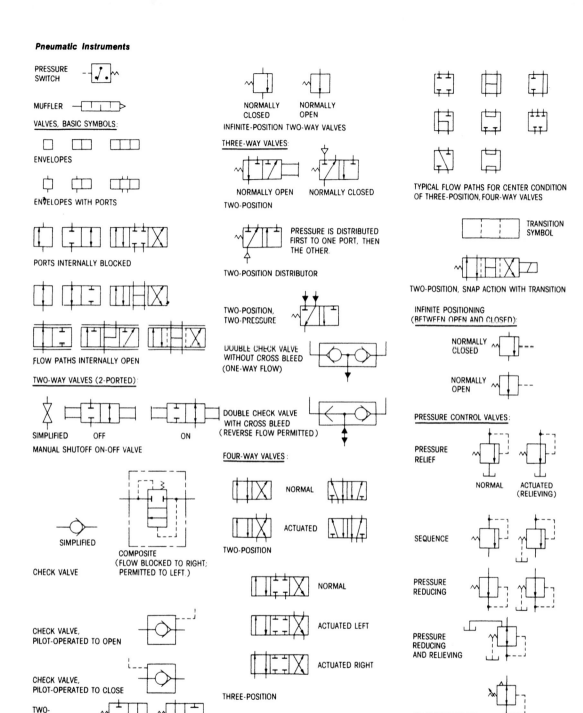

PRESSURE SWITCH

MUFFLER

VALVES, BASIC SYMBOLS:

ENVELOPES

ENVELOPES WITH PORTS

PORTS INTERNALLY BLOCKED

FLOW PATHS INTERNALLY OPEN

TWO-WAY VALVES (2-PORTED):

SIMPLIFIED   OFF   ON

MANUAL SHUTOFF ON-OFF VALVE

SIMPLIFIED   COMPOSITE
(FLOW BLOCKED TO RIGHT;
PERMITTED TO LEFT.)
CHECK VALVE

CHECK VALVE,
PILOT-OPERATED TO OPEN

CHECK VALVE,
PILOT-OPERATED TO CLOSE

TWO-
POSITION TWO-
WAY VALVES
            NORMALLY   NORMALLY
            CLOSED   OPEN

NORMALLY   NORMALLY
CLOSED   OPEN

INFINITE-POSITION TWO-WAY VALVES

THREE-WAY VALVES:

NORMALLY OPEN   NORMALLY CLOSED
TWO-POSITION

PRESSURE IS DISTRIBUTED
FIRST TO ONE PORT, THEN
THE OTHER.

TWO-POSITION DISTRIBUTOR

TWO-POSITION,
TWO-PRESSURE

DOUBLE CHECK VALVE
WITHOUT CROSS BLEED
(ONE-WAY FLOW)

DOUBLE CHECK VALVE
WITH CROSS BLEED
(REVERSE FLOW PERMITTED)

FOUR-WAY VALVES:

NORMAL

ACTUATED

TWO-POSITION

NORMAL

ACTUATED LEFT

ACTUATED RIGHT

THREE-POSITION

TYPICAL FLOW PATHS FOR CENTER CONDITION
OF THREE-POSITION, FOUR-WAY VALVES

TRANSITION
SYMBOL

TWO-POSITION, SNAP ACTION WITH TRANSITION

INFINITE POSITIONING
(BETWEEN OPEN AND CLOSED):

NORMALLY
CLOSED

NORMALLY
OPEN

PRESSURE CONTROL VALVES:

PRESSURE
RELIEF

NORMAL   ACTUATED
            (RELIEVING)

SEQUENCE

PRESSURE
REDUCING

PRESSURE
REDUCING
AND RELIEVING

AIR LINE PRESSURE
REGULATOR (ADJUSTABLE, RELIEVING)

INFINITE POSITIONING VALVES:

THREE-WAY VALVES

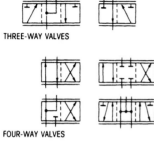

FOUR-WAY VALVES

FLOW-CONTROL VALVES:

ADJUSTABLE, NONCOMPENSATED
(FLOW CONTROL IN EACH DIRECTION)

ADJUSTABLE,
WITH BYPASS

FLOW CONTROLLED TO RIGHT,
FLOW TO LEFT BYPASSES CONTROL.

ADJUSTABLE AND PRESSURE
COMPENSATED, WITH BYPASS

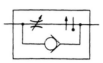

ADJUSTABLE, TEMPERATURE
AND PRESSURE COMPENSATED

AIR LINE ACCESSORIES:

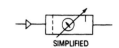

SIMPLIFIED

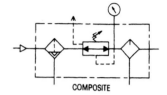

COMPOSITE

FILTER, REGULATOR, AND LUBRICATOR

424

INDEX

Page numbers in *italics* indicate diagrams and photographs. Tables are indicated by *tab.*